Klaus H. Lüdtke
Process Centrifugal Compressors

Springer

*Berlin
Heidelberg
New York
Hong Kong
London
Milan
Paris
Tokyo*

Engineering | ONLINE LIBRARY

springeronline.com

Klaus H. Lüdtke

Process Centrifugal Compressors

Basics, Function, Operation, Design, Application

With 211 Figures

 Springer

Dipl.-Ing. Klaus H. Lüdtke
Hedwigstrasse 10
D-13467 Berlin
Germany
luedtkek@t-online.de

ISBN 3-540-40427-9 Springer-Verlag Berlin Heidelberg New York

Library of Congress Cataloging-in-Publication-Data
A catalog record for this book is available from the Library of Congress.
Bibliographic information published by Die Deutsche Bibliothek.
Die Deutsche Bibliothek lists this publication in the Deutsche Nationalbibliographie;
detailed bibliographic data is available in the Internet at http://dnb.ddb.de

This work is subject to copyright. All rights are reserved, whether the whole or part of the material is concerned, specifically the rights of translation, reprinting, reuse of illustrations, recitations, broadcasting, reproduction on microfilm or in any other way, and storage in data banks. Duplication of this publication or parts thereof is permitted only under the provisions of the German copyright Law of September 9, 1965, in its current version, and permission for use must always be obtained from Springer-Verlag. Violations are liable for prosecution under the German Copyright Law.

Springer-Verlag is a part of Springer Science+Business Media
springeronline.com

© Springer-Verlag Berlin Heidelberg 2004
Printed in Germany

The use of general descriptive names, registered names trademarks, etc. in this publication does not imply, even in the absence of a specific statement, that such names are exempt from the relevant protective laws and regulations and therefore free for general use.

Typesetting: Protago-TeX-Produktion GmbH, Berlin
Cover design: medio Technologies AG, Berlin
Printed on acid free paper   62/3020/M - 5 4 3 2 1 0

Dedicated to my wife Marianne

and my children Susanne and Roland

Foreword

Throughout the last decades, centrifugal compressor research and development have been revolutionized. Computational fluid dynamics have provided a better understanding of the flow and physical phenomena, and the design of new centrifugal compressor components has been transformed from an "art" into a "science". New materials and manufacturing techniques now create new geometries that could only be dreamed of in the past, and new challenging applications have pushed the limits beyond what was considered the state of the art.

This new book presenting a comprehensive look at industrial compressors is therefore very timely.

Readers will find a large amount of information based on extensive experience, a clear and well-founded approach to real-gas handling and solutions to many practical problems. It will provide engineering contractors and users of industrial compressors with a better insight into the "how" and "why" of different design features thus allowing a more profound basis for discussions with manufacturers. It will also cast a light on the day-by-day design practice to academia by revealing the limitations and requirements of practical applications and economics.

This book combines a strict mathematical approach with practical experience and is illustrated with many examples. It fills in the gap between academic textbooks and encyclopaedic descriptions of industrial compressors. I have no doubt that this book, based on several decades of experience in the industry, both in the USA and Europe, will be well received by the centrifugal compressor community.

René Van den Braembussche
Professor
Turbomachinery and Propulsion Department of the von Karman Institute
Fellow of ASME

Preface

This book summarizes the author's experience of forty years work in the centrifugal compressor industry: in aerodynamic design, component development, and testing. The experience has resulted from a fruitful dialog with end users, contractors, cooperating manufacturers and research institutes.

The intention is to present information available only to an original equipment manufacturer (OEM) and to disseminate knowledge not included in any other book on centrifugal compressors.

This book can be regarded as a practical supplement to other works on centrifugal compressors. It offers help on many questions in the daily life of those who deal with centrifugal compressors, either as users, engineering contractors or compressor manufacturers. But it also addresses newly hired engineers in the industry making their first steps in an unaccustomed terrain and students who want to push the ivory tower doors open to the real world of turbomachinery.

Much of the author's experience stems from participating in bid clarification and order kick-off meetings with engineering companies and users, training of customer's representatives and regular lecturing to young engineers in the in-house training programs of well-known companies.

The author has been working with MAN Turbomaschinen AG, who took over the turbomachinery division of former Deutsche Babcock–Borsig AG in Berlin, Germany in 1996. Prior to joining Deutsche Babcock–Borsig he worked with Allis–Chalmers Corporation in Milwaukee, Wisconsin, USA. The author received his Diplom-Ingenieur degree in mechanical engineering from the Technical University of Berlin in 1962. He has been a Consulting Senior Engineer since 2000.

During his professional career the author was exposed to more or less the same questions and suggestions of customers, colleagues and his own management so that, after so many years, the wish matured to put the answers down in writing to prevent at least some of this experience from totally dissipating. Consensual (and controversial) dialog outcomes, problem solutions, evolutionary improvements, revolutionary developments and, last, but not least, hard evidence results from development and performance verification tests thus found their way, in the form of presented papers, into the Proceedings of the ASME Turbo Expo, the American A&M Turbomachinery Symposium, the German VDI Thermal Turbomachinery Meeting, the Polish Symkom Compressor and Turbine Conference, the British IMechE Fluid Machinery Congress and others. The most significant of these papers in turn form part of this book in conjunction with numerous other publications. So, in one way or another, this book constitutes the author's professional memoirs.

The term "process compressor" covers the typical multistage single-shaft centrifugal compressor, as it is predominantly used in the oil, gas, chemical and petrochemical industries, and in refineries. Exempt are centrifugal compressors in water chilling, plant air, and turbocharger units as well as in gas turbines and jet engines, where stage pressure ratios and Mach numbers are much higher and the numbers of stages per unit are lower, leading to totally different geometric configurations.

This book is addressed primarily to rotating-equipment specialists, project engineers, process engineers, process-engineering advisors in engineering contractors and compressor user companies, operations managers and operations support engineers in oil and gas field operations, natural gas processing, petroleum refining, petrochemical processing and chemical industries. In other words, all engineers are addressed who deal with planning, sizing estimates, writing specifications, sending out enquiries, evaluating bids, assessing new compressor designs and uprates, and appraising technical compressor-related documentation from the OEM.

Intended secondary addressees are educational (e.g., junior staff of compressor users, students of fluid flow machinery) and aerodynamic design engineers in compressor manufacturing companies.

The book will help the reader to understand the terminology and to speak the same language within the turbocompressor community and it will enable the compressor user to enter into a meaningful discussion with the manufacturer.

The contents of the book resulted from a four decade-long relationship between the compressor OEM and the user and can help the reader to select the right compressor for the particular application (is it middle-of-the-ballpark, borderline or prototype?), because it clearly defines the constraints which influence the compressor design to a much greater extent than anything else.

However, the philosophy of designing and assessing compressors presented in this book does not lay claim to exclusiveness, and it should be pointed out that it is not intended to devalue other philosophies that are equally successful. There are probably as many schools of thought as there are compressor manufacturers. There are highly empirical philosophies using huge data bases, and there are others based on scientific principles supported by computational models; and there are many more in between. Empirical or scientific – both approaches require decade-long experience that could best be defined as the "unconscious integration of all influencing quantities that cannot be ascertained by calculation".

All formulae are presented as "true physical equations" describing the pure physical facts without being intermingled with numerical values solely coming from the system of units used. So-called "unity braces" serve to convert any physical equation into the unit-dependent form that the reader is used to, be it the SI- system or the foot/inch- system.

Here are some of this book's features:

- the aero-thermodynamics is strictly based on real-gas formulae,
- the compressor similarity characteristics are interpreted by their implications rather than by their well-known formal structure,

- the gas density-dependent lateral rotor stiffness ratio is brought together with the flow coefficient, Mach number and number of stages to form a powerful interdependent aero-rotordynamic interface,
- five compressor regulation methods are systematically explained and compared,
- surge is broken down into the system surge caused by the compressor/system volume incompatibility and the stage surge triggered by the diffuser rotating stall,
- the discussion of compressor choke is supplemented by interpreting test results in the expansion mode beyond choke,
- pressure measurements in the impeller/diaphragm gaps are disclosed and analyzed permitting an improved thrust and leakage calculation,
- various measures of improving efficiency and operating range are described,
- the phenomenon of aerodynamic interstage mismatching is explained, and
- three common compressor uprating methods, i.e. higher speed, wider impellers and suction boosting, are outlined in detail.

Acknowledgements

Before I come to that, a word to readers whose mother tongue is English. Please excuse my mediocre and simplistic English; my time spent in America was just not long enough to acquire all the subtleties of the language. And I want to apologize to German readers for having written the book in English–I trust they will understand why. On the other hand, I am convinced that my German professional colleagues are well familiar with the termini technici in the compressor business. For those of them who want to preserve the traditional hundred years old German compressor vocabulary, I offer help in the form of a list of corresponding English–German compressor nomenclature which I guarantee to be authentic.

The experience laid down in this book would not have been possible without all the people who accompanied me during my professional life and to whom I would like to express my sincere gratitude: first of all to my wife Marianne for her moral support, patience and understanding during the long years of manuscript writing; to my former superiors Norbert Wemhöner and Rolf Göldner, who taught me the first steps in the world of industrial turbocompressors when I was still wet behind the ears; to Ludwig Kortenkamp, Dieter Jacobs, Klaus Förster and Peter Kazik for providing the atmosphere that is a sine qua non to be creative and to find fulfillment in the engineering work; to my former colleagues at Allis–Chalmers in Milwaukee, Ernst Henne, Russ John and Frank Rassman, for teaching me to "think compressors" the American way so that German systematism and American pragmatism could ideally supplement each other; to Dr. Hans-O. Jeske, member of the Executive Board of MAN Turbomaschinen AG for permission to publish this book and to Dr. Mathias Stark, Manager Design/Customer Support for paving the way to get it started; to my colleagues at MAN Turbo, who accompanied me during my last professional decade, for the excellent working atmosphere and team work, especially to Andreas Lempart, Manfred Moritz, Manfred Reichard, Vladimir Zarybnicky, Dr. Robert Ehrich, Dr. Detlef Bosin and Peter Sziedat for their professional cooperation, and to Marianne Quednau for contributing the compressor cross-sections.

Special thanks are given to Springer-Verlag for the pleasant synergy and the excellent design of this book.

Last but not least I want to thank the group of people to whom our company owes its existence: our long-term and new customers, who let us know by awarding us with contracts that they accept our technology, for many fruitful and stimulating discussions, resulting quite a number of times in fresh replies to old questions and thus initiating some of the chapters of this book.

Berlin, November 2003   Klaus Lüdtke

Contents

1 Basic Compressor Aero-Thermodynamics 1
  1.1 The Beginnings of Turbomachinery 1
    1.1.1 Process Centrifugal Compressor 1
    1.1.2 Turbocharger 2
    1.1.3 Centrifugal Refrigeration Machine 2
    1.1.4 Turbojet Aircraft Engine 2
    1.1.5 Axial Compressor 5
    1.1.6 Steam Turbine 5
    1.1.7 Stationary Gas Turbine 6
    1.1.8 Centrifugal Pump 6
    1.1.9 Rocket Engine 6
  1.2 Processes 8
  1.3 Definitions and Terminology 12
    1.3.1 Compressor 12
    1.3.2 Impeller, Stage, Section 13
  1.4 Method of Presenting Equations 14
  1.5 The First Law of Thermodynamics 16
  1.6 Reference Compression Processes 18
    1.6.1 Heat in the Temperature–Entropy Diagram 18
    1.6.2 Enthalpy Difference in the Temperature–Entropy Diagram 19
    1.6.3 Work in Reference Processes 20
    1.6.4 Application of Reference Work 25
    1.6.5 Head/Efficiency, Practical Formulae, Perfect Gas Behavior 25
  1.7 The Euler Turbomachinery Equation 30
    1.7.1 Change of State in the Mollier Diagram 32
  1.8 Velocity Triangles 33
    1.8.1 Inlet Triangles 35
    1.8.2 Exit Triangles 35
  1.9 Laws of Energy Transfer, Summary of Inferences 36
  1.10 Compressor Characteristics and their Significance 37
    1.10.1 Flow Coefficient 37
    1.10.2 Polytropic Head Coefficient 40
    1.10.3 Work Input Factor 40
    1.10.4 Polytropic Efficiency 40
    1.10.5 Machine Mach Number (Tip Speed Mach Number) 41
    1.10.6 Machine Reynolds Number 43
    1.10.7 Specific Speed 44
  References 44

2 Thermodynamics of Real Gases ... 47
2.1 Gas Thermodynamics ... 48
2.1.1 Polytropic and Isentropic Exponents ... 48
2.1.2 Summary of Formulae for Head and Discharge Temperature ... 54
2.2 Real-Gas Equations of State ... 58
2.2.1 BWRS Equation of State ... 59
2.2.2 RKS Equation of State ... 61
2.2.3 LKP Equation of State ... 61
2.2.4 PR Equation of State ... 62
2.2.5 Computation of Isentropic Exponents ... 63
2.2.6 Recommendations for the Use of Real-Gas Equations ... 66
2.3 Handy Formulae for Humid Gases ... 66
2.3.1 Relative Humidity, Volume Fraction, Humid Molar Mass ... 66
2.3.2 Absolute Water Content, Humid Mass, Volume Flow ... 67
2.3.3 Calculation of Saturation Pressure ... 68
2.3.4 Calculation of Dew Point ... 68
References ... 69

3 Aero Components – Function and Features ... 71
3.1 Impeller ... 71
3.1.1 Centrifugal Impeller Types ... 71
3.1.2 The Shrouded 2D-Impeller ... 72
3.1.3 The Shrouded 3D-Impeller ... 72
3.1.4 The Semi-Open Impeller ... 73
3.1.5 Impeller Family ... 74
3.1.6 Impeller Application ... 75
3.1.7 Impeller Manufacture ... 75
3.1.8 Impeller Stress Analysis ... 77
3.2 Inlet Duct ... 78
3.3 Adjustable Inlet Guide Vanes ... 79
3.4 Annular Diffuser ... 81
3.5 Vaned Diffuser ... 84
3.6 Return (Crossover) Bend ... 86
3.7 Return Vane Channel ... 87
3.8 Side-Stream Inlet ... 88
3.9 Volute ... 90
References ... 92

4 Compressor Design Constraints ... 93
4.1 Intrinsic Parameter Constraints ... 93
4.1.1 Flow Coefficient ... 93
4.1.2 Rotational Speed ... 94
4.1.3 Impeller Tip Speed ... 94
4.1.4 Mach Number ... 95
4.1.5 Discharge Temperature ... 97
4.1.6 Impeller Exit Width ... 98
4.1.7 Impeller Outer Diameter ... 98

    4.1.8 Shaft Stiffness 98
  4.2 Derived Compound Constraints 99
    4.2.1 Stage Head and Stage Pressure Ratio 99
    4.2.2 Maximum Number of Impellers per Rotor 100
    4.2.3 Maximum Volume Flow per Frame Size 104
    4.2.4 Minimum Discharge Volume Flow of Turbocompressors 108
    4.2.5 Impeller Arrangement 108
  References 111

5 Compressor Off-Design Operation 113
  5.1 Regulation Methods and Nomenclature 113
    5.1.1 Variable Speed 116
    5.1.2 Suction Throttling 117
    5.1.3 Adjustable Inlet Guide Vanes 118
    5.1.4 Adjustable Diffuser Vanes 120
    5.1.5 Bypass Regulation 122
  5.2 Customary Performance Maps for Multistage Process Compressors 123
    5.2.1 Variable Speed 126
    5.2.2 Suction Throttling 128
    5.2.3 Adjustable Inlet Guide Vanes 130
  5.3 Performance Curves for Variable Conditions 133
  5.4 In Search of an Invariant Performance Curve 134
  5.5 Performance-Curve Limits 140
    5.5.1 Surge 140
    5.5.2 Choke 148
    5.5.3 Beyond Choke (Expansion Mode in the Fourth Quadrant) 151
  References 156

6 Aerodynamic Compressor Design: Case Study 157
  6.1 Approximate Compressor Sizing 158
  6.2 Detailed Compressor Design 165
    6.2.1 Impeller Inlet 165
    6.2.2 Impeller Exit 174
    6.2.3 Stage Efficiency 184
    6.2.4 Head, Pressure, Temperature, Power 193
    6.2.5 Diffuser Aerodynamics 196
    6.2.6 Return Vane Channel 201
    6.2.7 Volute and Conical Exit Diffuser 202
  6.3 Impeller Blade Design 208
  6.4 Axial Rotor Thrust 210
  6.5 Leakage Flow through Labyrinth Seals 216
  6.6 Final Compressor Design 217
  6.7 Rotordynamics: Lateral Vibration Analysis 218
    6.7.1 Lateral Critical Speeds 218
    6.7.2 Stability 220
    6.7.3 Logarithmic Decrement 222
  6.8 Performance Map 224

6.9 Performance Test 227
References 229

7 Application Examples 231
7.1 Hydrogen-Rich Gas Compressors 232
7.2 Natural-Gas Compressors 236
7.3 Air Compressors 244
7.4 Refrigeration Compressors 247
7.5 Chlorine Compressors 251
7.6 Methanol Synthesis Gas Compressors 255

8 Improving Efficiency and Operating Range 259
8.1 Increase of Efficiency 259
    8.1.1 Optimization of Aero-Thermodynamics 259
    8.1.2 Increased Diffuser Diameter Ratio 261
    8.1.3 Improved Volute Geometry 264
    8.1.4 Impeller-Blade Cut-Back 266
    8.1.5 Cover-Disk Labyrinth Seal 266
    8.1.6 Reduction of Plenum Inlet Loss 267
    8.1.7 Reduction of Flow-Channel Surface Roughness 267
    8.1.8 Minimizing Impeller/Diaphragm Clearance 268
    8.1.9 Reduction of Recirculation Losses 270
8.2 Increase of Operating Range 271
    8.2.1 Delay of System Surge 271
    8.2.2 Delay of Stage Stall 273
References 279

9 Rerate of Process Centrifugal Compressors 281
9.1 Introduction 281
    9.1.1 Definition of Rerates 281
    9.1.2 Increase of Mass Flow 282
9.2 New Wider Impellers and Diaphragms 283
9.3 Higher Rotational Speed 285
    9.3.1 Additional Losses 287
    9.3.2 Case Study 1: Wider Impellers and Speed Increase Combined 287
9.4 Supercharging with a Booster Compressor 294
9.5 Rerate Methods: Advantages/Disadvantages 305
9.6 Anticipating Rerates 305
9.7 Conclusions 306
References 307

10 Standardization of Compressor Components 309
10.1 Meridional Impeller Blade Cutoff 309
10.2 Modular Design System 311
    10.2.1 Aerodynamic Stage Rating Range 312
    10.2.2 Impeller Family 313
    10.2.3 Frame-Size Standardization 314

10.2.4 Diaphragms 315
10.2.5 Discharge Volutes 315
10.2.6 Systematic Tests of Modular Design Stages 316

Appendix **317**
  Unit Equations 317
  English–German Glossary on Compressors 319

Subject Index **325**

1 Basic Compressor Aero-Thermodynamics

1.1 The Beginnings of Turbomachinery

1.1.1 Process Centrifugal Compressor

Late in the 19th century a French genius, Professor Auguste Rateau, invented the centrifugal compressor. By 1899 he had a single-stage prototype on his test stand compressing 0.5 m^3/s of atmospheric air to a discharge pressure of 1.5 bar (absolute) at a rotational speed of 20,000 rpm (Engeda 1998). In 1903 he had his first industrial unit operating in a steel works and by 1905 he had built the first five-stage single-shaft centrifugal compressor delivering 0.7 m^3/s of air with a 1.45 : 1 pressure ratio.

In 1905 his newly founded company, called S. A. Rateau, built the first multistage unit with five stages on one shaft. In the same year already a three-casing (!) experimental Rateau compressor operated at the Societé des Turbomoteurs à Combustion in Paris compressing 1 m^3/s of air to a discharge pressure of 4.5 bar at a rotational speed of 4,000 1/min and a power input of 350 hp. It was driven by the world's first operable gas turbine of Armengaud and Lemale of France, delivering (a) the combustion air to this novel turbine and (b) compressed air to a public utility.

Fig. 1.1 One of the world's first multistage centrifugal compressors, built by Brown Boveri & Cie, Switzerland in 1906 according to the patents of Rateau. It provided combustion air for the world's first operable gas turbine of Armengaud and Lemale and had three casings, volume flow 1.0 m^3/s, pressure ratio 4.5 : 1, speed 4,000 1/min and power 250 kW (Brown Boveri 1928).

This fascinating and versatile fluid flow machine with its pulsation-free flow and low weight/power ratio was to penetrate almost all branches of the fluid-processing industry and became the most widely spread compressor being able to handle practically all industrial gases and gas mixtures from the lowest to the highest molar mass and topping pressures of 800 bar.

In 1906 Brown Boveri & Cie (1928) of Switzerland acquired a licence from Rateau and built its first (experimental) centrifugal compressor as a very similar three-casing unit for gas turbine combustion air, as shown in Fig. 1.1. Since the test results were promising, they started full-scale production mainly for utilities, blast furnaces, steelworks, mining and chemical industries.

So the first centrifugal compressors are closely linked to the newly developed first gas turbines.

1.1.2 Turbocharger

The first turbocharger (an exhaust radial inflow gas turbine driving a centrifugal air compressor) was invented in 1905 (date of patent) by Alfred Büchi, chief engineer for Sulzer Brothers, Switzerland. In 1915 he introduced the prototype of a turbocharged diesel engine.

In World War I Auguste Rateau developed a turbocharged aircraft engine. In the late 1910s the General Electric Company developed a turbocharger and tested it together with the engine on Pikes Peak, Colorado at a height of 4,300 m; in 1920 an American biplane was equipped with the unit, and set a new altitude record of 11,000 m (Stodola 1924).

1.1.3 Centrifugal Refrigeration Machine

In 1921 Willis Carrier patented the centrifugal chiller, which comprised a single-stage centrifugal compressor operating between a (suction side) evaporator and a (discharge side) condenser in a closed refrigeration cycle. In the 1920s the new air-conditioning device spread from industry to department stores and theatres. In the 1930s, in the southern United States, private houses, office buildings, restaurants and hotels began to be equipped with air conditioners. They had an enormous impact on living and working conditions, transforming small communities into thriving metropolises.

1.1.4 Turbojet Aircraft Engine

In the beginning of a new era in aviation were two personalities of great inventive ability: Hans-Joachim Pabst von Ohain[1] in Germany and Frank Whittle in Britain.

Von Ohain was the first to design and develop a turbojet engine to power an aircraft. In February 1937 the prototype of the jet engine called He S-1 was suc-

---

[1] Last name: Pabst von Ohain; first name: Hans-Joachim.

cessfully tested at the Heinkel aircraft works. Impressed with the tests, von Ohain and Ernst Heinkel began to develop the He S-3 jet engine together with the experimental (non-military!) aircraft He-178. In August 1939, taking off from the Heinkel's works airstrip, Rostock-Marienehe in North Germany, the He-178 was the world's first jet powered aircraft and became the trailblazer of the jet age. The small plane (length 7.5 m, wing span 7.2 m) reached a speed of 700 km/h with an engine thrust of 4,890 N.

Fig. 1.2 The turbojet engine He S-3B developed by Pabst von Ohain to power the world's first jet aircraft He-178 in 1939 (Colliat 2001).

The engine consisted of an axial impeller with axial intake (Fig. 1.2) followed by a row of low solidity stationary vanes and by a high-flow coefficient semi-open centrifugal impeller with radial blades and vaned diffuser. The turbine was a single-stage radial inflow design. The fuselage was built around the von Ohain engine. The high-flow coefficient brought about a small outer diameter, making it easier to be accommodated in the air frame.

Frank Whittle developed a similar engine in Britain. In April 1937, just two months later than von Ohain, he started the very first run of his experimental jet engine, but it was not until May 1941 that the first British jet-propelled aircraft, the Gloster E.28/39, took off powered by Whittle's turbojet engine W.1.

The military aircraft (length 7.7 m, wing span 8.8 m) reached a speed of 545 km/h with 3,800 N thrust (later with second version W.2: 720 km/h at 7,550 N). The engine consisted of a semi-open double-flow centrifugal impeller with radial blades including a rudimentary inducer followed by a vane-island diffuser (Fig. 1.3 shows the variant W.1U). The double-flow design required a radial rather than an axial inlet with fixed inlet guide vanes to deflect the flow by 90°; but the design of course enabled a comparatively small outer diameter. The air frame was built around the engine as well.

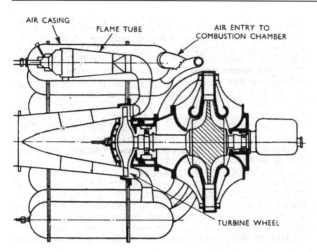

Fig. 1.3 The turbojet engine W.1U developed by Whittle in 1939 to power the jet aircraft Gloster E.28/39 of 1941 (Lloyd 1945).

Pabst von Ohain and Whittle are recognized as the co-inventors of the jet engine. Each did his development work separately and, although they definitely knew nothing of the other's achievements, both engines resemble each other not only in the general concept but also in many details. Both were awarded engineering's highest honor, the American "Charles Stark Draper Prize", in 1992 by the National Academy of Engineering for their contribution to aviation and humanity.

A speed increase of the aircraft could only be made possible by changing over completely to axial engine technology. The considerably smaller outer diameter permitted mounting two (instead of one) engines below the wings. That was achieved by the world' first production jet aircraft, the Me 262, designed by Willi Messerschmitt, of which a total number of 1,400 were manufactured. Its legendary jet engine, the Jumo 004 B, conceived by Anselm Franz of the Junkers Motorenwerke, had eight axial compressor stages and one axial turbine and developed a thrust of $2 \times 8,800$ N, pushing the speed of the 10.6 m long and 12.6 m wide fighter plane up to 870 km/h. The first flight was in July 1942.

Although the Germans led the world in jet-propelled airplanes, the United States was astonishingly slow to enter this new field of technology (Smithsonian Institution 2000). However, right after the British success with the Gloster E.28/39, the efforts were accelerated to design a jet aircraft. In September 1941 the Bell Aircraft Corporation was selected to build the fighter, and the General Electric Company was awarded the contract to duplicate the Whittle engine W.2B as a twin-engine configuration under the name of I-A, with Whittle's personal help, in 1942. In October 1942 the first flight of the Bell XP-59A Airacomet (length 12 m, wing span 14.9 m) took place from the Muroc Dry Lake in California reaching a speed of 628 km/h with a thrust of $2 \times 5,560$ N. The engines were mounted to the sides of the airframe. Although the plane did not see combat, it gave the US the necessary experience to pave the way to more advanced designs.

1.1.5 Axial Compressor

It is astonishing that all other fluid-flow machines had left behind their initial development for decades before the axial compressor began to be applied for high and very high volume flows and moderate pressure ratios. The idea, though, goes back to 1897 when Sir Charles Parsons in Britain had already conceived a multistage axial compressor; and also Auguste Rateau in France built one around the turn of the century. However, these first machines were inferior to centrifugal designs. In the 1920s no text book and other literature were found to refer to axial compressors. It was not until the great advances in airfoil aerodynamics were made that the advantages of the axial design became obvious: much lower weight at a given volume flow and pressure ratio and higher efficiency than the centrifugal compressor for flows > 30 m^3/s. In the 1930s Brown Boveri & Cie of Switzerland did pioneering work in axial technology, opening up the typical applications that are still valid today for the axial compressor.

1.1.6 Steam Turbine

There were two co-inventors who independently of each other developed the steam turbine: Carl Gustaf de Laval in Sweden and Sir Charles A. Parsons in England. Both presented their prototypes in the same year (!) to the public, namely in 1884.

De Laval's turbine was an impulse type design with four free steam jets directed onto the blades of a single-stage wheel (Fig. 1.4). This type is still used in modern steam turbines as a nozzle-group regulated control stage with variable, but partial admission. In 1896 to 1898 the American Charles G. Curtis equipped the de Laval turbine disk with one and two more blade rows separated by stationary deflection vanes ("Curtis turbine").

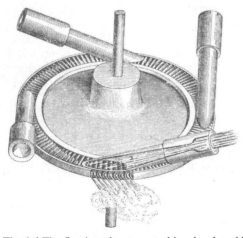

Fig. 1.4 The first impulse steam turbine developed by de Laval in 1884 (Stodola 1924).

Completely different in principle was Parson's turbine, which consisted of a double-flow 15-stage reaction drum type unit with steam admission in the middle of the two halves. The power output was 4 kW. The Parsons turbine evolved into one of the most used types in power plants and ship propulsion.

1.1.7 Stationary Gas Turbine

The first experimental gas turbine was that of R. Armengaud and C. Lemale of France in 1905, consisting of a two-stage Curtis wheel. Mainly due to the difficulty of designing a satisfactorily operating centrifugal compressor, the thermal efficiency of this first gas turbine was a meager 2 to 3%; in other words, the compressor consumed almost as much energy as the turbine was able to generate. Beginning in 1908, H. Holzwarth of Germany tried to circumvent this problem by using a discontinuous working principle by achieving the pressure rise in a deflagrating chamber that was alternatively closed on all sides, thus operating with alternating pressure (constant volume combustion). The first Holzwarth production turbines operated in the 1910s with a power output of up to 20,000 kW and a thermal efficiency of 13%. The first generator-drive gas turbine with constant pressure combustion[2] was manufactured by Brown Boveri & Cie of Switzerland in 1939, generating 4,000 kW at an efficiency of 18% without recuperator. By 1952 180 stationary gas turbines were in operation worldwide and by 1957 there were already 3,614 units with power outputs of between 10 kW and 30,000 kW.

1.1.8 Centrifugal Pump

According to Engeda (1998) the commercial production of the oldest turbomachine, the centrifugal pump, was started in the United States at the Massachusetts Pump Factory as early as 1818 based on the principles of French inventor Papin of 1689 (!). W.H. Andrews invented the volute casing and the shrouded impeller in 1839 and 1846, respectively. During England's Great Exhibition at the Crystal Palace, John Appold presented his pump having curved blades achieving an efficiency three times as high as with straight blades. In 1887 the first pumps were built with vaned diffusers, patented by the British physicist Osborne Reynolds in 1875 (better known for the "Reynolds number"). After 1893 centrifugal pumps were manufactured by Mather & Platt and Sulzer Brothers on a large scale.

1.1.9 Rocket Engine

On October 3, 1942 at 3:58 pm a man-made object, the liquid fuel rocket A4[3] reached for the first time a height of 85 km. It was launched from Testsite VII at

---

[2] This is according to the patent of Stolze of 1899, who conceived the right turbine but built the wrong compressor in Berlin-Weissensee/Germany in 1902.

[3] "A" for aggregate; later the A4 was dubbed "V2" by state propaganda; "V" for Vergeltung = revenge, retaliation.

Peenemünde on the island of Usedom in North Germany. After more than twelve long years of research, development, design, testing and production with more setbacks than successes this was a tremendous triumph for the innovative scientist Wernher von Braun and the enthusiastic engineer and chief of the Peenemünde Army Research Institute, Walter Dornberger.[4] Their dreams of eventually conquering outer space seemed to come true; dreams that they had pursued since Hermann Oberth, the German rocket pioneer, had published his book "Rocket to Interplanetary Space" in 1923[5]. They had a dream, but they awoke in the morning and found that they were working on one of the deadliest mass destruction weapons of World War II. As a matter of fact, the Rubicon had already been crossed in the mid-1930s, and long before 1942 they had been corrupted by one of the worst dictatorships that this earth has ever seen.

The technological success of the A4 was made possible by a turbomachine group, made by KSB, Germany (Fig. 1.5). It consisted of a two-blade-row Curtis turbine moving a centrifugal pump from each shaft end: one for 72 kg/s of liquid oxygen at 23 bar and the other for 58 kg/s of the 75% ethyl alcohol/25% water mixture at 18 bar. These constituted the two fuels that were injected into the burn chamber, where at 15 bar the temperature was brought to 2,500 °C. The 420 kW/3,800 rpm turbine was powered by an auxiliary fuel, a mixture of hydrogen peroxide, sodium permanganate and water that was expelled from a 130 liter tank by compressed nitrogen contained in bottles. The vapor at the turbine inlet had a pressure of 32 bar at 385 °C. This system generated 245,000 N thrust at the start, which increased to 712,000 N upon reaching the maximum speed of Mach 5.

Fig. 1.5 Cutaway view of original turbine/centrifugal pump unit of A4 rocket. From *left*: alcohol centrifugal pump, 2-stage turbine, liquid oxygen centrifugal pump. *Left* of turbine wheel: nozzle ring with circular inlet manifold; all three turbomachines have an exit volute casing (machine group on display in Peenemünde 2003, photo Lüdtke).

---

[4] Von Braun flew his first rockets on the "rocket airstrip" Berlin-Reinickendorf, a mere stone's throw away from the author's long-term company Borsig, where von Braun was a student trainee in 1931.

[5] The poetic German title was: "Die Rakete zu den Planetenräumen".

A total of 5,975 V2s were produced within 16 months; and about 4,300 were fired in all, leaving 35,000 dead, including slave laborers in Germany. In spite of all this the allies started secret missions right after the war to get hold of German rocket scientists, engineers, engineering archives and, of course, complete rockets and parts thereof in order to gain a lead over the next enemy.

The British, in their operation "Backfire", recruited 600 Germans, who fired three V2s in Cuxhaven, Germany in October 1945. The Soviets employed 5,000 Germans putting together twelve V2s in Bleicherode, Germany and sending them skyward in Kapustin Jar/Soviet Union (1st shot in October 1947), after having transferred hundreds of German specialists to the USSR by their organization "OSOAVIACHIM" (cooperative society in aviation and chemistry). The Sputnik was shot into orbit with a Wostok rocket based on V2 technology. The Americans, naming their operation "Overcast/Paperclip", started their first German rocket in April 1946 from White Sands, New Mexico, followed by 70 more captured V2s. And 118 transferred German scientists paved the way to the "Redstone", which is a direct V2 outgrowth, to the "Jupiter C" and to the "Saturn 5".

So did their dreams of sending rockets to the planets finally come true? A haunting question, however, remains: was it right or wrong that the WW II allies have utilized that bloodstained weapon as a progenitor for their future rocket developments? It is an exciting question as to what answer responsible historians would come up with.

This story tells us that almost every invention that *Homo sapiens* has made, ever since this innovative but aggressive novel species began roaming the earth 100,000 years ago, has been either good or bad for humanity. And of course turbomachinery has not been exempt from such development.

1.2 Processes

Chapter 1 covers in compact form the basics of process centrifugal compressors, i.e. the definitions, fundamental principles and reference processes of compression, including an overview of processes in gas field operations, hydrocarbon and chemical processing where centrifugal compressors are used as workhorses.

Process centrifugal compressors constitute only one out of the nine turbomachine species; but they are certainly the workhorses operating in many of the processes, shown in Fig. 1.6, which can be classified as follows:

- gas field operations,
- hydrocarbon processing (natural gas processing, petroleum refining, petrochemical processing), and
- chemical processing.

The end products, shown in Fig. 1.6, derive from petrochemical as well as from chemical processing. So approximately 15% of detergents are of petrochemical and 85% of chemical origin; for inorganic chemicals the percentages are 30% : 70%, for paints 50% : 50%, for fertilizers 85% : 15% and for plastics 100% : 0%.

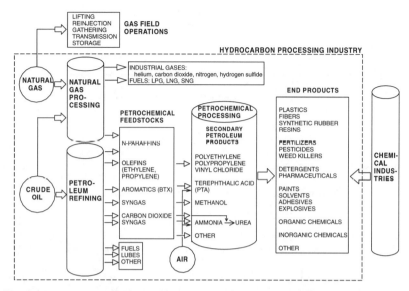

Fig. 1.6 Processes: gas field operations, hydrocarbon and chemical processing.

The compressors within these processes serve the following purposes:

- *Category I*: increasing (or reducing) gas flow pressure levels required for processing,
- *Category II*: providing pressure differences to overcome system resistances, thus enabling gas flows through reactors, heat exchangers and pipes, and
- *Category III*: refrigerating gas flows for cooling and liquefaction.

The following is a short journey through the world of process compressors along the path of rising molar mass.

As demonstrated in Tables 1.1 and 1.2, process compressors embrace a broad application range, starting with the lightest industrial gas having a molar mass of 2.2–3 in *gasoline production*, characterized by pressure ratios of as low as 1.2 for a six-stage machine but discharge pressures of as high as 200 bar.

Next on the list, as part of petrochemical processing, is the *ammonia synthesis gas* compressor handling a gas of 75% H_2 and 25% N_2, i.e. a molar mass of 8.5, which enables ammonia synthesis at pressures of up to 200 bar.

The *methanol synthesis gas* compressor, as part of petrochemical processing, handles a hydrogen-rich gas as well, with a molar mass of 11, and achieves a pressure of 85 bar.

A compressor for *ammonia refrigeration* with a molar mass of 17 is a component within the ammonia synthesis process and reaches an evaporation temperature (at the suction side) of around −30°C and lower.

An interesting application is *gas lifting*, with the object of injecting the associated gas ($CH_4 + C_nH_{2n+2}$, molar mass 18 to 26) into the oil column of the well at the deepest point in order to increase the net oil production rate by reducing the fluid density. Pressures are generally between 60 and 220 bar.

10 1 Basic Compressor Aero-Thermodynamics

Table 1.1 Some application examples of single-shaft process compressors

No	Process		Product	Gas handled by compressor		
				Name	Constituents	Molar mass
1	Petroleum refining	hydrotreating cat reforming isomerization hydrocracking ...	fuels (gasoline)	hydrogen-rich gas	70–93% H_2 +C_nH_{2n+2}	2.2–9
2	Petro-chemical processing	ammonia syngas	ammonia	syngas	75% H_2 +25% N_2	8.5
3		methanol syngas	methanol	syngas	69% H_2 +21% CO+CO_2	11.3
4		NH_3 refrigeration	ammonia	ammonia	NH_3	17
5	Gas field operations	gas lift	oil production increase	natural gas +...	60–90% CH_4 +C_nH_{2n+2}	18–26
6		gas reinjection		natural gas + ...	60–90% CH_4 +C_nH_{2n+2}	18–26
7	Petro-chemical processing	ammonia syngas	ammonia	air	N_2+O_2	28.8
8		ethylene	ethylene	ethylene	C_2H_4	30
9		ethylene	ethylene	cracked gas	C_nH_m	28
10		ethylene	ethylene	propylene	C_3H_6	42
11		urea	urea	carbon dioxide + ...	93–100% CO_2	42–44
12	Chemical industries	chlorine	chlorine	chlorine + ...	97–100% Cl_2 + air	68–71

Table 1.2 Process-related compressor parameters (No. corresponding to Table 1.1)

No.	Pressure ratio	Discharge pressure [bar]	Suction temp. [°C]	Compressor, number of				Max. polytr. stg. head [kJ/kg]	Category
				stages	sections	intercoolers	casings		
1	1.05–1.5	10–200	30–50	2–6	1	0	1	30–50	II
2		150–180	40	16	3	2	2		I
3	3–6	50–83	40–50	7–15	2–3	0–2	1–3	50	I
4	5–18	15–18	−5 to −40	7–12	2–3	0–1	1–2	54	III
5	20–40	60–220	30–60	13–17	2–4	1–3	2	45	I
6	60–120	300–600	30–60	18–26	4–6	3–5	2–4	35	I
7	35–45	35–45	30–50	10	3–4	2–3	2	55	I
8	30	12	−101	8	2	0	1	40	III
9	18–20	30	30	10	4–5	3–4	2	40	I
10	12–15	17	−43	6–7	2	0	1	26	III
11	155–200	100–200	40–50	12–18	4	3	2–3	30	I
12	3.4–13.2	5–13	−32 to +45	3–8	3–4	2–3	1	27	I

Reinjection of petroleum gas into the oil field ground requires still higher discharge pressures in the range of 300 to 600 bar, depending on the conditions of the particular oil field. These pressures are achievable with compressor trains consisting of up to four casings arranged in series.

A two-casing *air compressor* also forms part of the ammonia synthesis plant; the pressure ratio is around 35. Air compressors in terephthalic acid (PTA) plants have a pressure ratio of 20 to 25.

The perhaps most important petrochemical process is ethylene requiring a two-casing *cracked gas compressor* with four to five sections generating a discharge pressure of approximately 30 bar with a mixture of various hydrocarbons and hydrogen with a molar mass of around 25 to 30.

The *propylene refrigeration* compressor achieves a suction temperature of $-43°C$ and the *ethylene refrigeration* compressor achieves $-101°C$.

One of the great challenges of centrifugal compressor technology is the carbon dioxide compressor needed for *urea production*. The pressure ratio of as high as 200 bar is achieved in two to three casings arranged in series with various intercoolers.

These are just a few examples and there are many more where compressing, transport and cooling of gases are required, e.g. in *chlorine* (molar mass 69) and *hydrogen chloride* (molar mass 36.5) plants or in other industrial refrigeration branches using the newly introduced refrigerant R134a (molar mass 102) substituting the environmentally damaging refrigerant R12.

The total rating range of process centrifugal compressors is shown in Fig. 1.7.

All of these process compressors have a number of criteria in common:

- They require an unlimited design variability to cope with the wide ranges of industrial gases, pressures and temperatures, resulting in up to 26 stages and in up to four casings per compressor.

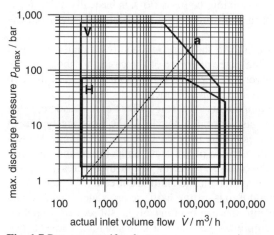

Fig. 1.7 Process centrifugal compressors: approximate rating ranges. V vertically split casing; H horizontally split casing; line a p_{dmax} for a suction pressure of $p_s = 1$ bar.

- In many applications, especially in the oil and gas industry, existing compressors have to cope with variations of the gas constituents, which is not an easy task for turbocompressors, since, very often, specified discharge pressures cannot even be maintained if the molar mass is reduced by as little as 5%.
- For most multistage compressors the variation of flow, suction temperature, molar mass, or speed causes an internal aerodynamic stage mismatching, which can reduce head, efficiency and operating range substantially.
- Almost all of the gases mentioned show an explicit real-gas behavior with compressibility factors as much as 20% lower than unity, which means that the volume flow from the first to the last compressor stage reduces even more than by increasing pressure alone, with the adverse effect of a reduced efficiency. Another real-gas influence is the split-up of the heat capacity ratio c_p/c_v into an isentropic volume exponent and an isentropic temperature exponent, whereby compression work and temperature increase follow different exponential laws.

1.3 Definitions and Terminology

1.3.1 Compressor

Centrifugal compressors are fluid-flow machines for the compression of gases according to the dynamic principle. The bladed impeller with its continual through-flow transfers the mechanical shaft energy into enthalpy, i.e. gas energy. Thus, the pressure, temperature and velocity of the gas leaving the impeller are higher than at the impeller inlet. The annular diffuser downstream of the impeller delays the gas velocity, thus providing a further pressure and temperature increase.

Fluid-flow machines are categorized in axial and radial designs by means of the main direction of flow in the meridional plane, i.e. a plane containing the axis of rotation. This meridional flow, determining the mass flow, is basically axial in axial machines and basically radial in centrifugal machines with the flow from the inside to the outward direction.

The radially outward flow direction in the impeller requires radially arranged annular diffusers which increase the outer diameter of the casing to about double the impeller diameter. The crossover duct directs the flow radially inward through the deswirling vanes before it is led into the next impeller.

The aerodynamically active components of a single-shaft compressor are displayed in Fig. 1.8. Suction of the gas is through nozzle A; the flow accelerates in plenum inlet B by two to three times; in the impeller C transfer of the mechanical energy to the gas takes place with increase of pressure, temperature and velocity; flow deceleration takes place in the diffuser, resulting in a further increase of pressure and temperature. A further impeller on the shaft requires a 180° change of flow direction in the crossover bend E, deswirling in the return vane channel F and a 90° turn to achieve axial inlet into the eye opening of the subsequent impeller. As illustrated in Fig. 1.8, the compressor can have a side-stream J between two stages with an additional suction nozzle H. If the gas is to be led out of the casing, either to be intercooled or at the compressor rear end, the diffuser D is followed by a volute K, which is a flow duct whose area increases along the circumference in the direction of rotation, serving as a gas-collecting chamber. The volute is followed by a conical exit diffuser L, where the gas is subjected to a further pressure increase. The final component is the discharge nozzle M.

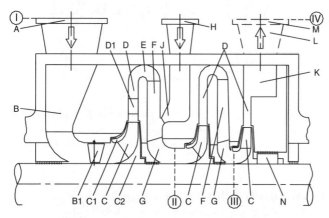

Fig. 1.8 Schematic of compressor aero components. *Inlet stage*: plane I to II; *middle stage*: plane II to III; *final stage*: plane III to IV. A suction nozzle, B plenum inlet, B1 adjustable inlet guide vanes (IGV), C1 cover disk (shroud), C impeller, C2 hub disk, D vaneless diffuser, D1 vaned diffuser, E return bend, F return vane channel, G return inlet, H side-stream nozzle, J side stream, K volute (scroll), L conical exit diffuser, M discharge nozzle, N balance piston.

1.3.2 Impeller, Stage, Section

The inlet compressor stage comprises: suction nozzle, plenum inlet, impeller, annular diffuser, crossover bend and return vane channel.

The middle stage comprises: return inlet, impeller, annular diffuser, crossover bend and return vane channel.

The final stage comprises: return inlet, impeller, annular diffuser, volute, conical diffuser and discharge nozzle.

The compressor section comprises all the compressor stages between the two nozzles, being characterized by the same mass flow and the same molar mass for all stages within this section; it can consist of one stage or up to ten stages. This section is also called the "stage group" or "process stage". The term "stage" instead of "section", often utilized by end users and contractors, can lead to misunderstandings and should therefore be avoided. In this terminology the compressor in Fig. 1.8 has three stages in two sections: one stage in section # 1 and two stages in section # 2. The single-flow compressor in Fig. 1.9a has three stages in section # 1 and four stages in section # 2, whereas the double-flow compressor in Fig. 1.9b, with a mirror-image group of stages, has three stages, albeit six impellers.

14   1 Basic Compressor Aero-Thermodynamics

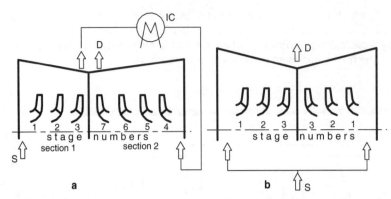

Fig. 1.9a, b Definitions: impellers, stages, sections. **a** single-flow compressor (sections in series): 7 impellers, 7 stages, 2 sections, 4 nozzles; **b** double-flow compressor (groups in parallel): 6 impellers, 3 stages, 1 section, 3 nozzles, S suction, D discharge, IC intercooler.

1.4 Method of Presenting Equations

The presented equations are of three types:

1. "Physical equations" clearly reveal the physical relationships, which they describe and are valid regardless of the system of units employed. In a physical equation the symbols represent physical magnitudes, which are fundamentally the product of the numerical value and the unit:

   physical magnitude = numerical value × unit.

   If one chooses a unit that is n times greater, the numerical value becomes one nth, i.e. the physical value is characterized by being invariant with regard to a change of the unit. When applying an equation, it is always advisable to replace the symbol in question by the number and its corresponding unit.

2. "Numerical equations" normally describe empirical relationships based on experiments rather than on theory. However, quite a number of numerical equations in the daily life of an engineer are just based on practical conventions with regard to the usage of customary units in different countries. This type requires employing predestined units only.

3. "Unit equations" numerically relate different units to each other, and thus serve to convert numerical values to arbitrary units. For example: 1 in = 25.4 mm; such an equation is made dimensionless having the value of 1.0:

$$1 = \left\{ \frac{1 \text{in}}{25.4 \text{mm}} \right\}$$

These so-called "unity braces" constitute a ratio of different units with a value of 1.0 and serve as handy multipliers in equations to bring about any desired end unit without changing the physical magnitude and validity of the equation. This method of presenting equations was first suggested by Baehr (1962) and is also described by Gieck (1989).

Example: Peripheral speed of a 500 mm disk at 8,000 rpm (= 1/min):

1. Physical equation:

   $u = \pi N d$.

   With the SI-units "1/s" for speed and "m" for length the peripheral speed is

   $$u = \pi \times 133.3 \, \frac{1}{s} \times 0.5 \, m = 209.4 \, \frac{m}{s}$$

2. Unit equation: if one is more accustomed to "1/min" for speed and "mm" for length, the pertaining unit equations as multipliers for the equation are

   $$1 = \left\{ \frac{1 \, min}{60 \, s} \right\} \quad \text{and} \quad 1 = \left\{ \frac{1 \, m}{1{,}000 \, mm} \right\}.$$

3. Numerical equation:

   $$u = \pi \times 8{,}000 \, \frac{1}{min} \left\{ \frac{1 \, min}{60 \, s} \right\} \times 500 \, mm \left\{ \frac{1 \, m}{1{,}000 \, mm} \right\} = 209.4 \, \frac{m}{s},$$

   since "min" and "mm" cancel each other out. So under these assumptions the final numerical equation becomes

   $$u = \frac{\pi N d}{6 \times 10^4} \, \frac{m}{s},$$

   with the mandatory units "1/min" for speed N and "mm" for diameter d.

4. Alternate unit equation: "1/min" for speed, "in" for length and "ft/s" for peripheral speed are to be used (for readers who have grown up with the US customary system of units). Hence,

   $$1 = \left\{ \frac{1 \, ft}{12 \, in} \right\}.$$

5. Alternate numerical equation:

   $$u = \pi \times 8{,}000 \, \frac{1}{min} \left\{ \frac{1 \, min}{60 \, s} \right\} \times 19.68 \, in \left\{ \frac{1 \, ft}{12 \, in} \right\} = 687 \, \frac{ft}{s},$$

   since "min" and "in" are cancelled out.

6. Under these assumptions the final numerical equation becomes

$$u = \frac{\pi N d}{720} \frac{\text{ft}}{\text{s}},$$

with the mandatory units "1/min" (= rpm) for speed N and "in" for diameter d.

So the simple rules are:

- Apply the physical equations in this book by replacing the symbols with the number and the usual (arbitrary) unit.
- Multiply any such part of the equations by the appropriate unity brace(s) (see Appendix).
- Cancel out units if equal in numerator and denominator much in the same way as numbers in an ordinary fraction.
- Use as many unity braces as necessary to bring about a result in the desired unit.

1.5 The First Law of Thermodynamics

The general form of the first law of thermodynamics is

$$P + \dot{Q} = \dot{m}\left[h_2 - h_1 + \frac{(c_2^2 - c_1^2)}{2} + g(z_2 - z_1)\right], \tag{1.1}$$

or in specific, i.e. mass-referenced terms and in differential form:

$$\mathrm{d}y + \mathrm{d}q = \mathrm{d}h + \mathrm{d}\left(\frac{c^2}{2}\right) + g\,\mathrm{d}z, \tag{1.2}$$

i.e. the work plus the heat transfer into a thermodynamic system increase its enthalpy, kinetic and potential energy from inlet to exit.

In Fig. 1.10 the thermodynamic system "compressor" is shown with the energies and mass flows crossing the system boundary.

The energy added to the system comprises:

- P mechanical power supplied via the drive shaft
- $\dot{m}h_{\text{stat\_s}}$ static enthalpy at the suction nozzle
- $\dot{m}c_s^2/2$ kinetic energy at the suction nozzle
- $\dot{m}g z_s$ potential energy at the suction nozzle

The energy leaving the system comprises:

- $\dot{Q}$ heat flow from the hot body to the environment
- $\dot{m}h_{\text{stat\_d}}$ static enthalpy at the discharge nozzle
- $\dot{m}c_d^2/2$ kinetic energy at the discharge nozzle
- $\dot{m}g z_d$ potential energy at the discharge nozzle

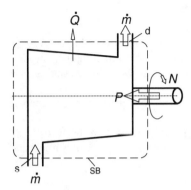

Fig. 1.10 The thermodynamic system "centrifugal compressor". s suction, d discharge, SB system boundary, $\dot{m}$ mass flow, $\dot{Q}$ heat flow, P mechanical power input, N rotation.

The conservation of energy in this open thermodynamic system leads to the first law of thermodynamics applied to turbocompressors (note: in contrast to general thermodynamics the heat flow from the hot compressor body to the environment is defined as positive):

$$P + \dot{m} h_{\text{stat\_s}} + \dot{m} \frac{c_s^2}{2} + \dot{m} g\, z_s = \dot{Q} + \dot{m} h_{\text{stat\_d}} + \dot{m} \frac{c_d^2}{2} + \dot{m} g\, z_d . \qquad (1.3)$$

The potential energy difference between the inlet and the exit is regarded as negligible for gases and the static enthalpy is combined with the kinetic energy into the total energy:

$$h = h_{\text{stat}} + \frac{c^2}{2} \qquad (1.4)$$

Since the heat loss $\dot{Q}$ of a compressor section due to convection and radiation amounts to only approximately 1% of the energy input, it is justified to regard the compressor as adiabatically enclosed. So with these simplifications the first law of thermodynamics for centrifugal compressors becomes very simple:

$$P = \dot{m}(h_d - h_s) = \dot{m}\Delta h , \qquad (1.5)$$

i.e. the specific work input equals the specific total enthalpy difference of the gas flow between the compressor inlet and exit. The term "specific" denotes "mass referenced".

1.6 Reference Compression Processes

In order to define the quality of a compression process, ideal (i.e. lossless) reference processes are defined. The specific work of the selected reference process serves as the criterion of assessment for the real specific work input, stating how far the real process is successful in avoiding losses.

For turbocompressors there are three common no-loss processes, as listed in Table 1.3.

All of the processes mentioned can be graphically presented in the form of areas in the temperature-entropy diagram. Before doing so, we recall some basic principles from thermodynamics.

Table 1.3 Reference processes

	Change of state	Thermodynamic system	Quantity remaining constant during change of state
Reference process	(1) polytropic	irreversible, adiabatic	polytropic efficiency
	(2) isentropic	reversible, adiabatic	entropy
	(3) isothermal	reversible, diathermic	temperature
Actual process	actual	irreversible, adiabatic	–

1.6.1 Heat in the Temperature–Entropy Diagram

According to the second law of thermodynamics the entropy change of a thermodynamic system is proportional to the reversibly transmitted heat across the boundary of the system (i.e. no dissipation caused heat is involved), with the absolute temperature as the proportionality factor:

$$dQ_{rev} = T dS, \qquad (1.6)$$

or in mass referenced terms (= specific) and integrated:

$$dq_{rev} = T ds, \qquad (q_{12})_{rev} = \int_1^2 T ds, \qquad (1.7)$$

and in compressor technology as per convention:

$$(q_{12})_{rev} = -\int_1^2 T ds. \qquad (1.8)$$

In other words, the area under any line of a reversible change of state in the T, s-diagram represents the reversible heat transfer to or from the system regardless of any power input (Fig. 1.11). Reversibly adding heat to the system ($Q < 0$) increases the entropy, and reversibly removing heat from the system ($Q > 0$) reduces the entropy.

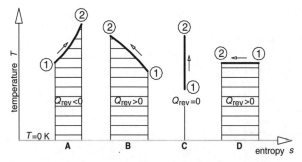

Fig. 1.11 T, s-diagram: reversible changes of state. Heat transfer areas are as follows. **A** $Q_{rev} < 0$, heat added; **B** $Q_{rev} > 0$, heat removed; **C** $Q_{rev} = 0$, adiabatic; **D** $Q_{rev} > 0$, heat removed.

1.6.2 Enthalpy Difference in the Temperature–Entropy Diagram

From Eq. (1.3) in mass-referenced and differential form we obtain

$$dq = dy - dh. \tag{1.9}$$

On the other hand, the specific compressor work input is

$$dy = vdp, \tag{1.10}$$

which is one of the fundamental equations of thermodynamics.

From Eqs. (1.8) and (1.9) we obtain

$$-Tds = vdp - dh. \tag{1.11}$$

For an isobar (p = constant; $dp = 0$) the enthalpy difference is

$$h_2 - h_1 = \int_1^2 T\,ds. \tag{1.12}$$

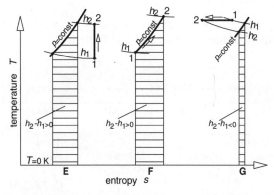

Fig. 1.12 T, s-diagram: arbitrary changes of state. The areas constitute enthalpy differences between points 2 and 1. **E, F** positive enthalpy differences, **G** negative enthalpy difference.

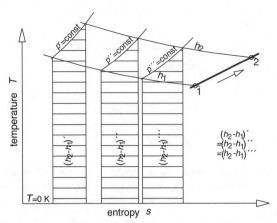

Fig. 1.13 T, s-diagram: identical areas of enthalpy difference.

This means the enthalpy difference of two arbitrary points of state 1 and 2 in the T, s-diagram becomes visible as an area under an arbitrary isobar section limited by the two lines of constant enthalpy (i.e. isenthalpic lines) through the points of state 1 and 2; see Fig. 1.12. The rule is: draw an isenthalpic line through point 1 until it intersects with an arbitrary isobar; then draw an isenthalpic line through point 2 and have it intersect with the same isobar. The area below this isobar cut-off down to $T = 0$ K represents the enthalpy change $h_2 - h_1$. Since any isobar can be selected, all areas shown in Fig. 1.13 are identical. Note that for real gas behavior the isenthalpic lines are inclined against the lines of constant temperature (i.e. isothermal lines), with increasing slope at higher pressures, whereas for ideal gas behavior both assume the more "familiar" form of horizontal lines in the T, s-diagram.

1.6.3 Work in Reference Processes

With these basics the various reference works and the actual work can be displayed in the form of areas in the T, s-diagram (Fig. 1.14 and Table 1.4).

Table 1.4 Graphical display of compressor work in the form of areas

Perfect gas,	Fig.	1.14a	1.14b	1.14c	1.14d
Real gas,	Fig.	1.14e	1.14f	1.14g	1.14h
		Reference process			*Actual*
		Isentropic	Polytropic	Isothermal	*process*
Perf. gas reference work		A–2T–2s–1–B	C–2T–2–1–D	E–2T–1–F	
Perfect gas actual work					G–2T–2–H
Real gas reference work		J–2h–2s–1–K	L–2h–2–1–M	N–2T–1–1p–O	
Real gas actual work					P–2h–2–Q

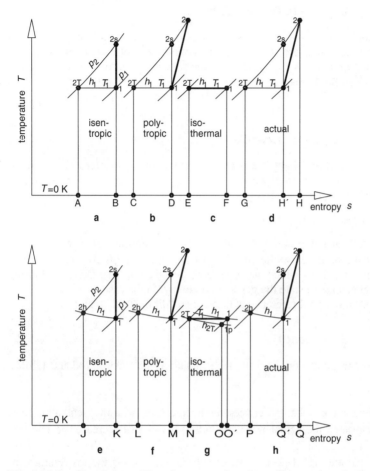

Fig. 1.14 *T, s*-diagram: specific reference work and specific actual work. *Top*: perfect gas behavior; *bottom*: real gas behavior.

The isentropic and polytropic reference processes are adiabatic since no heat crosses the system boundary. The actual process is adiabatic as well because the heat is irreversibly generated within the system through friction and does not cross the system boundary either.

Isentropic Work (Head)

The isentropic work (i.e. isentropic "head") 1–2s consists merely of the enthalpy difference $h_{2s} - h_1$. In Figs. 1.14a and 1.14e, the isenthalpics through points 1 and 2s intersect with the isobar p_2 at 2T and 2s (2h and 2s respectively). Therefore, the described area represents the isentropic work of reference:

$$y_s = h_{2s} - h_1.\qquad(1.13)$$

Isothermal Work (Head), Perfect Gas

The isothermal reference process is diathermic since heat crosses the system boundary. In Fig. 1.14c, since perfect gas behavior does not involve any enthalpy change, the compressor work input equals the reversibly removed heat, which in turn equals $-Tds$. So the rectangle represents the isothermal work:

$$y_T = Q. \tag{1.14}$$

Isothermal Work (Head), Real Gas

Real gas behavior means that there is an isothermal enthalpy difference according to Eq. (1.9). So the isothermal work is the reversible heat removed plus the isothermal enthalpy change (Fig. 1.14g):

$$y_T = Q + (\Delta h)_T, \quad \text{with} \quad (\Delta h)_T < 0. \tag{1.15}$$

The reversible heat is the area below the line 1–2T and the isothermal enthalpy difference is the area below the line 1–1p (the isobar cutoff by the enthalpics through points 1p and 1), which has to be deducted since it is negative, i.e. during the isothermal change of state the gas loses enthalpy. In the case of real gas behavior the isothermal work of reference is therefore a "diminished" rectangle.

Actual Work (Euler Head)

The actual compression work is irreversible and adiabatic (Figs. 1.14d and 1.14h):

$$\Delta h = h_2 - h_1. \tag{1.16}$$

So the area below the isobar p_2 between the intersections with the isenthalpic lines h_2 and h_1 represents the actual work, including losses that increase the entropy. Since the area G–2T–2–1–H' (respectively P–2h–2–1–Q') is the lossless polytropic work, the area H'–1–2–H (respectively Q'–1–2–Q) is the irreversible energy dissipation through internal friction, turbulence, flow separation, etc. (=loss). The area 1–2s–2–1 is no dissipative loss but the preheat, i.e., the heating expenditure or additional energy consumption due to the preheat of each consecutive compression step by the preceding step, which really constitutes the difference between isentropic and polytropic compression.

Polytropic Work (Head)

The polytropic work of reference is explained in Fig. 1.15. The change of state between suction condition 1 and discharge condition 4 in the T, s-diagram is broken down into three compression steps: 1–2, 2–3, and 3–4. The isentropic discharge conditions of steps 1, 2 and 3 are 2s, 3s and 3″, respectively. In accordance with the above, Table 1.5 describes the relationships.

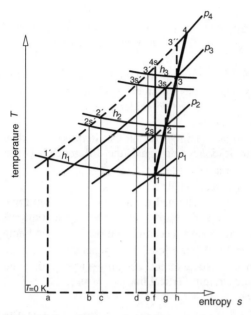

Fig. 1.15 T, s-diagram: clarification of the specific polytropic work of a three-stage compressor. Isentropic compressor work: a–1′–4s–f; sum of work in isentropic stages: a–1′–3″–3–3s–2–2s–f; polytropic compressor work: a–1′–4–1–f; actual change of state: 1–2–3–4.

Table 1.5 Explanation of polytropic work

	Specific isentropic work (head)	Area in Fig. 1.15
Compression step 1	y_{s1}	a–1′–2s′–b
Compression step 2	y_{s2}	c–2′–3s′–d
Compression step 3	y_{s3}	e–3′–3″–h
Compression steps 1+2+3	$\Sigma y_{si} = y_{s1} + y_{s2} + y_{s3}$	a–1′–3″–3–3s–2–2s–f *
Compressor	y_s	a–1′–4s–f
Heating expenditure, 3 steps	$e_3 = \sum_{i=1}^{3} y_{si} - y_s$	2s–4s–3″–3–3s–2
Heating expenditure, infinite number of compression steps	$e_\infty = \sum_{i=1}^{\infty} y_{si} - y_s = y_p - y_s$	1–4s–4–1

* Since area f–2s–2–g is identical with b–2s′–2–c, and g–3s–3–h is identical with d–3s′–3′–e.

This implies the following:

- The sum of the work in the isentropic stages (the area engulfed by the dashed lines) is greater than the isentropic compressor work by the heating expenditure.
- By increasing the number of stages for a given pressure ratio p_4/p_1 the staircase curve 1–2s–2–3s–3–3'' gradually approaches the line 1–2–3–4, being the line of constant polytropic efficency η_p = const and coincides with it for an infinite number of infinitesimal compression steps.
- The preheat (sometimes referred to as "heating loss" in the literature) has then reached its maximum for a given efficiency and equals the difference between polytropic and isentropic head: the area 1–4s–4–1.
- So the polytropic head equals the area a–1'–4–1–f.
- The specific polytropic work (head) y_p is defined as the sum of the isentropic heads for an infinite number of compression steps y_s along the actual compression path (defined by η_p = const), with each succeeding compression step being separated from the foregoing step by the frictional heat generated h_f. Figure 1.16 displays this type of change of state in the enthalpy–entropy diagram (note: here enthalpy differences are straight vertical lines!):

$$y_p = \lim_{z \to \infty} \sum_{i=1}^{z} y_{si}, \tag{1.17}$$

with z being the number of compression stages.

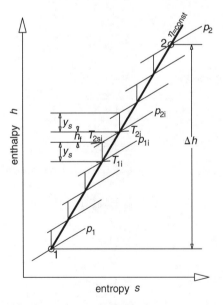

Fig. 1.16 h, s-diagram: explanation of specific polytropic work. y_s infinitesimal isentropic work, h_f irreversible (frictional) heat, 1-2 actual change of state.

- y_p is lossless because all frictional heat h_f is excluded by definition. Therefore it is suitable as a reference head.
- However, as odd as it may sound, y_p is at the same time irreversible, because the change of state in this adiabatically enclosed system cannot return on its own from point 4 to the point of origin 1, e.g. by expansion and work output due to the friction caused entropy increase, which is a priori irreversible.
- The polytropic head must not be confused with the actual work:

$$y_p \neq \Delta h \,. \tag{1.18}$$

- For a given pressure ratio the polytropic head increases when the efficiency reduces due to a rising suction temperature of the rear stages.

1.6.4 Application of Reference Work

Isentropic head: Advantages: clear definition, lossless, reversible, visible in the h, s-diagram, unaffected by the efficiency. Disadvantage: the isentropic head of a multistage compressor is smaller than the sum of stage isentropic heads; this means that in spite of unaltered stage efficiencies the isentropic compressor efficiency decreases when the pressure ratio increases.

Polytropic head: Advantage: polytropic compressor head is equal to the sum of stage polytropic heads and the polytropic efficiency is thermodynamically independent from the pressure ratio in case the compressibility factor stays constant for all compressor stages, i.e. for perfect gas behavior, many low pressure and low Mach number applications (e.g. H_2-rich gases). For real gas behavior the compressibility factor changes from suction to discharge; in most cases it decreases towards the rear compressor stages so that in turn the polytropic heads of these decrease as well; therefore in general the sum of stage heads differs from the compressor head:

$$\sum y_{pi} \neq y_{p\,\text{compressor}} \,. \tag{1.19}$$

In most cases, however, the deviation is only around 1% or less. Disadvantages: the definition of the polytropic head is ambiguous; it is not visible in the enthalpy-entropy diagram and, due to the interstage heat-up, it depends to some degree on the efficiency. In spite of these shortcomings this specific reference work is the most used in the industry.

Isentropic head: only used for multistage intercooled air compressors.

1.6.5 Head/Efficiency, Practical Formulae, Perfect Gas Behavior

The general law governing changes of state is:

$$pv^n = const \,. \tag{1.20}$$

The laws describing perfect gas behavior are:

$$pv = RT \tag{1.21}$$

(the so-called thermal equation of state) and

$$\kappa = \frac{c_p^0}{c_v^0}; \quad c_p^0 - c_v^0 = R; \quad c_p^0 = R\frac{\kappa}{\kappa-1}. \tag{1.22}$$

Derivation of pressure–temperature relationship: from Eq. (1.21) and combining with Eq. (1.20) we obtain

$$v^n = \text{const}\frac{T^n}{p^n}; \quad p\frac{T^n}{p^n} = \text{const}; \quad \frac{p^n}{pT^n} = \text{const}; \quad \frac{p^{n-1}}{T^n} = \text{const}. \tag{1.23}$$

This means that for an arbitrary compression between state 1 and 2 the last quotient in Eq. (1.23) remains constant:

$$\frac{p_2^{\frac{n-1}{n}}}{T_2} = \frac{p_1^{\frac{n-1}{n}}}{T_1}. \tag{1.24}$$

Table 1.6 Reference processes and exponent n

Change of state	Quantity remaining constant	Formula	Exponent	Pressure–temperature relationship, perfect gas behavior
Isentropic	entropy s	$pv^\kappa = \text{const}$	$n = \kappa$	$T_2/T_1 = (p_2/p_1)^{\frac{\kappa-1}{\kappa}}$
Polytropic	efficiency η_p	$pv^n = \text{const}$	$n = n$	$T_2/T_1 = (p_2/p_1)^{\frac{n-1}{n}}$
Isothermal	temperature T	$Pv = \text{const}$	$n = 1$	$T_2/T_1 = 1$
Isobaric	pressure p	$pv^0 = \text{const}$	$n = 0$	temp. and pressure unrelated
Isochoric	volume v	$pv^\infty = \text{const}$	$n = \infty$	$T_2/T_1 = p_2/p_1$

This results in the pressure–temperature relationship:

$$\left(\frac{p_2}{p_1}\right)^{\frac{n-1}{n}} = \frac{T_2}{T_1}, \tag{1.25}$$

with n assuming the values in Table 1.6.

Isentropic Head

$$y_s = h_{2s} - h_1 = c_p(T_{2s} - T_1) = c_pT_1\left(\frac{T_{2s}}{T_1} - 1\right) = c_pT_1\left[\left(\frac{p_2}{p_1}\right)^{\frac{\kappa-1}{\kappa}} - 1\right]. \tag{1.26}$$

Final isentropic head formula for perfect gas behavior:

$$y_s = RT_1 \frac{\kappa}{\kappa-1}\left[\left(\frac{p_2}{p_1}\right)^{\frac{\kappa-1}{\kappa}} - 1\right], \quad \text{with} \quad \kappa = \frac{\kappa_1 + \kappa_{2s}}{2}, \quad (1.27)$$

where κ_{2s} is defined at the isentropic discharge.

Isentropic Efficiency

$$\eta_s = \frac{y_s}{\Delta h} \quad (1.28)$$

Polytropic Head

The polytropic head formula for perfect gas behavior is derived differently. The polytropic change of state, the most important in compressor technology, is in general defined for an *infinitesimal* compression step as the constant ratio of the isentropic and actual enthalpy difference; this constant ratio is called the polytropic efficiency (Fig. 1.17):

$$\frac{v\,\mathrm{d}p}{\mathrm{d}h} = \eta_p = \text{const}. \quad (1.29)$$

The integration between suction state 1 and discharge state 2 yields:

$$\Delta h = \int_{p_1}^{p_2} \frac{v\,\mathrm{d}p}{\eta_p}. \quad (1.30)$$

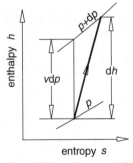

Fig. 1.17 h, s-diagram: infinitesimal compression step.

Normally the efficiency is not precisely constant along the complete path of compression. But it is assumed to be sufficiently constant in order to stay with the polytropic head as a reference work for comparison purposes. Then the equation can be written as

$$\Delta h = \frac{1}{\eta_p} \int_{p_1}^{p_2} v \, dp = \frac{y_p}{\eta_p}. \tag{1.31}$$

So the polytropic head is:

$$y_p = \int_{p_1}^{p_2} v \, dp. \tag{1.32}$$

From Eq. (1.20) we obtain

$$pv^n = p_1 v_1^n \rightarrow v = v_1 p_1^{1/n} \frac{1}{p^{1/n}}. \tag{1.33}$$

From Eq. (1.33) we obtain

$$y_p = v_1 p_1^{1/n} \int_{p_1}^{p_2} \frac{dp}{p^{1/n}} = v_1 p_1^{1/n} \frac{n}{n-1} \left[p_2^{\frac{n-1}{n}} - p_1^{\frac{n-1}{n}} \right]$$

$$= p_1 v_1 \frac{n}{n-1} \left[\left(\frac{p_2}{p_1} \right)^{\frac{n-1}{n}} - 1 \right]. \tag{1.34}$$

The polytropic head formula for perfect gas behavior is

$$y_p = RT_1 \frac{n}{n-1} \left[\left(\frac{p_2}{p_1} \right)^{\frac{n-1}{n}} - 1 \right], \quad \text{with} \quad n = \frac{n_1 + n_2}{2}. \tag{1.35}$$

The polytropic exponent n can be expressed in terms of κ and η_p as follows. The isentropic head for a small finite compression step i in Fig. 1.16 is

$$y_{si} = T_{1i} c_p \left[\left(\frac{p_{2i}}{p_{1i}} \right)^{\frac{\kappa-1}{\kappa}} - 1 \right]. \tag{1.36}$$

Replacing the pressure ratio by the polytropic temperature ratio yields

$$y_{si} = T_{1i} c_p \left[\left(\frac{T_{2i}}{T_{1i}} \right)^{\frac{\kappa-1}{\kappa} \frac{n}{n-1}} - 1 \right]. \tag{1.37}$$

As stated above, the transition to an infinite number of compression steps (i.e. $T_{2i} \to T_{1i}$) has the isentropic convert into the polytropic efficiency.

Differentiating y_{si} with respect to (T_{2i}/T_{1i}), we obtain

$$y'_{si} = \frac{d\, y_{si}}{d\left(\dfrac{T_{2i}}{T_{1i}}\right)} = T_{1i} c_p \frac{\kappa-1}{\kappa} \frac{n}{n-1} \left(\frac{T_{2i}}{T_{1i}}\right)^{\frac{\kappa-1}{\kappa}\frac{n}{n-1}-1}. \tag{1.38}$$

For $T_{2i} \to T_{1i}$:

$$y'_{si} = T_{1i} c_p \frac{\kappa-1}{\kappa} \frac{n}{n-1}. \tag{1.39}$$

The Euler head for the same small finite compression step i in Fig. 1.16 is

$$\Delta h_i = T_{1i} c_p \left(\frac{T_{2i}}{T_{1i}} - 1\right). \tag{1.40}$$

Differentiating Δh_i with respect to (T_{2i}/T_{1i}) we obtain

$$\Delta h' = \frac{d(\Delta h_i)}{d\left(\dfrac{T_{2i}}{T_{1i}}\right)} = T_{1i} c_p. \tag{1.41}$$

Dividing Eq. (1.39) by Eq. (1.41) yields the polytropic efficiency, since both heads were converted for an infinite number of compression steps:

$$\eta_p = \frac{y'_{si}}{\Delta h'} = \frac{T_{1i} c_p \dfrac{\kappa-1}{\kappa} \dfrac{n}{n-1}}{T_{1i} c_p} = \frac{\kappa-1}{\kappa} \frac{n}{n-1}. \tag{1.42}$$

With $\dfrac{n-1}{n} = \dfrac{\kappa-1}{\kappa \eta_p}$, the final polytropic head formula for perfect gas behavior is

$$y_p = RT_1 \frac{\kappa \eta_p}{\kappa-1}\left[\left(\frac{p_2}{p_1}\right)^{\frac{\kappa-1}{\kappa \eta_p}} - 1\right], \quad \text{with} \quad \kappa = \frac{\kappa_1 + \kappa_{2s}}{2}, \tag{1.43}$$

where κ_{2s} is defined at the isentropic discharge.

Polytropic Efficiency

$$\eta_p = \lim_{T_{2i} \to T_{1i}} \eta_s = \frac{y_p}{\Delta h}. \tag{1.44}$$

1.7 The Euler Turbomachinery Equation

This well-known equation interrelates thermodynamics with aerodynamics, describing the aerodynamic processes within the impeller brought about by the increase of total enthalpy. The single compressor stage is hereby regarded as adiabatically isolated from the environment.

The input torque has to effectuate the change of the angular momentum of the compressed mass flow plus leakage and has to additionally overcome the frictional torque of the cover and hub disk:

$$T = (\dot{m} + \Delta\dot{m})(c_{u2}r_2 - c_{u1}r_1) + T_F. \tag{1.45}$$

Introducing leakage efficiency and disk friction efficiency we write

$$\eta_L = \frac{\dot{m}}{\dot{m} + \Delta\dot{m}} \quad ; \quad \eta_F = \frac{T - T_F}{T}, \tag{1.46}$$

with $\Delta\dot{m}$ being the leakage through the cover disk seal, increasing the mass flow through the impeller (leakage and disk friction are frequently omitted by authors of traditional text books).

With gas power = torque × angular velocity, and $r\omega = u$ (blade speed):

$$P_i = T\omega = \dot{m}(c_{u2}u_2 - c_{u1}u_1)\frac{1}{\eta_L}\frac{1}{\eta_F} = \dot{m}\Delta h. \tag{1.47}$$

hence, the specific total enthalpy difference is

$$\Delta h = (c_{u2}u_2 - c_{u1}u_1)\frac{1}{\eta_L}\frac{1}{\eta_F} = (c_{u2}u_2 - c_{u1}u_1) + h_L + h_F, \tag{1.48}$$

which can also be expressed by the change of angular momentum *plus* leakage and disk friction loss. These are in terms of enthalpy:

$$h_L = (1 - \eta_L)\Delta h \quad ; \quad h_F = (1 - \eta_F)\eta_L \Delta h. \tag{1.49}$$

Equation (1.45) is known as the first form of the Euler turbomachinery equation. The polytropic head, using the hydraulic efficiency, is

$$\eta_p = \eta_h \eta_L \eta_F, \tag{1.50}$$

$$y_p = (c_{u2}u_2 - c_{u1}u_1)\frac{\eta_p}{\eta_L \eta_F} = (c_{u2}u_2 - c_{u1}u_1)\eta_h. \tag{1.51}$$

The second form of the Euler turbomachinery equation describes the energy transfer in terms of the static and kinetic enthalpy rise. Applying the cosine rule to the velocity triangles in Fig. 1.18 we obtain

$$w_2^2 = u_2^2 + c_2^2 - 2u_2c_2\cos\alpha_2 = u_2^2 + c_2^2 - 2c_{u2}u_2, \tag{1.52}$$

$$w_1^2 = u_1^2 + c_1^2 - 2u_1c_1 \cos\alpha_1 = u_1^2 + c_1^2 - 2c_{u1}u_1. \tag{1.53}$$

Subtracting from Eq. (1.52) yields

$$w_2^2 - w_1^2 = u_2^2 - u_1^2 + c_2^2 - c_1^2 - 2(c_{u2}u_2 - c_{u1}u_1). \tag{1.54}$$

Hence the second form of the Euler turbomachinery equation is

$$\Delta h = c_{u2}u_2 - c_{u1}u_1 + h_L + h_F$$

$$= \frac{u_2^2 - u_1^2}{2} + \frac{w_1^2 - w_2^2}{2} + \frac{c_2^2 - c_1^2}{2} + h_L + h_F. \tag{1.55}$$

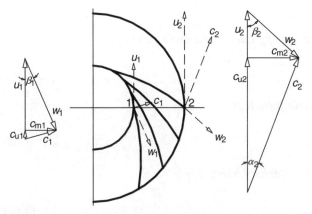

Fig. 1.18 Velocity triangles at the impeller inlet and exit. 1/2 impeller inlet/exit, u peripheral (metal) velocity, c absolute velocity, c_u peripheral component, c_m meridional component, w relative velocity, α absolute gas flow angle, β relative gas flow angle.

The first two terms of the equation constitute the static and the third the dynamic enthalpy rise. This leads to the interesting thermodynamic property of the fluid within the impeller, the so-called "rothalpy":

$$\Delta h_{stat} = h_{2\,stat} - h_{1\,stat} = \frac{u_2^2 - u_1^2}{2} + \frac{w_1^2 - w_2^2}{2}, \tag{1.56}$$

$$h_{2\,stat} + \frac{w_2^2}{2} - \frac{u_2^2}{2} = h_{1\,stat} + \frac{w_1^2}{2} - \frac{u_1^2}{2}. \tag{1.57}$$

So the energy transfer takes place in such a way that the combination of static and dynamic energy terms, called "rothalpy", is an invariant thermodynamic property throughout the impeller:

$$h_{\text{rot}} = h_{\text{stat}} + \frac{w^2}{2} - \frac{u^2}{2} = h^* - \frac{u^2}{2}. \tag{1.58}$$

The rothalpy remains constant in the impeller much in the same way as the total enthalpy remains constant in all of the stationary fluid ducts. Both facts are helpful for design and test analysis.

Specific rothalpy:

$$h_{\text{rot}} = h_{\text{stat}} + \frac{w^2}{2} - \frac{u^2}{2}. \tag{1.59}$$

Specific total enthalpy:

$$h = h_{\text{stat}} + \frac{c^2}{2}. \tag{1.60}$$

Specific total relative (in rotating frame of reference) enthalpy:

$$h^* = h_{\text{stat}} + \frac{w^2}{2}. \tag{1.61}$$

Specific energy of the centrifugal field:

$$\frac{u_2^2}{2} = h^* - h_{\text{rot}}. \tag{1.62}$$

Specific dynamic energy, stationary frame of reference:

$$\frac{c^2}{2} = h - h_{\text{stat}}. \tag{1.63}$$

Specific dynamic energy, rotating frame of reference:

$$\frac{w^2}{2} = h^* - h_{\text{stat}}. \tag{1.64}$$

1.7.1 Change of State in the Mollier Diagram

The change of state in a stage consisting of an impeller and a diffuser is illustrated in the h, s-diagram, Fig. 1.19, where all the specific energies described above are vertical lines. Thus, the energy transfer in the impeller with its constant rothalpy and the energy transformation in the diffuser with its constant total enthalpy become understandable. The thick solid line constitutes the change of the static enthalpy taking place in the impeller *and* diffuser. The dashed line represents the change of the total enthalpy taking place in the impeller only and remaining constant in the diffuser. The dash-dotted line is the change of the total enthalpy in the rotating frame of reference defined in the impeller only.

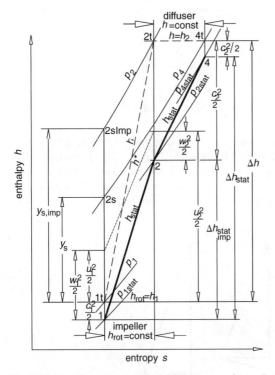

Fig. 1.19 h, s-diagram: static and total change of state in impeller and diffuser. h_{stat} course of static enthalpy, h course of total enthalpy in stationary frame of reference, h^* course of total enthalpy in rotating frame of reference, 1 impeller inlet, 2 impeller exit, 4 diffuser exit.

1.8 Velocity Triangles

The velocity triangles or vector diagrams are formed by the vectors of the gas velocities and the peripheral speed (tip speed) at the inlet and exit of the impeller.

The flow channel locations are indicated in Fig. 1.20.

The inlet triangle, Fig. 1.21, describes the state of the flow infinitesimally downstream of the blade leading edge (location "1"), i.e. within the blade channel constricted by the blade thickness.

The exit triangle, Fig. 1.22, is valid infinitesimally downstream of the blade trailing edge (location "3"), i.e. in the angular channel unrestricted by the impeller blades. Velocities in the stationary frame of reference, designated "c", are referred to as "absolute" and velocities in the rotating frame of reference, designated "w", are referred to as "relative".

Much in the same way as the cyclist sees rain drops slanting towards him, although they fall vertically, an impeller-mounted instrument realizes the relative velocity, whereas a stationary instrument perceives the absolute velocity.

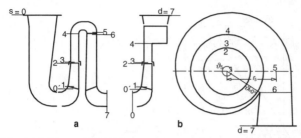

Fig. 1.20 Stage nomenclature. **a** inlet stage with return vane channel, **b** final stage with volute; s suction nozzle, d discharge nozzle; 0 stage inlet, 7 stage exit, 0´ upstream blade, 1 downstream leading edge, 2 blade exit (metal), 3 downstream blade exit (infinitesimal), 4 diffuser exit, 5 upstream deswirl vane and exit volute curved section, 6 downstream return vane inlet and exit volute, r_s center of gravity of area 5, ϑ wrap angle (azimuth), volute togue at $\vartheta = 0°$.

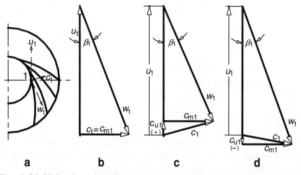

Fig. 1.21 Velocity triangles at impeller inlet (station 1). **a** velocity vectors downstream of blade inlet, **b** no prewhirl, **c** positive prewhirl (head decrease), **d** negative prewhirl (head increase), c absolute velocity, w relative velocity, subscript m meridional component, subscript u peripheral component.

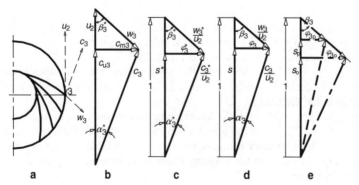

Fig. 1.22 Velocity triangles at impeller exit (station 3). **a** velocity vectors downstream of blade exit, **b** true dimensional triangle, **c** true dimensionless triangle, **d** practical triangle (test deduced, $s > s^*$ and $\varphi_3 < \varphi_3^*$), **e** off-design triangles: p partload, o overload.

Both gas velocities are coupled to each other via the peripheral (metal) speed "u" by the vectorial summation:

$$c = u + w \tag{1.65}$$

The absolute velocity c is split up into a meridional component c_m and a peripheral component c_u. c_m at any point of the flow path lies in the plane that contains the rotor axis; c_u is always perpendicular to c_m and lies in the plane that contains the vector of the peripheral velocity u. Note that all gas velocities shown are mean velocities, i.e. volume flows divided by the channel throughflow areas at the pertaining locations.

1.8.1 Inlet Triangles

The incoming velocity c_1 normally does not have any prewhirl ($\alpha_1 = 90°$). However, it can be influenced in its direction by fixed or adjustable inlet guide vanes called prerotation, producing either a rotationwise (positive) component $+c_{u1}$ (thus reducing the head as per Eq. (1.51)) or a counter-rotationwise (negative) component $-c_{u1}$ (thus increasing the head by a limited amount). The head variation at constant prewhirl α_1, however, is not uniform along the performance curve since the difference $c_{u3}-c_{u1}$ decreases at partload and increases at overload so that a rotationwise prewhirl does not cause a plain downward shift but a substantial leftward shift of the performance curve towards lower volume flow including surge and choke. For off-design operation at zero or constant prerotation the vector tip of w_1 slips along the vector of c_1.

1.8.2 Exit Triangles

The hard facts of fluid dynamics are depicted by the impeller exit triangle, Fig. 1.22: **b** shows the "true" triangle, i.e. c_{u3} is the actual peripheral projection of c_3 and c_{m3} contains the contributing cover disk leakage; **c** is the true dimensionless triangle normalized by u_2 with s^* being the impeller slip and φ_3^* the impeller exit flow coefficient. β_3^* is the relative flow exit angle which, due to the action of the counter rotative channel vortex, is some 10 to 15° smaller than the blade exit angle. α_3^* is the absolute exit angle decisive for the diffuser performance. **d** is a very handy version of the exit triangle. Since the impeller slip s^* is replaced by the work input factor s and the exit flow coefficient is freed of the tiresome cover disk leakage flow, it can very easily be deduced from PTC10-type compressor acceptance tests, which serve as an excellent analytical tool to detect any performance deficiencies:

$$s = \frac{\Delta h}{u_2^2} \qquad \varphi_3 = f\varphi \frac{\dot{V}_3/\dot{V}_{0t}}{4(b_2/d_2)} . \tag{1.66}$$

All parameters are calculable: Δh is calculated from the real gas equations of state, f is the recirculation loss from the balance piston and $\dot{V}_3/\dot{V}_{0t}$ is the Mach-number-dependent impeller volume ratio. So on the basis of s and φ_3 a fictitious triangle can be established which differs only insignificantly from the actual one but with the advantage of simplicity and meaningfulness.

The vector s is proportional to the gas power (if there is no prewhirl) and the vector φ_3 is proportional to the volume flow:

$$P_i = fs\Delta h, \tag{1.67}$$

$$\dot{V} = \varphi_3 \pi b_2 d_2 \frac{u_2}{f(\dot{V}_3/\dot{V}_{0t})}. \tag{1.68}$$

The absolute angle α_3 plays an important role for the diffuser performance, especially the minimum angle causing reverse flow in the diffuser inlet area triggering stage surge, as will be shown in Chap. 5. Figure 1.22e shows partload and overload triangles, whereby the tip of the (c_3/u_2)-vector slips along the vector of (w_3/u_2) at constant angle β_3, which is the case for the entire performance map.

1.9 Laws of Energy Transfer, Summary of Inferences

The first law of thermodynamics and the Euler turbomachinery equation define what happens to the energy input at the compressor shaft:

- The first law of thermodynamics, being a disclosure for the compressor as a whole, establishes that, in specific terms, the mechanical work input to the compressor corresponds on the gas side to an enthalpy increase plus the heat removed from the compressor across the boundary (Fig. 1.10 and Eq. (1.9)):

$$y = \Delta h + q. \tag{1.69}$$

- In an adiabatically enclosed compressor or compressor section the heat is $q = 0$ so that the enthalpy rise is the only gas energy term bringing about an increase of pressure and temperature. In reality, due to convection and radiation, the heat is: $q/y \approx 0.01$; so in most cases it is negligible.
- In an isothermally conducted compression of a perfect gas the enthalpy difference is $\Delta h = 0$, so that the removed heat is the only gas energy term bringing about an increase of pressure only. For real-gas behavior there is a small isothermal enthalpy reduction $\Delta h < 0$, which reduces the actual work input.
- The Euler turbomachinery equation is a statement about the impeller.
- The first form, Eq. (1.45), states that the lion's share of the energy input is used to increase the angular momentum of the gas, with the circumferential component of the absolute exit velocity, c_{u2}, being the most important energy parameter. The rest is leakage and disk friction loss comprising a few percent.
- The second form, Eq. (1.55), describes the breakdown of the work input into a valuable static and a not-so-beneficial dynamic enthalpy rise (plus leakage and disk friction). A considerable amount of the static rise is brought about by the change of the peripheral speed (u_2-u_1) which is characteristic of centrifugal compressors.

- The rothalpy is an invariant thermodynamic property of the change of state in the impeller.

1.10 Compressor Characteristics and their Significance

Table 1.7 Characteristics and their approximate ranges for single-shaft compressors

Characteristic	Meaning	Definition		Range (approx.)
Flow coefficient	Dimensionless actual suction volume flow	$\varphi = \dfrac{\dot{V}_s}{\dfrac{\pi}{4} d_2^2 u_2}$	(1.70)	$\varphi \approx 0.01\text{--}0.15$
Polytropic head coefficient	Dimensionless specific total work of compression of reference process (here: polytropic)	$\psi_p = \dfrac{y_p}{u_2^2/2}$ $\psi_p = 2\eta_p s$	(1.71) (1.72)	$\psi_p \approx 0.8\text{--}1.1$
Work input factor	Dimensionless specific total work of compression	$s = \dfrac{\Delta h}{u_2^2}$	(1.73)	$s \approx 0.57\text{--}0.66$
Polytropic efficiency	Total work of ref. process (here: polytropic) over total actual work	$\eta_p = \dfrac{y_p}{\Delta h}$	(1.74)	$\eta_p \approx 0.6\text{--}0.87$
Machine Mach number	Impeller tip speed over inlet sonic velocity	$M_{u2} = \dfrac{u_2}{\sqrt{\kappa_v Z_s R T_s}}$	(1.75)	$M_{u2} \approx 0.2\text{--}1.25$
Machine Reynolds number	Mass force over viscous force	$Re_{u2} = \dfrac{u_2 b_2 \rho_s}{\eta_s}$	(1.76)	$Re_{u2} \approx 10^5\text{--}10^8$
Specific speed	Dimensionless speed	$\sigma = \dfrac{\varphi^{1/2}}{\psi_p^{3/4}}$	(1.77)	$\sigma \approx 0.12\text{--}0.42$

1.10.1 Flow Coefficient

The flow coefficient is probably the characteristic number with the most far reaching implications: it determines how great an actual volume flow is achieved by an impeller of a given diameter rotating at a given tip speed. A small φ-value therefore implies small inlet and exit areas, i.e. narrow axial inlet and exit widths. The lower limit of φ is dictated by a dramatic increase of the friction loss at the flow channel walls due to the reduction of the hydraulic diameter and consequentially a breakdown of efficiency and head.

The upper limit of φ for centrifugal impellers arranged on a beamtype rotor (impellers arranged between two journal bearings) is dictated by:
- the through-drive shaft having a finite diameter,
- the maximum eye diameter which has to be smaller than the outer diameter to assure a centrifugal effect,
- the axial extension of the complete stage including return vane channel or volute,
- the increase of the impeller stress under centrifugal load.

So, an overhung type impeller (no through-drive shaft!) can be designed with a flow coefficient higher than 0.15. And it can be increased further if the axial space permits a mixed-flow impeller with an inclined diffuser, which is an hybrid between centrifugal and axial design.

Significance
1. The greatest significance of the flow coefficient is its influence on the *efficiency* as can be seen in Fig. 1.23. The stage efficiency grows with increasing flow coefficient because the channel width increases and the frictional forces lose their significance. The efficiency of a stage equipped with an impeller with single-curved blades (so-called 2D-blades) reaches a maximum at around $\varphi = 0.06$ and then decreases, since the growing impeller leading-edge length still has one constant blade inlet angle, causing incidence losses. To avoid these one has to extend the leading edge upstream into the eye opening and at the same time twist the blade in order to adapt to the changing flow angle along the blade height. Thus, the incidence losses are minimized and the longer blade channel gets better adapted to the flow diffusion. These so-called 3D-impellers push the stage efficiency further up until it reaches a maximum at around $\varphi = 0.09$. At flow coefficients of $\varphi > 0.1$ the stage efficiency sees another downturn mainly due to secondary flows and flow separation in the impeller blade channel and due to deteriorating diffuser performance.

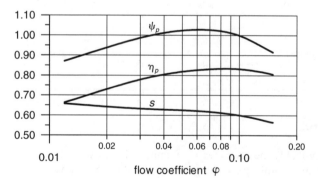

Fig. 1.23 Rough guidelines for basic values of polytropic head coefficient ψ_p, polytropic stage efficiency η_p, and work input factor s vs. flow coefficient φ, for impellers with backward curved blades, medium Mach and Reynolds numbers and medium impeller diameters. These must be corrected for size, shroud leakage, Mach number, Reynolds number, surface roughness, diffuser ratio, inlet loss, exit loss and geometry variants.

2. The *Impeller diameter and rotational speed* can be influenced by the flow coefficient, as demonstrated by the following formulae:

$$d_2 = \sqrt{\frac{\dot{V}_s}{\frac{\pi}{4} u_2 \varphi}} \qquad (1.78)$$

and

$$N = \frac{u_2}{\pi d_2}. \qquad (1.79)$$

So, for a given volume flow, an increase of the flow coefficient results in a smaller impeller outer diameter, which is favorable as far as cost and compressor size are concerned; but it also results in an increase of the rotational speed which may not be a desired secondary effect. So it may not always be feasible to stretch φ to the upper limit.

3. The *lateral critical* speed is influenced by the flow coefficient as follows: a higher flow coefficient results in an increased axial length of the stage L and thus in an increased bearing span. And since, on the other hand, a higher flow coefficient in most cases brings about a reduced hub diameter D_H (and thus shaft diameter) under the impeller, this double effect reduces the critical speed, questioning the advantageous aerodynamic design with a favorable efficiency. The geometric differences between a medium and a low-flow-coefficient stage are demonstrated in Fig. 1.24.

The flow coefficient is to a great extent an independent tool for the designer to determine the size and speed of the machine and is definitely the most influencing parameter for the efficiency and the head coefficient.

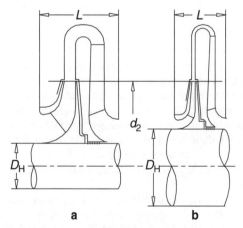

Fig. 1.24 Basic stage geometry differences. **a** stage with medium-flow-coefficient impeller: large axial length and thin shaft; **b** stage with low-flow-coefficient impeller: small axial length and thick shaft, D_H hub diameter, L axial stage length.

1.10.2 Polytropic Head Coefficient

It defines the reference head as a portion of the fictitious (metal) rim-speed kinetic energy $u_2^2/2$. So it is rather convenient to come up with the stage head once the tip speed is known. As an "artificial" compound characteristic ψ_p is a product of work input factor s and efficiency η_p. For "normal" process compressor impellers with backward curved blades having exit angles of roughly between 40° and 50° (from tangent) the polytropic head coefficients assume values that depend on φ, as shown in Fig. 1.23.

1.10.3 Work Input Factor

This is a "natural" inherent characteristic exclusively of the impeller (*not* the stage) selected. If one stays with the "work horses" of process compressors, i.e. backward-curved bladed impellers with exit angles of between 40° (low φ) and 50° (high φ) having between 18 and 20 blades, s is in the range shown in Fig. 1.23 with the average being around 0.63. If one disregards impeller leakage and disk friction losses as well as inlet prewhirl, the work input factor equals the impeller slip:

$$s \approx s^* = \frac{c_{u3}}{u_2}. \tag{1.80}$$

So, an impeller "slip" of, for example, $s = 0.6$ indicates that the absolute gas velocity in the circumferential direction lags behind the peripheral speed by 40%. The closest approach of circumferential gas velocity and peripheral metal speed is achieved by impellers with radially oriented blades at the exit (exit angle 90°) and is 13%.

1.10.4 Polytropic Efficiency

As stated previously, the purpose of the efficiency is to inform about the quality of a compression insofar as it compares the lossless polytropic reference process with the actual work, i.e. how successful the designer has been in minimizing losses in the flow channels such as friction, separation, incidence, leakage, and secondary and reverse flow. For the sake of avoiding too high a rotational speed or of accommodating many impellers on one shaft it is not always possible to reach the best aerodynamic solution (i.e. the optimum flow coefficient), since the final compressor design is a compromise between aerodynamics, rotordynamics, structural mechanics and, last but not least, price.

It is true that by finalizing the flow coefficient distribution in the compressor, as a result of compromises, the basic efficiency, as shown in Fig. 1.23, is fixed. But it has to be corrected by effects such as size, Mach number, Reynolds number, surface roughness, diffuser ratio, cover disk labyrinth clearances, etc. So the values given in Fig. 1.23 can vary by up to approximately ±0.04.

1.10.5 Machine Mach Number (Tip Speed Mach Number)

The Mach number is responsible for the compressibility of a flow. The higher the Mach number the greater is the density variation, whenever a change is imposed on the flow, be it expansion, diffusion or power input. So, for instance, an increasing Mach number will result in:

- a greater inlet expansion from flange to impeller eye due to the acceleration,
- a higher impeller pressure ratio, and
- an increased pressure recovery in the diffuser due to the deceleration.

This characteristic is a reference, not a real Mach number, because it is the ratio of a (metal) rim speed and the sonic inlet velocity. But any desired Mach number, like the important relative inlet Mach number at the impeller eye, often referred to as the "eye Mach number", can easily be derived from M_{u2} by means of a simple multiplier:

$$M_{w1} = \frac{w_1}{a_0} = \frac{w_1}{u_2}\frac{u_2}{a_0} = M_{u2}\frac{w_1}{u_2}, \quad (1.81)$$

where w_1/u_2 is a fixed ratio for the design point of any given impeller.

Significance
1. The most obvious is the *pressure ratio* dependency of the Mach number. In order to show this, the perfect gas polytropic head (Eq. (1.43)) is rearranged:

$$\frac{p_2}{p_1} = \left[\frac{su_2^2}{RT_1}\frac{\kappa-1}{\kappa}+1\right]^{\frac{\kappa\eta_p}{\kappa-1}} = \left[s(\kappa-1)M_{u2}^2+1\right]^{\frac{\kappa\eta_p}{\kappa-1}}. \quad (1.82)$$

For simplicity reasons the isentropic exponent at the inlet, κ_{vs} is set equal to the average isentropic exponent of the compression path, $(\kappa_1+\kappa_{2s})/2$ (see Eq. (1.43)).

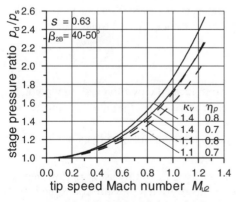

Fig. 1.25 Stage pressure ratio vs. tip speed Mach number. κ_v isentropic volume exponent, η_p polytropic stage efficiency, s work input factor, β_{2B} impeller blade exit angle.

As demonstrated in Fig. 1.25, the pressure ratio is indeed a strong function of the tip speed Mach number regardless of the gas; the heat capacity ratio and the efficiency play a secondary role only. Using the perfect gas formula does in no way restrict the general validity of this statement nor the above simplification for the isentropic exponent.

As a general rule it is worth keeping in mind:

- *the pressure ratio is a function of the tip speed Mach number,*
- *the head is a function of the tip speed.*

2. The Mach number/compressibility interdependance is also reflected in the *performance curve shape*, as shown in Fig. 1.26: In this example two compressor options were designed for a given specification: pressure ratio 4.67, molar mass 40.7 (hydrocarbons); the 2-stage version has an initial Mach number of 1.2 and the 5-stage alternative has one of 0.81.

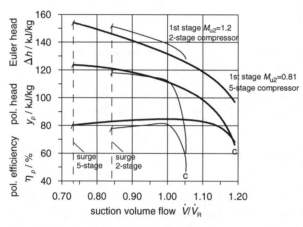

Fig. 1.26 Performance curve shape as function of the Mach number. Two compressor design options for a given specification: high Mach number 2-stage vs. medium Mach number 5-stage. Molar mass 40.7, pressure ratio 4.67, c choke point, $\dot{V}_R$ rated volume flow.

So, the higher Mach number leads to:

- a shorter curve with surge and choke point located closer to the design point,
- a flatter head curve at partload (lower head rise to surge) due to a greater decrease of the partload efficiency,
- a steeper curve at overload, mainly due to a progressive aerodynamic mismatch effect for each consecutive stage.

All this, of course, means restricted operational flexibility as far as flow and variable molar mass are concerned; it also means restricted pressure controllability.

3. A higher Mach number leads to a *reduced peak efficiency,* but also to a more pronounced efficiency peak.

4. As will be shown in Chap. 5, higher Mach numbers also bring about an increased aerodynamic *mismatching* potential, especially for higher compressor pressure ratios. This means essentially a progressive drift-off of each subsequent stage from its intrinsic best efficiency point once the first stage is operated at an off-design point. This is the case for changes of flow, gas constituents, temperature, speed, etc.

1.10.6 Machine Reynolds Number

The Reynolds number is by definition the ratio of mass force and viscous force; so it is an indicator of how much inertia prevails over viscosity:

$$Re = \frac{f_m}{f_v}. \tag{1.83}$$

According to the momentum equation, a mass flow with a through-flow area A and a velocity c exerts the mass force on a surface perpendicular to the direction of flow:

$$F_m = \dot{m}c = \rho A c^2. \tag{1.84}$$

The mass force per unit of area is

$$f_m = \rho c^2. \tag{1.85}$$

In a Newtonian fluid the viscous force per unit of area (i.e. the shear stress in the opposite direction of flow) is the product of the dynamic viscosity and the change of velocity perpendicular to the direction of flow:

$$f_v = \eta \frac{dc}{dy} \propto \eta \frac{c}{l} = \nu \rho \frac{c}{l}. \tag{1.86}$$

The reference areas that both forces act on are not identical; this, however, is not relevant for proportionality considerations. So the Reynolds number becomes

$$Re = \frac{cl}{\nu}. \tag{1.87}$$

For the "machine" Reynolds number c is replaced by u_2, length l by b_2 and the kinematic viscosity is defined at the suction side of the stage by the expression:

$$Re_{u2} = \frac{u_2 b_2}{\nu_s} = \frac{u_2 b_2 \rho_s}{\eta_s}. \tag{1.88}$$

Significance
1. In light of the above, the *efficiency* must definitely go up, if the Reynolds number increases (or, to put it more bluntly, if the pressure level increases). The machine Reynolds number, together with the relative surface roughness, is responsible for the friction at the compressor flow-channel walls. An equivalent pipe flow

can be defined having a friction factor similar to the compressor flow. Thus the well-known Moody diagram can be used to calculate the influence of both the machine Reynolds number and the relative arithmetic average roughness Ra/b_2 on the efficiency. The (iterative) Colebrook–White formula for the friction factor, reflecting the diagram, was published by Moody (1944):

$$\lambda = \frac{1}{\left[1.74 - 2\log_{10}\left(2\frac{Ra}{b_2} + \frac{18.7}{Re\sqrt{\lambda}}\right)\right]^2} .$$ (1.89)

Any change of the friction factor can be translated into an efficiency change by:

$$\Delta \eta_p = -\frac{k}{s}\Delta\lambda ,$$ (1.90)

where k is the experimentally determined proportionality factor.

2. *Reynolds number correction of test results*: Since the compressor performance is affected by the Reynolds number shop performance test results of efficiency and head coefficient have to be corrected if the test is conducted with another gas and/or at another pressure level. Correction formulae are given in the relevant test codes PTC-10-1997 and VDI 2045.

1.10.7 Specific Speed

This characteristic can be used instead of the flow coefficient and is an analogous influencing value for the efficiency. Although it is truly dimensionless by definition, it is used in some countries, especially in the USA, in the form of the numerical dimensional equation:

$$N_s = \frac{N\dot{V}_s^{1/2}}{y_p^{3/4}} ,$$ (1.91)

with speed N in rpm (= 1/min), volume flow $\dot{V}_s$ in cfm (= ft³/min) and polytropic head y_p in ft lb$_f$/lb$_m$ The resulting dimension of N_s cannot even be written down. The overall range is approximately $N_s \approx 300\text{–}1{,}100$ corresponding to the flow coefficient range of $\varphi \approx 0.01\text{–}15$. In this book neither σ nor N_s will be used further.

References

Baehr HD (1962) Thermodynamik, 1st edn. Springer-Verlag, Berlin, p 404
Brown, Boveri & Cie (1928) Turbokompressoren und Turbogebläse. Werkzeitschrift 1013, Baden, Switzerland (in German)
Colliat JP (2001) Images. http://jpcolliat.free.fr/he178/images/hes3b_01.jpg
Engeda A (1998) From the Crystal Palace to the pump room. ASME Paper 98-GT-22, International Gas Turbine & Aeroengine Congress, Stockholm 1998, Sweden; The American Society of Mechanical Engineers

Gieck KR (1989) Technische Formelsammlung/Technical Formulae, 29th German edn and 7th English edn. Gieck-Verlag, Germering, Germany

Historisch-Technisches Informationszentrum Peenemünde (2003). Documentation of development and application of the V-weapons: history, technology, fate of the slave laborers (Test site VII from where first A4/V2 was shot to 85 km height not yet open to the public). www-peenemuende.de

Lloyd P (1945) Combustion in the gas turbine. Proc. Inst. Mech. Eng. Vol 153, p 462 (War Emergency Issue No. 12), London, UK

Moody LF (1944) Friction factors for pipe flow. Trans. ASME, pp 671-684; The American Society of Mechanical Engineers

Smithsonian Institution, National Air and Space Museum (2002). www.nasm.edu

Stodola A (1924) Dampf- und Gasturbinen. Springer-Verlag, Berlin, Germany (in German)

2 Thermodynamics of Real Gases

Although the pertinent formulae for the compressibility factor, the polytropic volume and temperature exponent were published in the 1940s, the message has, even today, not reached all engineering contractors that different exponents for calculating head and temperature have to replace the traditional "ratio of specific heats", which is irrelevant for real process gases. In this chapter a systematic derivation and compilation of real-gas thermodynamics and the most common real-gas equations of state are presented along with the ready-to-use formulae with the purpose of establishing precise descriptions of the physical relationships and avoiding misunderstandings still widespread in the turbomachinery community.

As already mentioned, almost all of the gases handled by process compressors are characterized by a pronounced real-gas behavior. This means that three real-gas parameters have to be introduced to assure a proper compressor design and analysis of test findings:

- the compressibility factor Z required for the volume flow, head and power,
- the polytropic volume (change) exponent n_v required for the head, and
- the polytropic temperature (change) exponent n_T required for the discharge temperature.

All three depend on the gas composition, temperature and pressure, and are calculated from real-gas equations of state.

The k-value method is an older calculation procedure. Before relevant formulae and adequate equations of state and computers suited to applying them had become available, the compressibility factors were determined from individual charts of pure substances, including dubious mixture rules or (much better) from generalized graphs for gas mixtures, e.g. according to Nelson and Obert (1954), using so-called reduced pressure and reduced temperature calculated with the pseudocritical values (volume or mole fraction times critical value):

$$Z = f(p_r, T_r); \quad p_r = \frac{p}{\sum_{i=1}^{n}(r_i p_{ci})}; \quad T_r = \frac{T}{\sum_{i=1}^{n}(r_i T_{ci})}. \quad (2.1)$$

For head and temperature computation the perfect gas exponent is used:

$$\frac{n-1}{n} = \frac{\kappa - 1}{\kappa \, \eta_p} = f(T), \quad (2.2)$$

where κ is the ratio of the specific heat capacities for perfect gases:

$$\kappa = \frac{c_p^0}{c_v^0} = \frac{Mc_p}{Mc_p - \Re} = f(T), \qquad Mc_p = \sum_{i=1}^{n}(r_i Mc_{pi}) = f(T). \tag{2.3}$$

The molar isobaric heat capacity Mc_p for perfect gas behavior is compiled in tables for various gases as a function of temperature at pressures of approximately 1 bar. This so-called *k*-value or Mc_p-method operates without any pressure correction. The universal gas constant is

$$\Re = 8.31451 \pm 0.0007 \text{ kJ kmol}^{-1} \text{ K}^{-1} \tag{2.4}$$

This method can be used as an approximation with an accuracy of around ±3% for the head, especially in those cases where the software is not available to make a calculation on the basis of the real-gas equation (Nelson–Obert Chart in Fig. 6.1 and Mc_p-values in Fig. 6.2).

The advance of process compressors into higher pressure ranges (and also into lower-temperature ranges for refrigeration plants!) makes the exact knowledge of the relevant gas data an absolute necessity, since otherwise there would be a pronounced danger of aerodynamic stage mismatching for multistage machines, which would prevent the optimum efficiency, operating range and the desired prediction accuracy from being attained.

The aerodynamicist involved in compressor design cannot be spared the following lengthy excursion into the world of real-gas thermodynamics, since it was found that most authors of text books on turbomachinery are very tight-lipped on this subject. The aerodynamics are only as good as the proper gas thermodynamics is taken care of.

Before the real-gas equations are introduced, the thermodynamic expressions for the polytropic exponents, which deviate strongly from the familiar perfect-gas relations, will be derived.

2.1 Gas Thermodynamics

2.1.1 Polytropic and Isentropic Exponents

Volume Exponent

The starting point for deriving the exponents is the equation $pv^n = const$. The polytropic change of state is defined by

$$pv^{n_v} = const \rightarrow \ln p + n_v \ln v = const.$$

Therefore, $\quad n_v \, d(\ln v) = -d(\ln p) \rightarrow n_v = -\dfrac{d(\ln p)}{d(\ln v)}. \tag{2.5}$

The isentropic change of state is defined by

$$pv^{\kappa_v} = const \rightarrow \kappa_v = -\frac{d(\ln p)}{d(\ln v)}.\tag{2.6}$$

The *polytropic volume exponent* is defined at constant efficiency by

$$n_v = -\frac{v}{p}\left(\frac{\partial p}{\partial v}\right)_{\eta_p}.\tag{2.7}$$

This exponent is used for calculating the polytropic head:

$$y_p = Z_1 RT_1 \frac{n_v}{n_v - 1}\left[\left(\frac{p_2}{p_1}\right)^{\frac{n_v - 1}{n_v}} - 1\right], \quad \text{where } n_v = \frac{n_{v1} + n_{v2}}{2}.\tag{2.8}$$

The *isentropic volume exponent* is defined at constant entropy by

$$\kappa_v = -\frac{v}{p}\left(\frac{\partial p}{\partial v}\right)_s.\tag{2.9}$$

This exponent is used for calculating the isentropic head:

$$y_s = Z_1 RT_1 \frac{\kappa_v}{\kappa_v - 1}\left[\left(\frac{p_2}{p_1}\right)^{\frac{\kappa_v - 1}{\kappa_v}} - 1\right], \quad \text{where } \kappa_v = \frac{\kappa_{v1} + \kappa_{v2s}}{2}.\tag{2.10}$$

Temperature Exponent

Equation $pv^n = const$ is rearranged using $v = ZRT/p$ as follows

$$p\frac{T^{n_T}}{p^{n_T}} = const \rightarrow \frac{p^{n_T - 1}}{T^{n_T}} = const \rightarrow \frac{p^{\frac{n_T - 1}{n_T}}}{T} = \frac{p^m}{T} = const,\tag{2.11}$$

$$\rightarrow m\,d(\ln p) = d(\ln T), \quad \text{where } m = (n_T - 1)/n_T$$

$$\rightarrow m = \frac{n_T - 1}{n_T} = \frac{p}{T}\left(\frac{\partial T}{\partial p}\right)_{\eta_p}.\tag{2.12}$$

The compressibility factor Z is assumed constant for an infinitely small compression step.

The *polytropic temperature exponent* is defined at constant efficiency by

$$n_T = \frac{1}{1 - \frac{p}{T}\left(\frac{\partial T}{\partial p}\right)_{\eta_p}}.\tag{2.13}$$

This exponent is used for calculating the actual discharge temperature:

$$T_2 = T_1 \left(\frac{p_2}{p_1}\right)^{\frac{n_T-1}{n_T}}, \quad \text{where} \quad n_T = \frac{n_{T1}+n_{T2}}{2}. \tag{2.14}$$

The *isentropic temperature exponent* is defined at constant entropy ($\eta_p = 1.0$) by

$$\kappa_T = \frac{1}{1 - \frac{p}{T}\left(\frac{\partial T}{\partial p}\right)_s}. \tag{2.15}$$

This exponent is used for calculating the isentropic discharge temperature:

$$T_{2s} = T_1 \left(\frac{p_2}{p_1}\right)^{\frac{\kappa_T-1}{\kappa_T}}, \quad \text{where} \quad \kappa_T = \frac{\kappa_{T1}+\kappa_{T2s}}{2}, \tag{2.16}$$

and κ_{T2s} is defined at the isentropic discharge.

The terms $\left(\dfrac{\partial p}{\partial v}\right)$ and $\left(\dfrac{\partial T}{\partial p}\right)$ are transformed into calculable parameters as follows.

General Expression for the Polytropic Volume Exponent

The derivation is carried out in accordance with Dzung (1944) and was previously published by Beinecke and Lüdtke (1983).

It is indeed astonishing that these real-gas calculations, made known as early as 1944, went unnoticed in the turbomachinery industry until the 1970s or even 1980s, when adequate computerized real-gas equations of state became available. But even today the message has not got around to all engineering companies that different exponents for head and temperature replaced the (perfect gas) values $\kappa = c_p/c_v$, which have become irrelevant for process gases.

The total differential of the enthalpy as a function of the specific volume and pressure is

$$dh = \left(\frac{\partial h}{\partial p}\right)_v dp + \left(\frac{\partial h}{\partial v}\right)_p dv. \tag{2.17}$$

Substituting Eq. (1.29), we obtain

$$\left[\left(\frac{\partial h}{\partial p}\right)_v - \frac{v}{\eta_p}\right] dp = -\left(\frac{\partial h}{\partial v}\right)_p dv. \tag{2.18}$$

Since this is a polytropic change of state, the equation can be rewritten as

$$\frac{dp}{dv} = \left(\frac{\partial p}{\partial v}\right)_{\eta_p} = -\frac{\left(\frac{\partial h}{\partial v}\right)_p}{\left(\frac{\partial h}{\partial p}\right)_v - \frac{v}{\eta_p}} = -\frac{\left(\frac{\partial h}{\partial T}\right)_p \left(\frac{\partial T}{\partial v}\right)_p}{\left(\frac{\partial h}{\partial T}\right)_v \left(\frac{\partial T}{\partial p}\right)_v - \frac{v}{\eta_p}}.$$ (2.19)

The basic definition of the enthalpy is internal energy plus displacement energy (this is the so-called caloric equation of state):

$$h = u + pv \rightarrow u = h - pv.$$ (2.20)

The basic definition of the isochoric heat capacity is

$$c_v = \left(\frac{\partial u}{\partial T}\right)_v = \left(\frac{\partial h}{\partial T}\right)_v - v\left(\frac{\partial p}{\partial T}\right)_v.$$ (2.21)

The basic definition of the isobaric heat capacity is

$$c_p = \left(\frac{\partial h}{\partial T}\right)_p.$$ (2.22)

The real-gas equation of state (this is the so-called thermal equation of state) is

$$pv = ZRT.$$ (2.23)

Equation (2.23) produces three auxiliary partial derivatives:

$$\left(\frac{\partial p}{\partial v}\right)_T = -\frac{ZRT}{v^2}; \quad \left(\frac{\partial p}{\partial T}\right)_v = \frac{ZR}{v}; \quad \left(\frac{\partial T}{\partial v}\right)_p = \frac{T}{v},$$ (2.24)

where Z is assumed to be constant for these infinitesimals.
From Eq. (2.24) we obtain

$$\frac{(\partial p/\partial v)_T}{(\partial p/\partial T)_v} = -\frac{ZRT}{v^2}\frac{v}{ZR} = -\frac{T}{v} = -\left(\frac{\partial T}{\partial v}\right)_p.$$ (2.25)

From Eq. (2.21) we obtain

$$\left(\frac{\partial h}{\partial T}\right)_v \left(\frac{\partial T}{\partial p}\right)_v = c_v \left(\frac{\partial T}{\partial p}\right)_v + v.$$ (2.26)

By combining Eqs. (2.19), (2.22), (2.25) and (2.26) the partial derivative in Eq. (2.7) becomes

$$\left(\frac{\partial p}{\partial v}\right)_{\eta_p} = \frac{c_p \left(\frac{\partial p}{\partial v}\right)_T}{c_v + v\left(\frac{\partial p}{\partial T}\right)_v \left(1 - \frac{1}{\eta_p}\right)}.$$ (2.27)

Hence the general expression for the polytropic volume exponent is

$$n_v = -\frac{v}{p}\frac{c_p}{c_v}\frac{\left(\frac{\partial p}{\partial v}\right)_T}{1+\frac{v}{c_v}\left(1-\frac{1}{\eta_p}\right)\left(\frac{\partial p}{\partial T}\right)_v} \ . \tag{2.28}$$

For $\eta_p = 1.0$ the polytropic volume exponent changes into the isentropic exponent:

$$\kappa_v = -\frac{v}{p}\frac{c_p}{c_v}\left(\frac{\partial p}{\partial v}\right)_T \ . \tag{2.29}$$

Combining (2.28) and (2.29) produces

$$n_v = \frac{\kappa_v}{1+\frac{v}{c_v}\left(1-\frac{1}{\eta_p}\right)\left(\frac{\partial p}{\partial T}\right)_v} \ . \tag{2.30}$$

Before $\left(\frac{\partial p}{\partial T}\right)_v$ is resolved further, the temperature exponent has to be looked at in detail.

General Expression for the Polytropic Temperature Exponent

The total differential of the enthalpy as a function of pressure and temperature is, with Eq. (1.29):

$$dh = \left(\frac{\partial h}{\partial T}\right)_p dT + \left(\frac{\partial h}{\partial p}\right)_T dp = \frac{v}{\eta_p}dp \ . \tag{2.31}$$

Since this is a polytropic change of state, this equation can, with Eq. (2.22), be rewritten as

$$\frac{dT}{dp} = \left(\frac{\partial T}{\partial p}\right)_{\eta_p} = \frac{\frac{v}{\eta_p}-\left(\frac{\partial h}{\partial p}\right)_T}{\left(\frac{\partial h}{\partial T}\right)_p} = \frac{\frac{v}{\eta_p}-\left(\frac{\partial h}{\partial p}\right)_T}{c_p} \ . \tag{2.32}$$

We introduce the Clausius equation, which is a combination of the caloric with the thermal equations of state (Eqs. (2.20) and (2.23)); the derivation is described by Traupel (1977):

$$\left(\frac{\partial h}{\partial p}\right)_T = v - T\left(\frac{\partial v}{\partial T}\right)_p \ . \tag{2.33}$$

Combining Eqs. (2.32) and (2.33) yields:

$$\left(\frac{\partial T}{\partial p}\right)_{\eta_p} = \frac{v}{c_p}\left[\frac{1}{\eta_p} - 1 + \frac{T}{v}\left(\frac{\partial v}{\partial T}\right)_p\right]. \tag{2.34}$$

Thus Eq. (2.12) can be rewritten as:

$$m = \frac{p}{c_p}\left[\frac{v}{T}\left(\frac{1}{\eta_p} - 1\right) + \left(\frac{\partial v}{\partial T}\right)_p\right]. \tag{2.35}$$

For $\eta_p = 1.0$ the polytropic temperature exponent changes into the isentropic exponent:

$$m = \frac{n_T - 1}{n_T} = \frac{\kappa_T - 1}{\kappa_T} = \frac{p}{c_p}\left(\frac{\partial v}{\partial T}\right)_p \tag{2.36}$$

Final Formula for the Polytropic Temperature Exponent

Equations (2.35), (2.36) and (2.23) yield the final form:

$$m = \frac{n_T - 1}{n_T} = \frac{ZR}{c_p}\left(\frac{1}{\eta_p} - 1\right) + \frac{\kappa_T - 1}{\kappa_T}, \tag{2.37}$$

$$n_T = \frac{1}{1 - \frac{ZR}{c_p}\left(\frac{1}{\eta_p} - 1\right) - \frac{\kappa_T - 1}{\kappa_T}}. \tag{2.38}$$

Final Formula for the Polytropic Volume Exponent

The looked-for derivative $\left(\frac{\partial p}{\partial T}\right)_v$ is extracted from Eq. (2.25) as

$$\left(\frac{\partial p}{\partial T}\right)_v = -\left(\frac{\partial v}{\partial T}\right)_p\left(\frac{\partial p}{\partial v}\right)_T. \tag{2.39}$$

From Eqs. (2.36) and (2.29) we obtain

$$\left(\frac{\partial p}{\partial T}\right)_v = -\frac{c_p}{p}\frac{\kappa_T - 1}{\kappa_T}\left(\frac{\partial p}{\partial v}\right)_T = \frac{c_p}{p}\frac{\kappa_T - 1}{\kappa_T}\frac{pc_v}{vc_p}K_v = \frac{c_v}{v}\frac{\kappa_T - 1}{\kappa_T}K_v. \tag{2.40}$$

With Eq. (2.30) the *final formula for the polytropic volume exponent* becomes

$$n_v = \frac{\kappa_v}{1+\left(1-\dfrac{1}{\eta_p}\right)\dfrac{\kappa_T-1}{\kappa_T}\kappa_v}, \tag{2.41}$$

$$\frac{n_v-1}{n_v} = \frac{\kappa_v-1}{\kappa_v} + \frac{1-\eta_p}{\eta_p}\frac{\kappa_T-1}{\kappa_T}. \tag{2.42}$$

Isentropic Temperature and Volume Exponent, Real Gas Behavior

For a polytropic efficiency of $\eta_p=1.0$, i.e. an isentropic change of state, from Eqs. (2.38) and (2.41) we obtain the isentropic temperature exponent:

$$n_T = \kappa_T, \tag{2.43}$$

and the isentropic volume exponent:

$$n_v = \kappa_v. \tag{2.44}$$

Isentropic Temperature and Volume Exponent, Perfect Gas Behavior

The ratio of heat capacities, isentropic volume and temperature exponents are identical and a function of temperature only; the compressibility factor is unity:

$$\kappa = \frac{c_p^0}{c_v^0} = \kappa_v = \kappa_T = f(T), \quad \text{and} \quad Z=1. \tag{2.45}$$

The polytropic exponent for head and temperature, using Eqs. (2.37) and (2.24), assumes the familiar form:

$$\frac{n-1}{n} = \frac{n_v-1}{n_v} = \frac{n_T-1}{n_T} = \frac{\kappa-1}{\kappa\eta_p}. \tag{2.46}$$

2.1.2 Summary of Formulae for Head and Discharge Temperature

Isentropic head, perfect gas:

$$y_s = \int_1^{2s} v\,dp = RT_1\frac{\kappa}{\kappa-1}\left[\left(\frac{p_2}{p_1}\right)^{\frac{\kappa-1}{\kappa}}-1\right]. \tag{2.47}$$

2.1 Gas Thermodynamics

Isentropic head, real gas:

$$y_s = \int_1^{2s} v\,dp = Z_1 RT_1 \frac{\kappa_v}{\kappa_v - 1}\left[\left(\frac{p_2}{p_1}\right)^{\frac{\kappa_v-1}{\kappa_v}} - 1\right], \qquad (2.48)$$

where in (2.47): $\kappa = \dfrac{c_p{^0}_m}{c_p{^0}_m - R}$; and $c_p{^0}_m = \dfrac{1}{T_{2s} - T_1}\int_{T_1}^{T_{2s}} c_p{^0}\,dT$; (2.49)

and in (2.48): $\kappa_v = \dfrac{\kappa_{v1} + \kappa_{v2s}}{2}$.

Polytropic head, perfect gas:

$$y_p = \int_1^2 v\,dp = RT_1 \frac{\kappa\eta_p}{\kappa - 1}\left[\left(\frac{p_2}{p_1}\right)^{\frac{\kappa-1}{\kappa\eta_p}} - 1\right]. \qquad (2.50)$$

Polytropic head, real gas:

$$y_p = \int_1^2 v\,dp = Z_1 RT_1 \frac{n_v}{n_v - 1}\left[\left(\frac{p_2}{p_1}\right)^{\frac{n_v-1}{n_v}} - 1\right], \qquad (2.51)$$

where in (2.50): $\kappa = \dfrac{c_p{^0}_m}{c_p{^0}_m - R}$; and $c_p{^0}_m = \dfrac{1}{T_2 - T_1}\int_{T_1}^{T_2} c_p{^0}\,dT$; (2.52)

and in (2.51): $n_v = \dfrac{n_{v1} + n_{v2}}{2}$.

Isothermal head, perfect gas:

$$y_T = \int_1^{2T} v\,dp = RT_1 \ln\frac{p_2}{p_1}. \qquad (2.53)$$

Isothermal head (using the entropy), real gas:

$$y_T = \int_1^{2T} v\,dp = T_1(s_1 - s_{2T}) - (h_1 - h_{2T}). \qquad (2.54)$$

Actual (Euler) head, perfect gas:

$$\Delta h = c_p{^0}_m (T_2 - T_1). \qquad (2.55)$$

Actual (Euler) head, real gas:

$$\Delta h = h_2 - h_1. \qquad (2.56)$$

Discharge temperature, perfect gas:

$$T_2 = T_1 \left(\frac{p_2}{p_1}\right)^{\frac{\kappa-1}{\kappa\eta_p}}. \tag{2.57}$$

Discharge temperature, real gas:

$$T_2 = T_1 \left(\frac{p_2}{p_1}\right)^m. \tag{2.58}$$

Isentropic volume/temperature exponent, isobaric molar heat capacity, perfect gas:

$$\kappa = \kappa_v = \kappa_T = \frac{c_p^0}{c_v^0} = \frac{Mc_p}{Mc_p - \Re}; \quad Mc_p = M \times c_p^0; \tag{2.59}$$

Isentropic volume exponent, real gas:

$$\kappa_v = -\frac{v}{p}\left(\frac{\partial p}{\partial v}\right)_s. \tag{2.60}$$

Isentropic temperature exponent, real gas:

$$\kappa_T = \frac{1}{1 - \frac{p}{T}\left(\frac{\partial T}{\partial p}\right)_s}. \tag{2.61}$$

Polytropic volume/temperature exponent, perfect gas:

$$n_v = n_T = \frac{\kappa\eta_p}{1 + \kappa\eta_p - \kappa}; \quad \frac{n_v - 1}{n_v} = \frac{n_T - 1}{n_T} = \frac{\kappa - 1}{\kappa\eta_p}. \tag{2.62}$$

Polytropic volume exponent, real gas:

$$n_v = -\frac{v}{p}\left(\frac{\partial p}{\partial v}\right)_{\eta_p}; \quad \frac{n_v - 1}{n_v} = \frac{\kappa_v - 1}{\kappa_v} + \frac{1 - \eta_p}{\eta_p}\frac{\kappa_T - 1}{\kappa_T}. \tag{2.63}$$

Polytropic temperature exponent, real gas:

$$n_T = \frac{1}{1 - \frac{p}{T}\left(\frac{\partial T}{\partial p}\right)_{\eta_p}}; \quad m = \frac{n_T - 1}{n_T} = \frac{ZR}{c_p}\left(\frac{1}{\eta_p} - 1\right) + \frac{\kappa_T - 1}{\kappa_T}. \tag{2.64}$$

Isobaric specific heat capacity, perfect and real gas:

$$c_p = \left(\frac{\partial h}{\partial T}\right)_p. \quad (2.65)$$

Compressibility factor:

perfect gas: $Z = 1$; real gas: $Z = \dfrac{pv}{RT}$. (2.66)

Legend for Eqs. (2.47) to (2.66):
 M molar mass, s entropy, h enthalpy;
 subscripts: 1 suction, 2 actual discharge, 2s isentropic discharge, 2T isothermal discharge, s entropy constant, p pressure constant, η_p polytropic efficiency constant, m average;
 superscript: 0 perfect gas behavior.

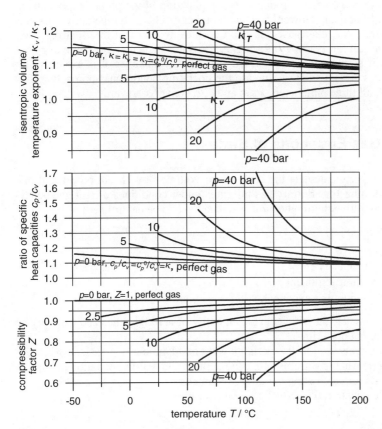

Fig. 2.1. Example for real gas behavior. Propane: isentropic exponents, ratio of specific heat capacities, compressibility factors; calculated with LKP-equation of state.

The simple perfect gas isentropic exponent, being identical with the ratio of specific heat capacities, splits up into an isentropic volume exponent and an isentropic temperature exponent once the gas increasingly differs from the perfect gas behavior, i.e. for increasing pressure or decreasing temperature. These exponents are both pronounced functions of temperature *and* pressure. However, one does not have to differentiate between the "two gas behaviors", since the availability of the comprehensive real-gas formulae assures the proper treatment of perfect gases automatically; so in this context perfect gas behavior is only a borderline case of real gas behavior. If the conditions of a real gas change towards more perfect conditions, κ_v and κ_T gradually appproach the perfect gas ratio of specific heat capacities $\kappa = c_p/c_v$ (and of course the compressibility factor Z approaches unity).

Propane is an excellent example to demonstrate this effect (Fig. 2.1). The higher the pressure and the lower the temperature, the more the isentropic volume and temperature exponents differ from each other and from the ratio of specific heat capacities (at 0 bar), which is solely a function of temperature.

By the way, in the case of propane, the application of the proper exponents instead of c_p^0/c_v^0 results in a lower head and a higher discharge temperature; this may not be consistent for other gases! It also becomes evident that the isentropic volume exponent can fall below 1.0 for higher pressures! The real-gas value of c_p/c_v is completely different from c_p^0/c_v^0 and from κ_v and κ_T; however, it is irrelevant for compressor design and analysis. In case the operating conditions approach perfect gas behavior, all three parameters κ_v, κ_T and c_p/c_v converge towards $\kappa = c_p^0/c_v^0$.

2.2 Real-Gas Equations of State

In order to properly calculate the compressor volume flow, head, power and discharge temperature for arbitrary gas mixtures under conditions of real gas behavior the following five thermodynamic gas data have to be determined:

- molar mass
- compressibility factor
- isobaric heat capacity
- isentropic volume exponent
- isentropic temperature exponent.

(The sixth parameter, the efficiency, is an entity of aerodynamics and can only be worked out in that discipline).

In the foregoing section the pertaining thermodynamic expressions were derived; however, what is still missing, is an appropriate real-gas equation of state to serve as a basis for calculating the above parameters.

The simple equation for perfect gas behavior is:

$$Z = \frac{pv}{RT} = 1. \tag{2.67}$$

There are two classical techniques to correct for real gas behavior.

2.2 Real-Gas Equations of State

The first was suggested by van der Waals (1873):

$$\left(p+\frac{a}{v^2}\right)(v-b)=RT, \quad \text{or} \quad Z=\frac{v}{v-b}-\frac{a}{RTv}, \tag{2.68}$$

with a, b accounting for the intermolecular adhesion and the molecules' volume. The second was introduced by Kammerlingh Onnes (1901) and has a virial form:

$$Z=1+\frac{B}{v}+\frac{C}{v^2}+\cdots \tag{2.69}$$

However, it was not until the 1940s that these formulae of the physics textbooks were modified to satisfy the needs of process engineering.

Benedict, Webb, and Rubin published a new virial equation (1940), which got an extension by Starling (1973), known as the *BWRS* equation of state.

Redlich and Kwong suggested a new van der Waals-type equation (1949) augmented by Soave (1972), known as the *RKS* equation of state.

Lee and Kesler proposed a novel virial equation (1975) modified by Plöcker, Knapp, and Prausnitz (1978), known as the *LKP* equation of state.

Peng and Robinson published their modification of the van der Waals equation (1976) known as the *PR* equation of state.

All of these are generalized equations of state, i.e. they are applicable for gas mixtures as they occur predominantly in process compressors; and it has become common practice, for the sake of consistency, to use them for single gases as well, although there might exist specially adapted equations for each individual gas.

The generalized equations are based on the principle of "corresponding states" for pure substances characterized by three parameters: critical values of temperature T_c, pressure p_c or specific volume v_c (or density ρ_c) plus the acentric factor ω, which represents the acentricity or nonsphericity of a molecule, expressing the complexity of a molecule with respect to geometry and polarity. ω ranges from zero for monatomic gases to approximately 0.9 for complex molecule structures. For mixtures pseudo-critical values are used: Σ (molar fraction $\times$ critical value).

The mixing rules of all equations incorporate so-called "binary interaction parameters" k_{ij} accounting for the intermolecular forces between any pair of gases that have been experimentally determined. The values of k_{ij} are differently defined for each equation and have therefore differing magnitudes. For the LKP equation the binary parameters are around 1.0 for binaries similar in molar mass, molecule structure or in the same group of gases; dissimilar binaries have k_{ij} values as low as 0.9 and as high as 3.0. Values of T_c, p_c, v_c and ω were published by Reid et al. (1977); values of k_{ij} were published by Knapp et al. (1982). Four real-gas equations, quoted from Knapp et al. (1982), are described below.

2.2.1 BWRS Equation of State

$$Z=1+B/v+C/v^2+D/v^5+(C'/v^2)(1+\gamma/v^2)\exp(-\gamma/v^2), \tag{2.70}$$

where $B = (1/RT)(B_0RT - A_0 - C_0/T^2 + D_0/T^3 - E_0/T^4)$,
$C = (1/RT)(bRT - a - d/T)$, $D = (1/RT)\alpha(a + d/T)$,
$C' = (1/RT)(c/T^2)$.

The coefficients of the BWRS equation are given in Table 2.1

Table 2.1 BWRS coefficients (quoted from Knapp et al. (1982))

Pure component coefficients	j	A_j	B_j	Coefficients for mixtures
$\rho_c B_0 = A_1 + B_1\omega$	1	0.44369	0.11545	$B_0 = \sum_i x_i B_{0i}$
$\rho_c A_0/(RT_c) = A_2 + B_2\omega$	2	1.28438	–0.92073	$A_0 = \sum_i \sum_j x_i x_j A_{0i}^{1/2} A_{0j}^{1/2}(1-k_{ij})$
$\rho_c C_0/(RT_c^3) = A_3 + B_3\omega$	3	0.356306	1.70871	$C_0 = \sum_i \sum_j x_i x_j C_{0i}^{1/2} C_{0j}^{1/2}(1-k_{ij})^3$
$\rho_c^2 \gamma = A_4 + B_4\omega$	4	0.544979	–0.2709	$\gamma = \left[\sum_i x_i \gamma_i^{1/2}\right]^2$
$\rho_c^2 b = A_5 + B_5\omega$	5	0.528629	0.34926	$b = \left[\sum_i x_i b_i^{1/3}\right]^3$
$\rho_c^2 a/(RT_c) = A_6 + B_6\omega$	6	0.484011	0.75413	$a = \left[\sum_i x_i a_i^{1/3}\right]^3$
$\rho_c^3 \alpha = A_7 + B_7\omega$	7	0.070523	–0.04445	$\alpha = \left[\sum_i x_i \alpha_i^{1/3}\right]^3$
$\rho_c^2 c/(RT_c^3) = A_8 + B_8\omega$	8	0.504087	1.32245	$c = \left[\sum_i x_i c_i^{1/3}\right]^3$
$\rho_c D_0/(RT_c^4) = A_9 + B_9\omega$	9	0.030745	0.17943	$D_0 = \sum_i \sum_j x_i x_j D_{0i}^{1/2} D_{0j}^{1/2}(1-k_{ij})^4$
$\rho_c^2 d/(RT_c^2) = A_{10} + B_{10}\omega$	10	0.073283	0.46349	$d = \left[\sum_i x_i d_i^{1/3}\right]^3$
$\rho_c E_0/(RT_c^5) = A_{11} + B_{11}\omega\exp(-3.8\omega)$	11	0.006450	–0.02214	$E_0 = \sum_i \sum_j x_i x_j E_{0i}^{1/2} E_{0j}^{1/2}(1-k_{ij})^5$

$k_{ii} = 0$ for $i = j$ (pure substance interaction); x_i, x_j are the molar fractions.

2.2.2 RKS Equation of State

The RKS equation is nearly a classical van der Waals equation:

$$Z = v/(v-b) - a/[RT(v+b)]. \qquad (2.71)$$

The coefficients of the RKS equation are given in Table 2.2.

Table 2.2 RKS coefficients (quoted from Knapp et al. (1982))

Pure component coefficients	Coefficients for mixtures
$a = 0.42747 \alpha R^2 T_c^2 / p_c$	$a = \sum_i \sum_j x_i x_j a_{ij}$
$\alpha = \left[1 + m\left(1 - \sqrt{T/T_c}\right)\right]^2$	$a_{ij} = \sqrt{a_i a_j}(1 - k_{ij})$
$m = 0.48 + 1.574\omega - 0.176\omega^2$	$k_{ij} = 0$ for $i = j$
$b = 0.08664 RT_c / p_c$	$b = \sum_i x_i b_i$

2.2.3 LKP Equation of State

$$Z = Z^{(0)} + (\omega/\omega^{(r)})(Z^{(r)} - Z^{(0)}) \qquad (2.72)$$

The compressibility factors of the "simple fluid" $Z^{(0)}$ and the "reference fluid" $Z^{(r)}$ are calculated separately by a modified BWR equation as functions of reduced variables $T_r = T/T_c$, $p_r = p/p_c$ and $v_r = p_c v/(RT_c)$:

$$Z = p_r v_r / T_r = 1 + B/v_r + C/v_r^2 + D/v_r^5 \\ + [C_4/(T_r^3 v_r^2)][(\beta + \gamma/v_r^2)\exp(-\gamma/v_r^2)], \qquad (2.73)$$

where $B = b_1 - b_2/T_r - b_3/T_r^2 - b_4/T_r^3$,

$C = c_1 - c_2/T_r + c_3/T_r^3$,

$D = d_1 + d_2/T_r$.

The coefficients of the LKP equation are given in Table 2.3.

Table 2.3 LKP coefficients (quoted from Knapp et al. (1982))

Constant	Simple fluid	Reference fluid	Mixing rules
b_1	0.1181193	0.2026579	
b_2	0.265728	0.331511	
b_3	0.154790	0.027655	$T_{cM} = (1/v_{cM})^{\eta} \sum_i \sum_j x_i x_j (v_{cij})^{\eta} T_{cij}$
b_4	0.030323	0.203488	$T_{cij} = (T_{ci} T_{cj})^{1/2} k_{ij}^*$
c_1	0.0236744	0.0313385	
c_2	0.0186984	0.0503618	$k_{ii}^* = k_{jj}^* = 1$
c_3	0.0	0.016901	$v_{cM} = \sum_i \sum_j x_i x_j v_{cij}$
c_4	0.042724	0.041577	
$d_1 \times 10^4$	0.155488	0.48736	$v_{cij} = (1/8)(v_{ci}^{1/3} + v_{cj}^{1/3})^3$
$d_2 \times 10^4$	0.623689	0.0740336	
β	0.65392	1.226	$v_{ci} = Z_{ci} R T_{ci} / p_{ci}$
γ	0.060167	0.03754	$Z_{ci} = 0.2905 - 0.085 \omega_i$
ω	0.0	0.3978	
			$p_{cM} = R T_{cM} Z_{cM} / v_{cM}$
			$Z_{cM} = 0.2905 - 0.085 \omega_M$
			$\omega_M = \sum_i x_i \omega_i$
			$\eta = 0.25$

p_{cM}, T_{cM}, v_{cM}, Z_{cM} are the pseudocritical values. The simple fluid constants were determined from the experimental data for argon, krypton and methane; the reference fluid constants were determined from the experimental data for n-octane.

2.2.4 PR Equation of State

$$Z = v/(v-b) - (v/RT)a/[v(v+b) + b(v-b)] \tag{2.74}$$

The coefficients of the PR equation are given in Table 2.4.

2.2 Real-Gas Equations of State

Table 2.4 PR coefficients (quoted from Knapp et al. (1982))

Pure component coefficients	Coefficients for mixtures
$a = 0.45724 R^2 T_c^2 \alpha / p_c$	$a = \sum_i \sum_j x_i x_j a_{ij}$
$\alpha = [1 + \chi(1 - \sqrt{T/T_c})]^2$	$a_{ij} = \sqrt{a_i a_j}\,(1 - k_{ij})$
$\chi = 0.37464 + 1.54226\omega - 0.26992\omega^2$	$k_{ij} = 0$ for $i = j$
$b = 0.07780 R T_c / p_c$	$b = \sum_i x_i b_i$

The double summations in the mixing rules for multi-component gases cover:
(a) the sum of the binary interactions of the first fraction x_1 with each of the others $x_2, x_3, \ldots$ (j-summation) and
(b) the sum of the binary interactions of each consecutive fraction with each of the others (i-summation).

Examples of some k_{ij}-values are given in Table 2.5.

Table 2.5 Typical examples of the binary interaction parameters k_{ij} (Knapp et al. (1982))

Pairs of gases	Equation			
	BWRS	RKS	LKP	PR
Inorganic/inorganic, similar molar mass CO_2/N_2O	−0.0037	0.0022	0.9944	0.0048
Inorganic/inorganic, different molar mass CO_2/H_2	0.0278	−0.3426	1.625	−0.1622
Organic/organic, similar molar mass C_2H_4/C_2H_6	0.0030	0.0089	0.9911	0.0089
Organic/organic, different molar mass CH_4/nC_6H_{14}	0.0764	0.0374	1.3089	0.0422
Heavy organic/light inorganic nC_6H_{14}/H_2	−0.0404	−0.0800	2.4530	−0.0300
Light organic/heavy inorganic CH_4/CO_2	0.0100	0.0933	0.9874	0.0919

The table conveys an impression of how much the LKP-k_{ij} values depart from 1.0 with growing dissimilarity between the components of the pair and of how much they depart from zero for the other equations.

2.2.5 Computation of Isentropic Exponents

As has been shown, the compressibility factor can be calculated directly from the described real-gas equations of state.

The isentropic volume exponent is calculated from Eq. (2.29):

$$\kappa_v = -\frac{v}{p}\frac{c_p}{c_v}\left(\frac{\partial p}{\partial v}\right)_T. \tag{2.75}$$

The isentropic temperature exponent is calculated from Eq. (2.40) rearranged:

$$\kappa_T = \frac{c_p}{c_p + p\frac{(\partial p/\partial T)_v}{(\partial p/\partial v)_T}}. \tag{2.76}$$

For the partial derivatives reduced variables have to be used and for mixtures pseudocritical values have to be used:

$$\left(\frac{\partial p}{\partial T}\right)_v = \left(\frac{\partial p_r}{\partial T_r}\right)_{v_r}\frac{p_{c(M)}}{T_{c(M)}}, \tag{2.77}$$

$$\left(\frac{\partial p}{\partial v}\right)_T = \left(\frac{\partial p_r}{\partial v_r}\right)_{T_r}\frac{p_{c(M)}^2}{RT_{c(M)}}. \tag{2.78}$$

Equations (2.75) and (2.76) require the heat capacities c_v and c_p calculated in the form of departures from the perfect gas values. Isochoric heat capacity departure:

$$\Delta c_v = c_v(T,v) - c_v^0(T) = T\int_\infty^v \left(\frac{\partial^2 p}{\partial T^2}\right)_v dv. \tag{2.79}$$

Isobaric heat capacity departure:

$$\Delta c_p = c_p(T,p) - c_p^0(T) = -T\int_0^p \left(\frac{\partial^2 v}{\partial T^2}\right)_p dp = \Delta c_v - R - T\frac{(\partial p/\partial T)_v^2}{(\partial p/\partial v)_T}. \tag{2.80}$$

For the LKP equation, Plöcker (1977) calculated these in detail:

$$\frac{c_v - c_v^0}{R} = \frac{2(b_3 + 3b_4/T_r)}{T_r^2 v_r} - \frac{3c_3}{T_r^3 v_r^2} - 6E, \tag{2.81}$$

$$\frac{c_p - c_p^0}{R} = \frac{c_v - c_v^0}{R} - 1 - T_r\frac{(\partial p_r/\partial T_r)_{v_r}^2}{(\partial p_r/\partial v_r)_{T_r}}. \tag{2.82}$$

The partial derivatives for the LKP equation are quoted from Reid, Prausnitz, Sherwood (1977):

$$\left(\frac{\partial p_r}{\partial T_r}\right)_{v_r} = \frac{1}{v_r}\left[1 + \frac{b_1 + b_3/T_r^2 + 2b_4/T_r^3}{v_r} + \frac{c_1 - 2c_3/T_r^3}{v_r^2}\right] + \tag{2.83}$$

$$+ \frac{1}{v_r} \left[\frac{d_1}{v_r^5} - \frac{2c_4}{T_r^3 v_r^2} \left\{ \left(\beta + \frac{\gamma}{v_r^2} \right) \exp\left(-\frac{\gamma}{v_r^2} \right) \right\} \right].$$

$$\left(\frac{\partial p_r}{\partial v_r} \right)_{T_r} = -\frac{T_r}{v_r^2} \left[1 + \frac{2B}{v_r} + \frac{3C}{v_r^2} + \frac{6D}{v_r^5} \right] \quad (2.84)$$

$$- \frac{T_r}{v_r^2} \frac{c_4}{T_r^3 v_r^2} \left\{ 3\beta + \left(5 - 2\left(\beta + \frac{\gamma}{v_r^2} \right) \right) \frac{\gamma}{v_r^2} \right\} \exp\left(-\frac{\gamma}{v_r^2} \right),$$

$$E = \frac{c_4}{2 T_r^3 \gamma} \left[\beta + 1 - \left(\beta + 1 + \frac{\gamma}{v_r^2} \right) \exp\left(-\frac{\gamma}{v_r^2} \right) \right]. \quad (2.85)$$

The special trick for using the LKP equation consists in calculating the heat capacity departures and the above derivatives separately for the simple and the reference fluid with the acentric factor serving as a means of interpolation:

$$X = X^{(0)} + \omega / \omega^{(r)} \left(X^{(r)} - X^{(0)} \right), \quad (2.86)$$

where $\quad X = \dfrac{\Delta c_v}{R}, \dfrac{\Delta c_p}{R}, \left(\dfrac{\partial p_r}{\partial T_r} \right)_{v_r}, \left(\dfrac{\partial p_r}{\partial v_r} \right)_{T_r}. \quad (2.87)$

Acentric factors ω (constant for each substance) and molar isobaric heat capacities for perfect gases $c_p^0 = c_p^0(T)$ can be found in numerous tables, e.g. in polynomial numerical form in the book of Reid, Prausnitz, Sherwood (1977) for 468 pure substances (temperature T in K):

$$c_p^0 = CPVAPA + (CPVAPB)T + (CPVAPC)T^2 + (CPVAPD)T^3 \quad \frac{\text{kcal}}{\text{kmol K}}. \quad (2.88)$$

Molar isobaric heat capacities for perfect gas mixtures are calculated with the molar fractions as follows:

$$c_{pM}^0 = \sum_i x_i c_{pi}^0 \quad \text{and} \quad \sum_i x_i = 1. \quad (2.89)$$

The molar isochoric heat capacity for perfect gas behavior (single gas and mixture) is given by

$$c_v^0 = c_p^0 - R; \quad c_{vM}^0 = c_{pM}^0 - R. \quad (2.90)$$

Thus all relevant parameters for volume flow, head, power and discharge temperature for mixtures with real gas behavior are calculable: compressibility factors with Eqs. (2.70)–(2.73); isentropic volume and temperature exponents with Eqs. (2.75) and (2.76) together with (2.79) through (2.90); polytropic volume and temperature exponents with Eqs. (2.42) and (2.37).

2.2.6 Recommendations for the Use of Real-Gas Equations

Compressibility factors were calculated for various pure substances with the BWRS, RKS and LKP equations of state within the relevant ranges of pressure and temperature for process compressors. The findings were compared with the gas property tables published by Canjar and Manning (1967), Vargaftik (1975), and Angus et al. (1976). The results for these selected frequently occurring gases were as follows:

- air: LKP best agreement (deviations < 0.2%); BWRS slightly worse (< 0.3%); RKS still acceptable (< 0.8%),
- hydrocarbons (paraffins, olefins, $C_1...C_3$): LKP best agreement (< 0.7%); RKS worse (< 1.2%); BWRS still acceptable (< 1.7%),
- carbon dioxide: BWRS best agreement for $p_{max} < 50$ bar (< 0.2%); LKP slightly worse (< 0.5%); LKP best agreement for $p_{max} = 50$–250 bar (< 2%); BWRS worse (< 5%); RKS not recommended
- hydrogen: RKS best agreement (< 0.5%); LKP still acceptable for $p < 200$ bar; BWRS not recommended,
- chlorine: RKS best agreement (< 0.5%); BWRS satisfactory (< 1%); LKP highest deviation but still acceptable (< 1.6%),
- ammonia: LKP and BWRS best agreement (however, deviation < 3%!); RKS not recommended.

It should be mentioned, though, that the question "which equation for which gas?" is discussed controversially and therefore answered differently. Reid, Prausnitz, Sherwood, for instance, recommend RKS and LK for hydrocarbons, N_2, CO_2, and H_2S and do not recommend the Soave equation if the mixture contains H_2.

2.3 Handy Formulae for Humid Gases

2.3.1 Relative Humidity, Volume Fraction, Humid Molar Mass

Very often the specified gas composition in an enquiry covers dry gas constituents only, and the moisture, if any, is stated in the form of the relative humidity.

The relative humidity is defined as the ratio of the water partial pressure over the saturation pressure:

$$\varphi = \frac{p_{H_2O}}{p_{sat}}. \tag{2.91}$$

Since the volume fraction of the water vapor is the ratio of the partial pressure over the total pressure, the combination with Eq. (2.91) results in

$$r_{H_2O} = \frac{p_{H_2O}}{p} = \varphi \frac{p_{sat}}{p}. \tag{2.92}$$

The molar mass of humid gas using the dry gas molar mass is given by

$$M_{hum} = r_{H_2O} M_{H_2O} + (1 - r_{H_2O}) M_{dry}, \quad (2.93)$$

where $M_{H_2O} = 18.016$ kg/kmol is the molar mass of water vapor, and M_{dry} is the molar mass of dry gas.

2.3.2 Absolute Water Content, Humid Mass, Volume Flow

The absolute water vapor content is given by

$$x = \frac{R_{dry}}{R_{H_2O}} \frac{\varphi p_{sat}}{p - \varphi p_{sat}} \quad / \text{ kg H}_2\text{O/kg dry gas}, \quad (2.94)$$

where $R_{H_2O} = 0.4615$ kJ kg^{-1} K^{-1} is the gas constant of water vapor and R_{dry} is the gas constant of dry gas.

The gas constant of humid gas is given by

$$R_{hum} = \frac{0.4615x}{1+x} + \frac{R_{dry}}{1+x} \quad / \text{ kJ kg}^{-1}\text{ K}^{-1}. \quad (2.95)$$

The humid mass flow is given by

$$\dot{m}_{hum} = \dot{m}_{dry}(1+x). \quad (2.96)$$

The molar isobaric heat capacity of humid gas is given by

$$Mc_{p\_hum} = r_{H_2O} Mc_{p\_H2O} + (1 - r_{H_2O}) Mc_{p\_dry}, \quad (2.97)$$

where $M_{cp\_H_2O} = 33.6$ (20 °C) – 33.9 (80 °C) kJ kmol^{-1} K^{-1} is the molar isobaric heat capacity of water vapor and Mc_{p_dry} is the molar isobaric heat capacity of dry gas.

This allows us to calculate the ratio of the heat capacities for a perfect gas as:

$$\kappa_{dry} = \frac{Mc_{p\_dry}}{Mc_{p\_dry} - \Re} \; ; \quad \kappa_{hum} = \frac{Mc_{p\_hum}}{Mc_{p\_hum} - \Re}, \quad (2.98)$$

where $\Re = 8.3145$ kJ kmol^{-1} K^{-1} is the universal gas constant.

In the case where the flow is specified in terms of the standard (dry) volume flow, the calculation of the actual humid volume flow is

$$\dot{V} = \dot{V}_0 \frac{T}{T_0} \frac{Z}{Z_0} \frac{p_0}{p_{dry}} = \dot{V}_0 \frac{T}{T_0} \frac{Z}{Z_0} \frac{p_0}{p - p_{H_2O}} = \dot{V}_0 \frac{T}{T_0} \frac{Z}{Z_0} \frac{p_0}{p - \varphi p_{sat}}, \quad (2.99)$$

where T_0, Z_0, p_0 are the standard conditions (SI-system: $T_0 = 273.15$ K, $p_0 = 1.01325$ bar); and T, Z, p are the actual conditions; p_{dry} is the partial pressure of dry gas.

The above equation reflects the fact that, in the presence of water vapor, the dry gas partial pressure decreases, which, in turn, increases the actual volume flow.

2.3.3 Calculation of Saturation Pressure

An interesting option for calculating the water saturation pressure p_{sat}, rather than using tables, is the Clausius–Clapeyron equation (see Baehr (1989)):

$$\frac{dp}{dT} = \frac{r_0 p}{RT^2}, \qquad (2.100)$$

where r_0 is the heat of evaporation, assumed to be constant (approximation).

This equation states that the natural logarithm of the saturation pressure $\ln(p_{sat})$ versus the inverse absolute temperature $(1/T)$ constitutes (nearly) a straight line, which allows us to formulate a very good approximation for the saturation pressure of water vapor: $\ln(p_{sat}) = A - B/T$. This yields a simple formula for the saturation pressure (A and B depend on the (accurate) straight line "anchor points"):

$$p_{sat} = e^{(A-B/T)} \; / \; \text{bar}, \qquad (2.101)$$

where $A = 14.2805$, $B = 5288.358$; Eq. (2.101) is valid for a temperature range of $t = 9\,°C - 47\,°C$ with a maximum error of $p_{sat} = \pm 0.5\%$; p_{sat} is accurate for $t = 15\,°C$ and $t = 40\,°C$ (anchor points); $T = t + 273.15$ K.

2.3.4 Calculation of Dew Point

If the water content is part of the specified gas constituents the dew point at the compressor suction flange can be checked by calculating the relative humidity:

$$\varphi = r_{H_2O} \frac{p}{p_{sat}} \quad \text{or:} \quad \varphi = m_{H_2O} \frac{M_{hum}}{M_{H_2O}} \frac{p}{p_{sat}}, \qquad (2.102)$$

where r_{H_2O} is the water vapor volume fraction, m_{H_2O} is the water vapor mass fraction, M_{hum} is the molar mass of the humid gas, $M_{H_2O} = 18.016$ kg/kmol is the molar mass of the water vapor.

If, at the preset conditions of pressure and temperature, the relative humidity equals or exceeds 1.0, the gas is at or below the theoretical dew point with the danger of liquid carryover, which may jeopardize the impellers by erosion. It is recommended, however, that, in the absence of a water separator, the relative humidity at the compressor inlet be adequately below unity (say 5–10%), since through possible cooling at the piping walls the gas may nevertheless locally drop below the dew point, leading to liquid carryover.

Of greater importance is the dew point calculation for intercooled compressors. This determines whether a water separator is necessary and, in the case of dropping below the dew point, how great the condensing mass flow is.

Using Eq. (2.101) the dew-point temperature can be calculated explicitly as

$$T_{\text{dew}} = \frac{B}{A + \ln\left(1 + \frac{1}{x_1} \frac{R_{\text{dry}}}{R_{H_2O}}\right) - \ln p_{\text{IC}}} \quad / \text{K}, \tag{2.103}$$

where x_1 is the ratio kg H_2O/kg dry gas (absolute water content upstream of intercooler), p_{IC}/bar is the intercooler pressure, $A = 14.2805$, $B = 5288.358$, $t = T - 273.15/°C$, R_{dry} is the gas constant of dry gas, and $R_{H_2O} = 0.4615$ kJ kg^{-1} K^{-1}.

In the case where the recooling temperature is lower than the dew-point temperature, there will be a certain amount of water drop out. The condensing mass flow can be calculated as follows. Using Eq. (2.94) and setting the relative humidity, $\varphi = 1.0$ yields the maximum absolute water content at the recooling temperature, T_R in question (downstream of the cooler):

$$x_{2\max} = \frac{R_{\text{dry}}}{R_{H_2O}} \frac{p_{\text{sat}}(T_R)}{p_{\text{IC}} - p_{\text{sat}}(T_R)}. \tag{2.104}$$

The water drop out in the first intercooler is the difference between x_1 and $x_{2\max}$:

$$\Delta x = \eta_{\text{sep}} \frac{R_{\text{dry}}}{R_{H_2O}} \left(\frac{\varphi p_{\text{sat}}(T_s)}{p_s - \varphi p_{\text{sat}}(T_s)} - \frac{p_{\text{sat}}(T_R)}{p_{\text{IC}} - p_{\text{sat}}(T_R)} \right) / \frac{\text{kg } H_2O}{\text{kg dry gas}}, \tag{2.105}$$

where p_s, T_s, φ are the compressor pressure, temperature, relative humidity at suction, p_{IC} is the intercooler pressure, T_R is the recooling temperature (gas temperature at the intercooler exit) and $\eta_{\text{sep}} = 0.85–0.98$ is the degree of separation, depending on the separator design. The water separation in the subsequent coolers can be calculated accordingly, accounting for the degree of separation.

References

Angus S, Armstrong B, Reuck KM (1976) International thermodynamic tables of the fluid state. Vol 3, carbon dioxide. IUPAC (International Union of Pure and Applied Chemistry); Pergamon Press, Oxford, UK

Baehr HD (1989) Thermodynamik. 7th edn. Springer, Berlin Heidelberg New York, pp 170–171 (in German)

Beinecke D, Lüdtke K (1983) Die Auslegung von Turboverdichtern unter Berücksichtigung des realen Gasverhaltens. In: VDI-Berichte 487, VDI-Verlag Düsseldorf, pp 271–279

Benedict M, Webb RB, Rubin L C (1940) J. Chem. Phys. 8

Canjar LN, Manning FS (1967) Thermodynamic properties and reduced correlations for gases. Gulf Publishing Co., Houston

Dzung LS (1944) Beiträge zur Thermodynamik realer Gase. Schweizer Archiv 10, pp 305–313 (in German)

Kammerlingh Onnes H (1901) Comm. Phys. Lab., Leiden, Netherlands

Knapp H, Döring R, Oellrich L, Plöcker U, Prausnitz JM (1982) Vapor-liquid equilibria for mixtures of low boiling substances. Chemistry Data Series, vol VI, Dechema, Frankfurt, Germany, pp 771–793

Lee BI, Kesler MG (1975) AIChE J. 21(3)

Nelson LC, Obert EF (1954) Trans. ASME 76, pp 1057

Peng DY, Robinson DB (1976) Ind. Eng. Chem. Fund. 15

Plöcker U (1977) Doctoral Thesis. Technical University of Berlin, Germany

Plöcker UJ, Knapp H, Prausnitz JM (1978) Ind. Eng. Chem. Proc. Des. Dev. 17

Redlich O, Kwong JNS (1949) Chem. Rev. 44

Reid RC, Prausnitz JM, Sherwood TK (1977) The properties of gases and liquids. 3rd edn. McGraw Hill, New York, pp 96 and 127

Soave G, (1972) Chem. Eng. Sci. 27

Starling KE (1973) Fluid thermodynamic properties of light petroleum systems. Gulf Publishing Co., Houston

Traupel W (1977) Thermische Turbomaschinen. 1st vol, 3rd edn. Springer, Berlin Heidelberg New York, p 8 (in German)

Van der Waals JD (1873) Dissertation, University of Leiden, Netherlands

Vargaftik NB (1975) Tables on the thermophysical properties of liquids and gases. 2nd edn. Hemisphere Publishing Co., Washington London

3 Aero Components – Function and Features

At least ten components with distinct functions make up the complete flow channel within the compressor, Fig. 1.8. The primary component responsible for the energy transfer is the impeller, and the stationary components serve the purpose of supplying the flow to the impeller or processing the flow leaving the impeller.

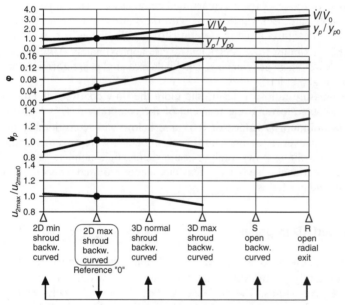

Fig. 3.1 Impeller types and properties. Arrows indicate impeller development starting from reference: volume flow, head, flow coefficient, head coefficient and tip speed.

3.1 Impeller

3.1.1 Centrifugal Impeller Types

The aero-thermodynamic and structural mechanical capabilities and limitations of this fast rotating and therefore highly stressed constituent determine the application potential of the compressor. Impeller sizing and setting of the rotational speed are determined by several disciplines: aero-thermodynamics must assure volume flow, head, efficiency and operating range; stress analysis must prove static and dynamic integrity; rotor dynamics must warrant smooth running; and production engineering should enable economic manufacture. Figure 3.1 is a summary of the different impeller designs and their most important parameters; the arrows indicate, starting from the traditional medium-flow-coefficient 2D-impeller, how the centrifugal wheel evolved into a high capacity, high head, high efficiency device, on the one hand, and also into a very narrow impeller for high-pressure applications, on the other hand.

3.1.2 The Shrouded 2D-Impeller

This is the impeller type which the first centrifugal compressor was equipped with shortly after 1900, and the 2D-impeller, although manufactured in a different way, is still the workhorse of today's process centrifugal compressors. This impeller has a cover disk; its blades are backward curved with the same curvature across the entire blade width, a design commonly referred to as "2D-blades" or, in short, "2D-impeller". The maximum flow coefficient φ_{max} is approximately 0.06; higher values will tend to lower stage efficiencies. The blade exit angles are normally between 40° and 50° (Fig. 3.2, left).

The higher the pressure ratio per casing the more the flow coefficients are reduced from stage to stage towards the rear end of the compressor. This leads to 2D-impellers with minimum φ values of 0.01 and below with considerably reduced efficiencies. However, it was not until such impellers were developed that compressors with low volume flows, high pressure ratios, high numbers of stages per casing and low rotational speeds were feasible, as are required, especially in refineries and in oil and gas fields.

3.1.3 The Shrouded 3D-Impeller

An increase of the volume flow at constant wheel diameter and speed leads to higher φ values (presently around 0.15) and consequently to wider eye openings and greater impeller exit widths. The blade exit angles are normally between 45° and 60°, from tangent. In order to attain high efficiencies the blade length has to be increased and the blade leading edge has to be adapted to the different flow angles at the hub and eye diameter. These stipulations lead to blades twisted in space with a leading edge pulled upstream into the eye opening with different curvatures across the blade width. They also result in a smaller hub diameter and an increased axial stage length (Fig. 3.2, right). These impellers, commonly referred to as 3D-impellers, are characterized by a comparatively high efficiency at a reduced outer diameter (for a given specification), a wide operating range, a high rotational speed and a reduced maximum number of stages per single-shaft compressor casing.

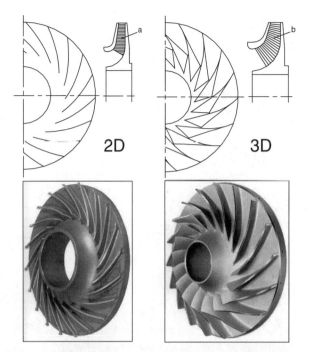

Fig. 3.2 Impeller types for single-shaft compressors. *Left*: 2D-impeller, shrouded with backward-curved blades. *Right*: 3D-impeller, shrouded with twisted backward-curved blades. a straight-line blade elements for 3-axis NC flank milling, b straight-line blade elements for 5-axis NC flank milling. Photos: hub disks with integral blades before brazing of shrouds.

3.1.4 The Semi-Open Impeller

A higher tip speed is a prerequisite for a further increase of volume flow and head. This requires elimination of the cover disk to get rid of its additional centrifugal load, optimization of the hub disk shape, further elongation of the blades into the eye (which then becomes the famous "inducer"), increase of the blade exit angle from tangent, and conical blades with a thickness distribution that minimizes blade vibration. The shroudless wheels operate with a narrow gap between the blade tips and the diaphragm and also have twisted blades.

The impeller with backward-leaning blades whose axial projection show a slight S-shape is referred to in short as S-impeller. The blade exit angle is mostly between 45° and 70° (Fig. 3.3, left).

The R-impeller, an abbreviation for a wheel with radially ending blades, with an exit angle of 90° enables the highest tip speeds and, hence, the highest heads of all impellers at considerably lower efficiency, smaller operating range and flatter performance curve, i.e. a lower head rise from normal to surge (Fig. 3.3, right).

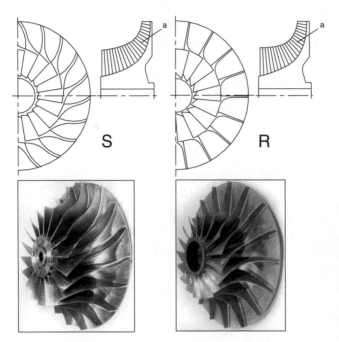

Fig. 3.3 Impeller types for integrally geared compressors. *Left*: S-type impeller, semi-open with backward-leaning blades. *Right*: R-type impeller, semi-open with radial exit blades. a straight-line blade elements for 5-axis NC flank milling.

3.1.5 Impeller Family

Compressor manufacturers do not invent the wheel anew from contract to contract but use instead a system of pre-engineered stages and impellers that have a more or less standardized geometry. At least the impeller blade shapes with inlet and exit angles are fixed and scalable with the other dimensions, like inlet and exit width, eye and hub diameter being adapted to the special application. Others have a narrowly staggered true modular design system with a completely fixed but scalable geometry, which has the advantage of a relatively high prediction accuracy since virtually all modular stages have test-derived performance curves for the complete range of industrial gases. So the components with known characteristics are selected for the special case rather than designed anew; and the compound compressor performance curve is determined by aerodynamically superimposing the known stage curves. Scaling permits adaptation to the contract conditions in question on the basis of limited deviations from the laws of aerodynamic similarity. The diversity and complexity in the process compressor world is obtained by using the same elements with as little modifications as possible for all applications.

An impeller family like this is shown in Fig. 3.4 for the complete range of flow coefficients. Impeller "convergency" variation for different molar masses and hub/tip ratio variation to satisfy rotordyamics are optional modifications of the base line.

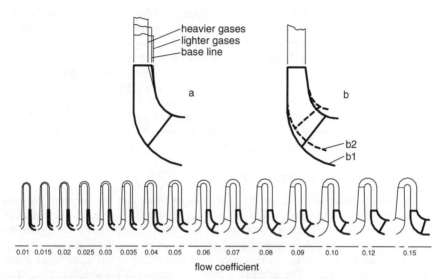

Fig. 3.4 Impeller family. *Bottom*: base line; *top*: meridional variants. *a* convergency variation, *b* hub/tip ratio variation, *b1* baseline, *b2* higher hub/tip ratio.

3.1.6 Impeller Application

Single-shaft compressors are almost exclusively equipped with shrouded 2D- and 3D-impellers. Before the advent of 3D-impellers multistage compressors in rare cases had a semi-open first-stage impeller. Semi-open impellers nowadays are predestined as overhung designs with axial suction nozzles in single or multistage integrally geared compressors, since the advantage of a high tip speed is not compromised by the rotordynamics of the short pinion and since the small tip clearance is under control. In many applications integrally geared compressors are also equipped with shrouded 3D-impellers and, at higher discharge pressures, with 2D-impellers as rear stages.

3.1.7 Impeller Manufacture

Depending on the compressor OEM, the user specification and the impeller size there are the following manufacturing methods.

Blade Milling

The flow channels are NC-milled from the solid hub forging. 2D-impellers require three translational axes. 3D-impellers as well as S- and R-impellers require five axes for the relative movements between the machine tool and the work piece: three translational x, y and z and two rotational movements α and β in two planes perpendicular to each other. In principle it does not matter whether the tool or the wheel or both perform these movements; in practice, however, it is advisable to distribute the five movements in such a way that the maximum rigidity is obtained for the milling procedure. Since both blade surfaces are made up of straight lines in space at any arbitrary point of the blade surface, which is not detrimental to aerodynamics at all (see framework of straight lines from hub to shroud in Fig. 3.2), flank milling is possible along these straight lines, whereby the milling tool mills along the entire blade height (in contrast to line or point milling). In the proper sense of the word flank milling approximates the straight line surface by a slight helix due to the actual thickness of the tool.

High-Temperature Vacuum Brazing

In the case of shrouded impellers the cover disk is either high-temperature vacuum brazed or welded to the blades. The brazing is performed for impeller diameters of approximately a maximum of 1,000 mm at present. For impellers with higher diameters the absolute deformation during the heat-up process falls above or below the allowable limits for the brazing gap, thus preventing a sufficient bond along the entire blade length. The brazing material is mostly a gold/nickel alloy. The brazing process is automated and time/temperature controlled and does not result in any obstruction of the flow channel. It is applicable also for extreme narrow impeller widths and restricts the tip speed only marginally, since the strength of the new alloy in the diffusion zone of the braze is only slightly lower than that of the base material.

Welding of the Cover Disk

Welding of the cover disk to the blades is performed by inserting the electrode into the flow channel from the inside or the outside. In the case where the actual width between disks is smaller than approximately 20 mm the shroud is welded from the outside through milled slots that have to be adapted exactly to the blade shapes (slot welding).

Separate Blade Manufacture

Blades may also be manufactured separately by die forging or casting and then welded to cover and hub disk. This is mostly applied with impeller diameters in excess of 1,300 mm.

Cast Impellers

One-piece sand casting is used for shrouded and semi-open impellers > 400 mm diameter. Inclusions or cavities, however, reduce the integrity of the casting and hence the maximum tip speed, and the pattern costs usually require more than three casts and prevent a quick and efficient geometry variation. Investment casting or the lost-wax method is used for smaller diameters and higher numbers of pieces. There are very high pattern costs with no options for varying the geometry.

Stereolithography

A quick and relatively low-cost procedure patented in 1986 is used for pattern manufacture; it is called 3D-stereolithography or rapid prototyping. An ultraviolet laser beam controlled by the impeller CAD data "constructs" the positive wheel pattern in a light-sensitive synthetic resin bath by successively hardening of thin horizontal layers. The model in many cases serves first of all for design verification or as a first prototype and can also be used in a cire perdu process to produce a cast impeller.

Riveted Impellers

The shroud is riveted to the blades; the hub disk, blades and cover disk have axial bores for the through rivets. This is applicable for 2D-blades only; thicker blades reduce the efficiency, and the strength of the rivets diminishes the maximum tip speed. Riveting is an outdated jointing technology.

Electroerosion

Electroerosion or electric discharge machining (EDM) is based on material removal by spark discharges in a dielectric fluid (usually a hydrocarbon) between the workpiece and the tool which has the shape of the flow channel. The method produces a single-piece impeller but is limited to small sizes and very simple channel shapes.

5-axis NC-milling of impellers with brazed or welded cover disks is regarded as the jointing technique with the highest quality as far as precision, strength, material integrity and erosion and corrosion resistance are concerned.

3.1.8 Impeller Stress Analysis

Stresses and strains are normally calculated according to the finite-element-method (FEM); see Fig. 3.5. Calculations were carried out by means of a simplyfied quasi-3D approach, applying angular elements with triangular cross-sections for the disks and quasi-flat radially oriented elements for the blades that transmit no tangential forces.

The method was checked by strain measurements on the running wheel and found to be satisfactory. Lines of constant combined stress in a 2D-impeller are shown.

Since material strengths are defined for mono-axial stresses only, the multi-axial stress field with its tangential, axial and radial normal and shear stress components has to be reduced to an appropriate equivalent stress termed "combined" stress. The hypothesis applied here is the "maximum shear strain energy criterion" or "von Mises" method (another one is the "maximum shear-stress criterion").

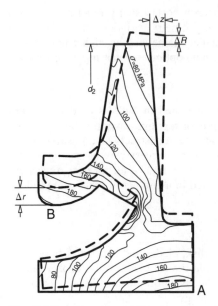

Fig. 3.5 Impeller stress and strain. Example: $d_2 = 400$ mm, $u_2 = 200$ m/s, $\Delta R = 0.06$ mm, $\Delta r = 0.11$ mm, $\Delta z = 0.09$ mm (distortion out of scale), σ/Mpa = N/mm^2 combined stress. A, B are regions of highest stress.

Tangential or hoop stresses are by far the dominant stress components in an impeller under centrifugal load.

In this 2D-impeller equally high stresses occur in the impeller bore on the rear side of the hub (location A) and in the inner rim of the cover disk at the impeller eye (location B). With increasing eye diameter (i.e. increasing flow coefficient) the cover disk stresses increase in absolute terms and in relation to the bore stresses. The impeller deforms by shrinking axially, widening radially (with the exception of the bore, where the shrink fit counteracts the centrifugal force) and by either tilting backward towards the hub side (2D-impeller) or tilting forward towards the shroud side (3D-, S- and R-impellers). Due to the fictitious blade elements in this simplified calculation model, the blade stresses shown are rough approximations only, which one does not have to be concerned about because blade stresses are not the determining factors for maximum impeller loads.

3.2 Inlet Duct

Single-shaft process compressors in almost all cases have a radial centerline intake nozzle with suction flange. As shown in the cross-sectional sketch, Fig. 1.8, the body head part contains the inlet duct from the flange to the first impeller eye.

This flow channel is sometimes referred to as the plenum inlet in contrast to the much narrower return channel inlet of the middle stage. It serves the purpose of guiding the incoming radial flow into the axial direction. After passing through the circular area of the inlet nozzle the flow is divided (Fig. 3.6a) by the intake rib, accelerated twofold to threefold and diverted into the eye while going around the shaft.

The shaped flow splitter or suction baffle prevents a collision of the two remaining partial flows. The gas leaves the plenum inlet through an annulus directly leading into the eye of the impeller. The overall stage performance requires, as far as possible, a vortex-free flow with uniform velocity profiles in the meridional and circumferential directions in this plane and a minimum pressure loss in the plenum inlet. If the inlet duct is not properly designed, it will not only result in increased duct losses but could adversely affect the impeller performance by a distorted state at the inlet.

With barrel compressors the radially oriented suction nozzle can be connected to the body under any desired angle. Horizontally split compressors require either a top or a bottom entry. The latter is the most used, since the removable top casing half is normally kept free of any pipe connections for the sake of easy maintenance. If the specification for a horizontally split compressor calls for a ground-level block rather than a table foundation plus a pipe-free upper casing half, the radial suction nozzle must be given up and replaced by a horizontal one, leading necessarily to a tangential connection with the lower body half, since the nozzle cannot interfere with the parting joint; see Fig. 3.6b. The suction nozzle is sized in such a way that the plenum inlet loss is minimized.

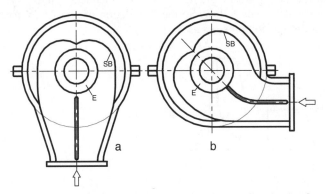

Fig. 3.6a, b Compressor inlet duct. **a** radial centerline intake, **b** tangential offset intake, *E* impeller eye, *SB* suction baffle.

3.3 Adjustable Inlet Guide Vanes

A row of movable axial inlet guide vanes upstream of the impeller (Fig. 3.7) is another (optional) device to provide a service to the impeller enabling it to vary its head at constant volume flow and speed, which otherwise would not be feasible. The vanes create gas prerotation, thus influencing primarily the head of the adjoining impeller according to Euler's equation, Eq. (1.51). Positive prerotation reduces the head, negative prerotation increases it within small limits. The vane operation is shown in Fig. 3.7.

The axial location is preferred over the radial arrangement because of its far greater potential to distribute the flow circumferentially evenly to the impeller. This is not necessarily warranted with the radial row when partially closed, since both halves have totally different flow patterns and resistances which deflect the flow in quite a different way, as can readily be seen in Fig. 3.8. This possible partial admission can lead to lower partload efficiencies and even to an adverse impeller performance.

In contrast to axial and integrally geared compressors, centrifugal single-shaft compressors cannot be equipped with adjustable IGVs ahead of each stage due to an extreme lack of interstage space dictated by the rotordynamic requirements. The maximum number of IGV rows is two for back-to-back arranged impellers.

Not only single-stage but also multistage process compressors are sometimes equipped with just one row of AIGVs upstream of the first stage, because the regulability, although diminished, is warranted and offers in most cases a wider operating range and an improved efficiency over suction throttled compressors. It is surprising: turndown and partload power consumption are improved relative to suction throttling, even for six- and seven-stage machines. The reason for that is that when the operating point approaches the surge at constant discharge pressure the pertaining IGV regulated performance curve becomes steeper through the influence of the modified first-stage curve. Thus the point on the pressure curve with the horizontal tangent (i.e. zero pressure gradient), which is identical with the system surge, shifts to a smaller volume flow. So the installation of one AIGV row may be justified for multistage compressors as well.

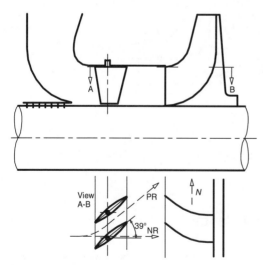

Fig. 3.7 Adjustable inlet guide vanes, axial positioning. Example: inducer-type impeller with vane position 39° from axial (positive prerotation), N impeller rotation, NR no flow prerotation, PR prerotation.

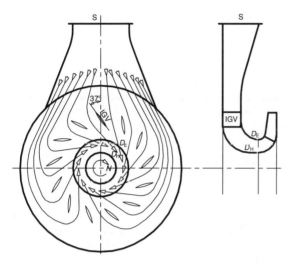

Fig. 3.8 Adjustable inlet guide vanes, radial positioning. Example: vane position 37° from radial (positive prerotation), S suction nozzle, N rotation, D_H hub diameter, D_E eye diameter, IGV inlet guide vanes.

3.4 Annular Diffuser

Unfortunately the kinetic energy at the impeller exit constitutes 35 to 40% of the total work input! Since the target of process gas compression is to achieve static pressure, this huge portion of this not readily usable dynamic enthalpy has to be converted through diffusion (deceleration) as far as possible to a static isentropic enthalpy rise, which in turn will lead to an increase of the static pressure. This is the role of the annular diffuser directly downstream of the impeller. It consists of two parallel or, in some cases, converging walls forming an open radial passage from the impeller exit to some defined outer radius. The most common diffuser design for process compressors is vaneless, radially oriented and parallel-walled with radius ratios of between 1.4 and 2.0; but there are also pinched (converging) and vaned diffusers.

The absolute gas velocity at the impeller exit reaches, at tip speeds of 320 m/s, values of 800 km/h, i.e. four times that of hurricanes. And, by the way, the flow pattern is not unlike that of a hurricane outside the eye insofar as the predominant circumferential component of the velocity basically follows the law of conservation of angular momentum of a non-viscous (frictionless) flow: rc_u=const, i.e. c_u is inversely proportional to the radius. In reality the reduction of the circumferential velocity is still stronger due to viscous effects:

$$c_{u4} = \frac{r_3}{r_4} c_{u3} - F_r ,\qquad(3.1)$$

where F_r represents the friction correction.

The reduction of the relatively small meridional component of the velocity follows the continuity equation:

$$c_m = \frac{\dot{m}}{2\pi r b \rho}. \tag{3.2}$$

Therefore, a diffuser with parallel walls handling an incompressible flow follows the law: $c_m r = \text{const}$, i.e. c_m is inversely proportional to the radius as well.

Hence the angle of the non-viscous flow is:

$$\alpha = a\tan\left(\frac{c_m}{c_u}\right) = a\tan\left(\frac{\dot{m}r}{2\pi r b \rho k}\right) = a\tan\left(\frac{k^*}{b\rho}\right). \tag{3.3}$$

This means that for a non-viscous ($F_r = 0$) and incompressible ($\rho = \text{const}$) flow in a parallel-walled diffuser ($b = \text{const}$) the flow angle remains constant, i.e. the streamlines are logarithmic spirals. So, to a large degree, this is still the case in a parallel-walled wide diffuser, where viscous effects are negligible, handling a low Mach number flow (e.g. H_2-rich gases).

Under the influence of pinching and friction the flow angle increases, and under the influence of compressibility it reduces.

The transformation of the velocity triangle in the diffuser is shown in Fig. 3.9.

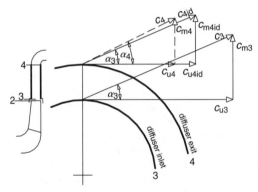

Fig. 3.9 Vaneless diffuser, parallel-walled, velocity triangle transformation. id: ideal = non-viscous, incompressible flow ($\alpha_{4id} = \alpha_3$); *dashed triangle*: viscous, compressible flow ($\alpha_4 > \alpha_3$).

So the recovery of the dynamic enthalpy in the vaneless diffuser is achieved by deceleration in two different ways:

- Conservation of angular momentum reduces the circumferential velocity component.
- Increase of annular throughflow area reduces the meridional velocity component.

3.4 Annular Diffuser

At the exit of a diffuser with a radius ratio of $r_4/r_3 = 1.65$ the kinetic energy is still 16 to 28% of the total work input, depending on the flow coefficient of the stage.

In Fig. 3.10 it is made clear that, in order to obtain a static pressure rise, it is essential to bring about a static isentropic enthalpy rise; an enthalpy rise solely based on entropy production does not create any pressure. A diffuser pertaining to a low flow coefficient stage reduces the kinetic energy much more than a high flow coefficient diffuser (at the same radius ratio), but mainly due to higher losses; so no adequate pressure recovery is actually brought about.

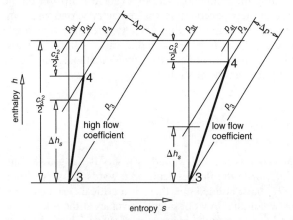

Fig. 3.10 Vaneless diffuser performance. *3* diffuser inlet, *4* diffuser exit.

The isentropic enthalpy recovery is defined as:

$$c_{hR} = \frac{\Delta h_s}{c_3^2/2}. \tag{3.4}$$

Applying perfect-gas relationships:

$$c_{hR} = \frac{T_{4s} - T_3}{T_{3t} - T_3} = \frac{\dfrac{T_{4s}}{T_3} - 1}{\dfrac{T_{3t}}{T_3} - 1} = \frac{\left(\dfrac{p_4}{p_3}\right)^{\frac{\kappa-1}{\kappa}} - 1}{\left(\dfrac{p_{3t}}{p_3}\right)^{\frac{\kappa-1}{\kappa}} - 1}. \tag{3.5}$$

If c_{hR} is zero, the diffuser pressure rise $\Delta p = p_4 - p_3$ is zero. One might as well omit the exponents in Eq. (3.5), leading to the pressure recovery:

$$c_{pR} = \frac{\dfrac{p_4}{p_3} - 1}{\dfrac{p_{3t}}{p_3} - 1} = \frac{p_4 - p_3}{p_{3t} - p_3} \tag{3.6}$$

The pressure recovery, which is more commonly used in the literature than the isentropic enthalpy recovery, states how much of the dynamic inlet pressure is converted into a static pressure rise by the diffuser. Both coefficients, numerically nearly identical, express the same facts of the matter. Typical values are 0.4 to 0.6, depending on width, length, surface roughness and, very important, inlet blockage by low energy material which is often present at the impeller exit, e.g. in the cover disk/blade suction side corner. The latter, especially, can have a severe adverse effect on the diffuser performance.

In 1960 Traupel (1962), in a seminar at the Technical University of Zürich, presented his approach of calculating the change of state in a vaneless diffuser on the basis of consecutive finite-radius steps, accounting for geometry variations, compressibility and viscous effects and coming up with velocity, pressure, flow angle and isentropic enthalpy recovery, depending on the radius ratio. Although below the level of present-day CFD (computational fluid dynamics) methods, his theory is still very valuable, especially for parameter studies.

3.5 Vaned Diffuser

One of the disadvantages of vaneless parallel-walled diffusers is the growing instability of the flow at the inlet once the flow is reduced towards the surge limit. As Senoo and Kinoshita (1977) have pointed out, there is a critical (vaneless) diffuser inlet flow angle, below which the flow onesidedly reverses at the inlet, causing rotating stall and eventually triggering the incipience of surge. This is because the drastically reduced meridional velocity component at the front wall of the diffuser does not have sufficient kinetic energy to maintain its forward movement to overcome the radial pressure gradient that has been forced upon the flow through the conservation of angular momentum.

Senoo et al. (1983) reported on the favorable effect of low-solidity diffuser (LSD) vanes located downstream of the impeller tip. These airfoil cascades with a large pitch/length ratio, having no throat whatsoever between adjacent vanes, stabilize the diffuser flow substantially by imposing non-critical angles on the diffuser flow at partload. In addition they avoid the adverse effects of conventional normal solidity diffuser vanes, whose throats have a strong tendency to choke at little more than the design flow. LSD vanes increase the flow angle (from tangent) at part load so that no surge-triggering reverse flow can take place where the vaneless diffuser normally stalls. Thus turndown, head rise to surge, partload efficiency and overall performance-curve extension can considerably be improved. The prerequisite is that for the normal operating range the vane incidence angle must be

$$i = \alpha_{\text{chord}} - \alpha_{\text{flow}} \geq 0 , \qquad (3.7)$$

where α_{flow} is the arithmetic average of the profile inlet and exit flow angle.

This assures a positive profile lift coefficient c_a (i.e. a pressure recovery c_p and hence an efficiency increase) and a sufficiently small drag coefficient in the operating range and requires positioning of the vanes accordingly. The chord angle is between the log spiral chord of the airfoil and the circumferential direction.

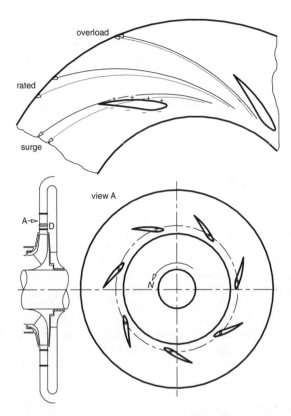

Fig. 3.11 Low-solidity diffuser vanes. *Dotted* streamlines: without LSD vanes (constant angle, logarithmic spirals), *solid* streamlines: with LSD vanes, N rotation, D dowel for vane setting, +++ pressure side, – – – suction side.

Typical LSD vanes are shown in Fig. 3.11. The dotted streamlines, calculated without LSD vanes for non-viscous conditions, are logarithmic spirals (constant angle); the solid streamlines show the amount of flow angle increase through the vanes. The results of an example calculation are displayed in Table 3.1.

Table 3.1 Example: diffuser calculation with LSD vanes

	Operating point		
	Partload	Rated	Overload
Flow angle for vaneless diffuser	11°	19°	26°
Flow angle with LSD			
Cascade inlet	11°	19°	26°
Cascade exit	15.7°	22.4°	26.3°
Airfoil chord angle	21°	21°	21°
Incidence i	7.6°	0.3°	-5.1°
Pressure recovery c_p, cascade	0.68	0.30	0

So the flow angle at partload is considerably increased, increasing the efficiency, which results in a higher curve slope and hence in an improved turndown of the compressor. The rated point head and efficiency are increased as well; the volumetric overload point, characterized by a flow angle of 26° remains as it is, whereas for flows greater than that the head and efficiency decrease, i.e. the choke point comes closer to the rated point.

Besides their use for operating range improvement, LSD vanes can also be used to increase the rated point efficiency, which of course is another objective; however, both goals can normally not be achieved simultaneously.

3.6 Return (Crossover) Bend

The normally vaneless return bend is the connecting member between the annular diffuser with its centrifugal flow and the deswirling apparatus with its centripetal flow, thus producing the typical meandering flow channel system of centrifugal compressors. Besides this function it does not have any aerodynamic significance. In other than process compressor applications (e.g. in some jet engine compressor designs) it may be integrated into the return vane channel for deswirling purposes. It has become good practice to increase the return bend throughflow area from the diffuser exit to the return vane inlet to compensate for the return vane blade blockage at position 6 (see Fig. 3.12). The inevitable flow separation at the inner rim disappears shortly after the acceleration due to the blade blockage.

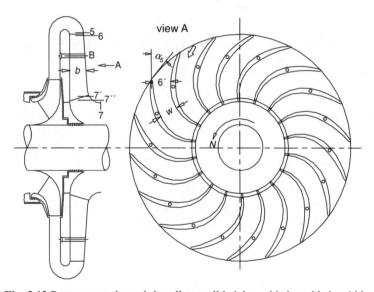

Fig. 3.12 Return vane channel. b wall-to-wall height, w blade-to-blade width, α_5 vane inlet angle, 5 and 6 vane inlet, 6´ throat, 7´and 7´´ vane exit, 7 stage exit, B diaphragm bolting, N rotation.

The pressure recovery of the return bend is

$$c_{pR\_45} = \frac{p_5 - p_4}{p_{4t} - p_4} \approx 0.00 \text{ to } -0.10, \tag{3.8}$$

i.e. the static pressure is constant or decreases slightly.
The loss coefficient is

$$c_{L\_45} = \frac{p_{4t} - p_{5t}}{p_{4t} - p_4} \approx 0.2 - 0.3. \tag{3.9}$$

3.7 Return Vane Channel

It is mandatory that the flow with its strong tangential velocity component is de-swirled before it enters into the next impeller. A swirling flow at the impeller inlet would reduce the head and efficiency substantially and cannot be tolerated. The impeller requires an orderly flow in the meridional plane only. Therefore, the return or deswirl vanes must be designed in such a way as to assure a uniform meridional velocity profile both radially and tangentially in the eye opening of the subsequent impeller.

From a radially open space, characterized by the conservation of angular momentum rc_u = const, the vortex flow enters a multichannel path consisting of a high-density cascade, where Bernoulli's equation $p/\rho + c^2/2$ = const prevails. Whereas the throughflow area in the vaneless diffuser is $A_d = 2r\pi b$, it is $A_{rv} = wb$ in the return vane channel (channel width × height). Each flow passage is formed by two adjacent vanes and the front and rear diaphragm walls.

The inlet flow angle is usually 26° to 32°, depending on the flow coefficient, impeller exit angle and diffuser radius ratio, and the outlet flow angle is supposed to be 90°. It goes without saying that the 60° flow deflection must proceed with as little loss as possible. So, due to the many return-vane-channel surfaces causing friction, the average velocity must be kept low, especially where the deflection is greatest.

Therefore, as can be seen from Fig. 3.12, there is an increase of the blade-to-blade width w in the first two thirds of the channel between position 6´ and 7´of some 30% and a decrease in the last third, where the vanes approach the desired flow direction of 90°. The meridional (or axial) wall-to-wall height b has to increase from b_6 (being determined by the foregoing diffuser) to b_7, which is determined by the eye opening of the following impeller insofar as there has to be an acceleration from the deswirl vane exit, position 7´´, to the impeller eye, position 7, in order to achieve a velocity profile that ought to be free of the vane trailing-edge wakes and separation that may develop in the 90° change of direction. So the increase of b along the deswirl vane is normally between 20% and 40%, depending on the two impellers to be connected. The flow-channel throughflow area from 6´ to 7´ increases by some 30% to 50%, which is utilized for diffusion, i.e. an increase of the static pressure, in exchange for velocity.

The pressure recovery of the return vane channel including the inlet bend to the next impeller is:

$$c_{pR\_57} = \frac{p_7 - p_5}{p_{5t} - p_5} \approx 0.4 - 0.5. \tag{3.10}$$

The pertaining loss coefficient is:

$$c_{L\_57} = \frac{p_{5t} - p_{7t}}{p_{5t} - p_5} \approx 0.25. \tag{3.11}$$

These are valid for the optimum performance point of the return vane channel, i.e. for zero vane incidence at the leading edge (approximately). c_{pR} decreases towards part and overload and c_L increases towards part and overload.

The centerline is superimposed by a thickness distribution (maximum 7% to 10% of blade length) mainly dictated by the diameter of the bolt which, at least in the designs of most manufacturers, is required to hold the diaphragm elements together (see Fig. 3.12).

There are a number of design philosophies for the channel geometry that result in different vane shapes, for instance a linear reduction of either the tangential component c_u or of the angular momentum rc_u. Tests revealed that the actual differences are only minor. However, simple circular arc blades and rectangular return bends should be avoided because of excessively high loss coefficients. Much more important is the radius ratio of the annular diffuser and hence the inlet radius to the return vane, which is normally identical with the diffuser exit radius. The longer the diffuser the higher the stage efficiency, because return vane channel losses reduce by roughly 2% for every percent of diffuser radius increase. So in spite of a longer flow path, the diffuser/return vane package with a greater radius ratio is energetically advantageous over a "compact" package with a small radius ratio.

3.8 Side-Stream Inlet

Industrial refrigeration compressors serve the purpose of achieving the specified evaporation (i.e. refrigerating) temperature of normally between +5°C and –105°C depending on the requirements of the cooling process. The evaporation temperature corresponds to the saturation pressure of the specific fluid which, accounting for some pressure loss, is the suction pressure of the compressor, whereas the discharge pressure of the compressor is the saturation pressure corresponding to the condensing temperature of the refrigeration cycle. In many of these cooling processes up to three refrigerating capacities at different temperature levels are specified; so the compressor has to handle up to two additional side streams plus possibly another side stream from a so-called economizer, i.e. a liquid/gas separator, improving the overall cooling cycle efficiency.

So, as an example, a seven-stage ethylene compressor with four sections, handling 100% mass flow at –101°C in the first section, takes 280% side mass flow at –70 °C in the second section, another 200% at –59°C in the third section and finally another 33% at −49 °C in the fourth section, so that the discharge mass flow is the fifteenfold greater suction mass flow.

The compressor has to be designed to meet not only the suction pressure at the required discharge pressure but also the various specified interstage pressures at the incoming secondary mass flows, i.e. the refrigerating temperatures at the pertaining refrigerating capacities. Flow mixing can of course be accomplished externally in a pipe, necessitating an intermediate discharge and suction nozzle. An attractive alternative is to let the two flows mix internally in the return channel of the foregoing stage, as shown in Fig. 3.13. This saves one nozzle with its associated loss and axial length thus contributing to a shorter, i.e. stiffer shaft.

So, a side-stream inlet is a flow element enabling the internal injection of a cold secondary mass flow to a hot primary mass flow. In the mixing zone the different temperatures and velocities of both flows adapt to each other. It is important that the element:

- has as little loss as possible,
- attains smooth velocity profiles radially and circumferentially in the adjoining impeller eye,
- eliminates angular momentum as far as possible.

The secondary flow, which can be up to three times greater than the primary flow, is preferably injected at the last third of the return-vane length, where the vanes are basically radially oriented, reducing the resistance of the side-stream channel. Injection near the vane leading edges would, due to the flat angle at this point, not warrant an even flow distribution from the centerline nozzle to both halves of the perimeter because blade blockage in one half would prevent an orderly annular admission.

An actual side-stream inlet for 50% additional mass flow is shown in Fig. 3.13.

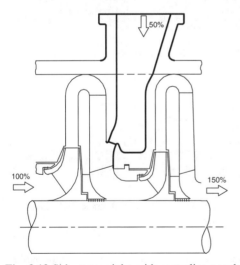

Fig. 3.13 Side-stream inlet with centerline nozzle. Example: 50% (additional) side stream mass flow.

3.9 Volute

If the gas has to be brought out of the casing, either to be intercooled, merged with an external side stream or at the final compressor discharge, the vaneless diffuser is followed by a volute or scroll. The latter is a flow-collecting duct whose through-flow area increases around the circumference as more mass fractions are added. The final component downstream of the scroll is the conical diffuser, often regarded as an integral part thereof, connecting the scroll exit to the discharge nozzle.

The other aero components described so far have very similar geometries with the different manufacturers. Not so the process compressor volutes, which exhibit an abundance of cross-sectional shapes (Fig. 3.14): circular, elliptical, quadratic, rectangular. All of these are either symmetrical, inclined or one-sided offset in relation to the radial direction of the diffuser exit. More important than shape, however, is how the scroll develops radially with the wrap angle (or azimuth). The classical circular offset scroll of the text books (the so-called "Oesterlen" scroll) is put on top of the diffuser, thus having an increasing center-of-gravity (CG) radius towards the exit; it is called the "external" scroll. The "central" scroll has a CG radius being equal to the diffuser exit radius; and the "internal" scroll is wrapped in such a way that its CG radius becomes smaller than the diffuser radius towards the exit.

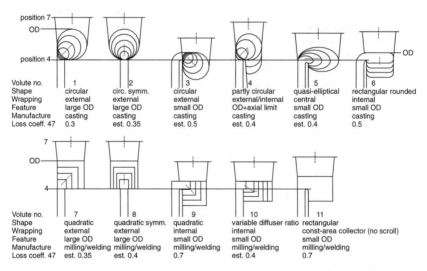

Fig. 3.14 Volutes including conical exit diffusers. OD scroll outer diameter, 4 volute inlet, 7 conical diffuser exit, *dotted line* course of center of gravity.

Whereas in the vaneless diffuser the meridional (radial) velocity component c_m is responsible for through-flow (continuity equation: $c_{m2} r \pi b \rho$ = const), this component is dissipated in the scroll because it is converted into a secondary vortex flow, causing the lion's share of the scroll losses and overlaying the basic rc_u = constant flow pattern throughout the scroll. So the large peripheral velocity component c_u has taken over the through-flow function (continuity equation: $c_u A \rho$ = const).

Since the volute flow regime is still governed by the law of conservation of angular momentum, the external volute achieves a static pressure increase and a smaller loss through a lower velocity level at the price of a larger outer diameter. In the internal volute the velocity is accelerated, leading to a static pressure reduction and a lower efficiency with the advantage of a smaller outer diameter, i.e. a more compact design. The central volute is a pure collector without accelerating or decelerating the flow.

It is essential that the volute is well matched to the impeller and diffuser at the best stage efficiency point, with the criterion being an even peripheral distribution of the static pressure at the annular diffuser exit. This, by retroaction, automatically assures an even peripheral distribution of the static pressure at the impeller exit as well, which in turn is a prerequisite for the impeller performance. A distorted pressure distribution would cause a periodic load change of the impeller channels with adverse effects on the efficiency and operating range. (An example of how the performance is influenced by scroll matching is shown in Fig. 6.25.)

In off-design conditions the scroll/impeller mismatching causes an acceleration in the scroll at overload (pressure reduction) and a diffusion at partload.

The pressure recovery and loss coefficients including conical exit diffuser are:

$$c_{pR\_47} = \frac{p_7 - p_4}{p_{4t} - p_4}, \qquad (3.12)$$

$$c_{L\_47} = \frac{p_{4t} - p_{7t}}{p_{4t} - p_4}. \qquad (3.13)$$

Guidelines for the loss coefficient were published by Mishina and Gyobu (1978, Fig. 3.14, volute no. 1, 9, 11), by Ayder and Van den Braembussche (1991, volute no. 6) and by the FVV (1998, volute no. 7, 9). Loss coefficients of volutes no. 2, 3, 4, 5, 8, 10 were estimated by "logical interpolation". Figure 3.14 also includes the course of the center of gravity along the azimuth. An increasing CG radius means deceleration achieving a positive pressure recovery in the volute (volute no. 7: c_{pR} = 0.4 as measured); decreasing radius means reacceleration and losing some of the static pressure obtained in the diffuser (volute no. 9: c_{pR} = 0.0).

The Oesterlen volute (Fig. 3.14, volute no. 1) loses much of its advantage since its only economic manufacture is casting with an inherently rough surface. So in many cases the milled square volute with its smooth surface is a better choice. The effect of its high loss coefficient can be eliminated by a long vaneless diffuser (reducing $c_4^2/2$) and wrapping the volute internally. Very high stage efficiencies are being achieved by such a diffuser/scroll combination, which offers the geometric flexibility often required for the various process compressor configurations. The absolutely worst collector is the one with a peripherally constant flow area. It leads to stage efficiencies that are 5–7% lower than normal.

References

Ayder E, Van den Braembussche R (1991) Experimental study of the swirling flow in the internal volute of a centrifugal compressor. ASME Paper 91-GT-7, Internat. Gas Turbine and Aeroengine Congress, Orlando, Florida; ASME, New York

FVV (Forschungsvereinigung Verbrennungskraftmaschinen) (1998) Spiralenströmungen. Vol. 659, FVV, Frankfurt, Germany (proprietory, in German)

Mishina H, Gyobu I (1978) Performance investigations of large capacity centrifugal compressors. ASME Paper 78-GT-3, Gas Turbine Conf., London, UK; ASME, New York

Senoo Y, Hayami H, Ueki H (1983) Low-solidity tandem cascade diffusers for wide-flow-range centrifugal blowers. ASME Paper 83-GT-3, Gas Turbine Conf., Phoenix, Arizona

Senoo Y, Kinoshita Y (1977) Influence of inlet flow conditions and geometries of centrifugal vaneless diffusers on critical flow angle for reverse flow. Trans. ASME, J. Fluids Eng., pp 98–103

Traupel W (1962) Die Theorie der Strömung durch Radialmaschinen. Verlag G. Braun, Karlsruhe, Germany (in German)

4 Compressor Design Constraints

4.1 Intrinsic Parameter Constraints

A word of caution is required before proceeding further: since the design limitations disclosed in this chapter, stem from the accumulated knowledge and practical experience of just one compressor manufacturer, they cannot be widely generalized. The differences of geometric impeller variants, for instance, will certainly lead to different tip-speed limits and diverse statements on the maximum number of impellers per rotor. Or other OEMs may have developed impellers with higher and lower flow coefficients or may simply have larger impeller outer diameters in their spectrum of frame sizes. In many aspects, however, the market has had quite a normative effect, as far as design limits are concerned: in some written or unwritten specifications rotational speeds above 15,000 rpm, or impeller-eye Mach numbers in excess of 0.85, or stage heads above 35 kJ/kg will not be accepted (without saying why). So, in a way, the market overrules some of the individual OEMs own limits, thus generalizing the technological constraints the various manufacturers have come up with on their own. With these exceptions in mind the following constraints ought to be regarded as subjective and should not be absolutized.

An in-depth knowledge of the aero-thermodynamics required to design a compressor would be insufficient without the experience of comprehending the limits of the various design parameters. Flow coefficient, rotational speed, impeller tip speed, Mach number, discharge temperature, impeller exit width, impeller diameter and rotor stiffness, with their constraints and powerful interdependencies, determine the feasibility of a compressor regarding the number of casings, impellers per casing, intercoolers, efficiency, operating range. In this design process the interfaces between the disciplines aerodynamics, rotordynamics and structural mechanics have to be observed.

The starting point is the intrinsic constraints of single parameters, which certainly vary from manufacturer to manufacturer.

4.1.1 Flow Coefficient

The limits $\varphi_{min} \approx 0.01$ and $\varphi_{max} \approx 0.15$ and the reasons for them have already been described in Sect. 1.9.

Although the highest feasible flow coefficient results theoretically in the smallest possible frame size and the highest possible efficiency, the flow coefficients have to be lower than maximum in most cases in order to reduce the rotational speed, increase the shaft stiffness ratio or enable a single-casing solution.

4.1.2 Rotational Speed

Many conservative engineering contractor and user specifications call for an upper ceiling of approximately 15,000 rpm; however, manufacturers' limits are around 20,000 rpm for single-shaft compressors with good, long-term running experience, and very high-speed compressors are available with speeds of up to 25,000 rpm. The determining criterion is the observance of the necessary separation margins between critical and operating speeds, which becomes more difficult at higher rotational speeds.

In the case of an electric-motor drive with a speed of 1,500 or 1,800 rpm via a single-stage speed-increasing gear, the maximum compressor speed is limited to approximately 15,000, or 18,000 rpm respectively, since the maximum speed ratio is approximately 10:1.

4.1.3 Impeller Tip Speed

The maximum combined stress of impellers of arbitrary material under centrifugal load is described by the numerical equation:

$$\sigma = u_2^2 \rho / C \ \text{N/mm}^2, \tag{4.1}$$

with tip speed u_2 in m/s and impeller material density ρ in kg/m^3. The maximum combined stress of *steel impellers* under centrifugal load is described by the numerical equation:

$$\sigma = u_2^2 / C^* \ \text{N/mm}^2. \tag{4.2}$$

The factors C and C^* depend on the impeller flow coefficient, since this characteristic has the greatest influence on the impeller geometry (see Table 4.1).

Table 4.1 Rough guidelines for proportionality factors C, C^* (single-shaft, multistage)

Flow coefficient φ	Constant C	Constant C^* [m^3/Gg]
0.01	$\approx 1.65 \times 10^6$	≈ 210
0.05	$\approx 1.50 \times 10^6$	≈ 190
0.12	$\approx 1.33 \times 10^6$	≈ 170

1 Gg (Gigagram) = 10^9 g

Density ρ (kg/m^3): steel 7,850; aluminum 2,700; titanium 4,500. C and C^* vary with the specific impeller design, i.e. disk shape, shaft and eye diameter, number and length of blades, etc. and may assume different values with other OEMs.

The allowable stress-related impeller tip speed for arbitrary materials is

$$u_{2\max} = \sqrt{\frac{C}{\rho} \frac{\sigma_{100}}{R_{p0.2}} R_{p0.2}} \quad \text{m/s}, \tag{4.3}$$

or, for steel impellers:

$$u_{2\max} = \sqrt{C * \frac{\sigma_{100}}{R_{p0.2}} R_{p0.2}} \quad \text{m/s}. \tag{4.4}$$

The ratio $\sigma_{100}/R_{p0.2}$ (stress at 100% speed/material yield strength) is often specified not to exceed 0.6–0.7. Standard material yield strengths are given in Table 4.2.

Table 4.2 Material yield strengths for two different alloyed steels for impellers

Steel	Mat. yield strength $R_{p0.2}$ [N/mm² = Mpa]	Application
X5CrNi13.4 (0.05% C, 13% Cr, 4% Ni)	950 (maximum)	High impeller tip speed
	830 (normal)	Normal impeller tip speed
	590 (low)	Resistance to stress corrosion cracking (gases with H_2S)
X8Ni9 (0.08% C, 9% Ni)	600	Low-temperature toughness (refrigerants at −150°C to +120°C)

Example: Maximum tip speed, Eq. (4.4): $\varphi = 0.05$; $C* \approx 190$ m³/Gg; $\sigma_{100}/R_{p0.2} = 0.6$; $R_{p0.2} = 830$ MPa. This yields $u_{2\max} \approx 308$ m/s.

Although the highest permissible stress-related tip speed results in the smallest possible number of stages and the smallest frame size, the tip speed in most applications is lower than the maximum in order to reduce the rotational speed, the Mach number, and the stage head or to increase the flow coefficient, shaft stiffness ratio and operating range.

4.1.4 Mach Number

The main criterion is the flexibility required by the operational needs of the process compressor, which has to cope with frequent changes of the external conditions such as the flow, temperature, gas composition and speed. So the compressor has to have a minimum performance curve width and a minimum curve slope (head rise to surge). Typical values, which are often specified, are:

$$\dot{V}_{\text{surge}}/\dot{V}_{\text{rated}} \leq 0.85; \quad \dot{V}_{\text{choke}}/\dot{V}_{\text{rated}} \geq 1.1; \quad y_{p\_\text{surge}}/y_{p\_\text{rated}} \geq 1.05 \tag{4.5}$$

where "surge" flow is defined on the unthrottled 100%-speed curve and "choke" flow is defined where the head approaches zero. All three values are a strong function of the Mach number and the number of compressor stages. In contrast to that the often-cited criterion of a loss-intensive local transsonic velocity near the blade suction side surface loses much of its significance, as can be seen from Fig. 4.1. This single stage with a radial inlet and a volute was designed for a tip-speed Mach number $M_{u2} = 1.12$ and a subsonic relative suction-side inlet Mach number; at $M_{u2} = 1.31$ the efficiency reduces by only two percentage points, although the relative inlet Mach number is locally around 1.17. At $M_{u2} = 1.4$ the efficiency is reduced by 5%. These losses, partly attributed to impeller inlet/exit and scroll mismatching, would still be tolerable. The very narrow operating range at these high Mach numbers, however, would normally not be acceptable with process compressors.

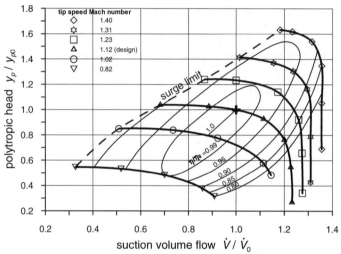

Fig. 4.1 Single-stage compressor performance map for variable Mach number. Measured with refrigerant R 12, flow coefficient 0.07, with plenum inlet and volute. Parameter: referenced polytropic efficiency.

So, with regard to the operating criteria, Eq. (4.5), the (highly empirical) approximate limits of the tip speed Mach number, derived from measurements of compressors with various numbers of stages, are given in Fig. 4.2. It has to be kept in mind, though, that these constraints depend on the OEMs particular method of stage design and geometry. These values can be exceeded if the compressor is not subject to variations of the operating parameters or with variable speed drives.

The relative inlet Mach number at the outer rim of the leading edge (blade-congruent or incidence-free), which is the actual main source for the performance curve shape, can easily be calculated from M_{u2}:

$$M_{w1} = \frac{w_1}{a_0} = \frac{u_1}{a_0 \cos \beta_1} = M_{u2} \frac{d_1}{d_2} \frac{1}{\cos \beta_1}. \quad (4.6)$$

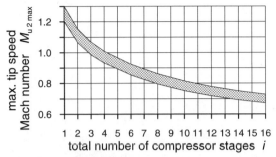

Fig. 4.2 Rough guidelines for the maximum tip-speed Mach number to achieve a satisfactory operating range.

This Mach number, often referred to as the "eye Mach number", is for reasons of simplicity referenced to the sonic velocity at total inlet conditions. The precise sonic velocity accounting for inlet expansion is

$$a_1 = \sqrt{\kappa_v Z_{0t} R T_{0t} \left(1 - \frac{\kappa_v - 1}{2} M_{w1}^2 \right)}. \tag{4.7}$$

This incidence-free Mach number has to be increased by some 20% to 30% to come up with the actual value near the suction side of the blade, as calculated by, for instance, CFD methods.

Independently of the above, the maximum eye Mach number is sometimes specified to be 0.75–0.85; this, of course, has a bearing upon the pertinent tip-speed Mach number limit.

4.1.5 Discharge Temperature

Approximate empirical rated point temperature limits are as follows:
- Limits imposed by the compressor construction as such, i.e. impeller shrink fits, proximity to seals and bearings:
  normal   200 °C
  maximum   250 (280) °C.
  Temperatures of 250 °C and higher necessitate heat shields at the bearing or back-to-back impeller arrangement, where the high temperature occurs at the rotor mid-span. These limits apply for air compressors, e.g. in ammonia synthesis plants where high discharge temperatures are required. In other cases where the gas would permit these high values in principle (e.g. nitrogen) one or more intercoolers are inserted into the compression process to reduce the power input.
- In many process compressors the gas itself limits the temperature in order to avoid dissociation, decomposition, fire hazards or in some cases polymerization, which clogs flow channels, labyrinth seals, or impeller–diaphragm gaps.

Here are some examples (guide lines):

ammonia	160 °C
chlorine	120 °C
polymerizating hydrocarbons	100–110 °C
acetylene	60 °C

- Applications in the oil and gas industry:

natural gas	170–180 °C
nat. gas with crudes susceptible to cracking	150 °C

4.1.6 Impeller Exit Width

It has been good practice to limit the impeller exit and diffuser widths to a minimum of 3 to 4 mm in general and especially to minimize the effect of build-up due to particle carryover in the gas flow. In narrow impeller channels a small congestion can already have a severe adverse effect on the compressor performance in the form that the volume flow, head and efficiency decrease and shaft vibrations increase, with the consequence of frequent time-consuming plant shutdowns to clean the compressor internals.

4.1.7 Impeller Outer Diameter

The established overall range of single-shaft process centrifugal impeller diameters is 150 to 1,500 mm. Other OEMs extend the range up to approximately 2,000 mm. The lower limit is determined by the described maximum rotational speed and by increased tolerances under normal manufacturing circumstances; the upper limit by dramatically increasing the outer dimensions and mass and the very high costs of foundations, transport and maintenance. For higher volume flows either a split-up into several units, or double-flow casings, or selection of an axial compressor would be better solutions.

4.1.8 Shaft Stiffness

The rotor stiffness ratio be defined as the first lateral critical speed on rigid-rotor supports divided by the maximum continuous speed:

$RSR = N_{c1\text{-rig}}/N_{max}$

The numerator corresponds to the rotor geometry and the denominator corresponds to the operating mode. If RSR drops below a certain value, i.e. the shaft is too long, or too thin or the speed is too high, the rotor will start to vibrate at high amplitudes with frequencies that are near the first rigid-support critical speed. These dangerous subsynchronous vibrations are excited by so-called cross-coupling effects through whirling flows in annular gaps that have become eccentric under the influence of a deflection. They can occur in the bearing oil film, in the impeller interference fits and in the labyrinth seals. Cross-coupling excitations in the shaft and impeller labyrinths are present in all turbomachines and become really relevant for high gas densities. They can be avoided by a sufficiently high rotor stiffness ratio. Fulton (1984) has defined an empirical stiffness/density relationship which is essential for compressor design insofar as it determines the maximum number of impellers that can be accommodated per rotor. The following equation was deduced from Fulton's findings:

$$RSR = \frac{N_{cl\_rig}}{N_{max}} \geq F_{min} = \frac{1}{3.24 - 0.36\ln\left(\frac{\rho_s + \rho_d}{2}\right)}. \tag{4.8}$$

$\rho_m = (\rho_s + \rho_d)/2$ is the average density of the gas between suction and discharge. The normal range is: $RSR = 0.4 - 0.6\ (0.7)$.

As described earlier, high flow coefficient impellers with their great axial length and their small shaft diameter might thus result in a rotor with insufficient stiffness. This can be remedied by reducing the flow coefficient and hence the efficiency level.

If $RSR < F_{min}$ is unavoidable, special measures must be provided to prevent flow rotation upstream of the labyrinths, e.g. vortex breakers or shunt holes.

4.2 Derived Compound Constraints

The described intrinsic parameter constraints form an interdependent meshwork with great implications on head and pressure ratio per stage, number of stages per casing, maximum volume flow per casing and, last but not least, efficiency. We present some of the most important constraints derived from the intrinsic limits in the following sections.

4.2.1 Stage Head and Stage Pressure Ratio

In order to achieve the highest possible stage pressure ratio the Mach number M_{u2}, as shown in Fig. 4.2, has to be as high as possible. For light gases like hydrogen-rich gases, however, this will result in tip speeds $u_2 = M_{u2} a_0$ exceeding the stress-related maxima (see Eqs. (4.3) and (4.4)) because the sonic velocity a_0 is high. So M_{u2} and, consequently, the stage pressure ratio p_2/p_1 has to be reduced.

On the other hand, in order to obtain the highest feasible stage polytropic head, the tip speed u_2, as follows from Eq. (4.4), has to be as high as possible. For heavy gases like propane, carbon dioxide and chlorine, this will result in tip-speed Mach numbers $M_{u2} = u_2/a_0$ exceeding the limits shown in Fig. 4.2 because the sonic velocity is low. So u_2 and, consequently, the stage head y_p has to be reduced (see Eq. (1.73)).

These interrelationships are displayed in the four-quadrant diagram, Fig. 4.3, where the sonic inlet velocity as a given specified parameter serves as a starting point (on the left abscissa). The limits of stress-related tip speed and of tip-speed Mach number guide the way to maximum stage heads and stage pressure ratios.

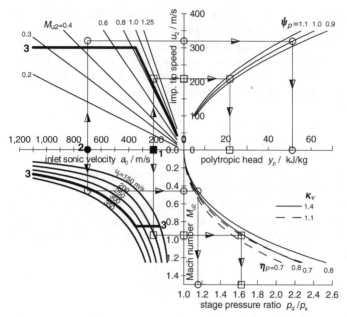

Fig. 4.3 Stage polytropic head and pressure ratio. *Example 1*: high molar mass, low head, high pressure ratio. *Example 2*: low molar mass, high head, low pressure ratio. *Line 3*: conservative upper limit of tip speed and tip-speed Mach number. κ_v isentropic volume exponent.

So the rule is:

low molar mass gas → high sonic velocity, high tip speed, high head, low pressure ratio

high molar mass gas → low sonic velocity, low tip speed, low head, high pressure ratio

4.2.2 Maximum Number of Impellers per Rotor

Figure 4.4 is a compilation of statistical data on compressors operating in the field without and with one intercooler running at tip-speed Mach numbers of between 0.2 and 1.0 and having up to nine stages with flow coefficients from 0.01 to 0.12. In order to be comparable, all of these rotors were theoretically adjusted to an average impeller tip speed of 270 m/s and a common rotor stiffness ratio of $RSR = N_{cl\_rig}/N_{max} = 0.42$ (being a conventional general minimum for process compressors). This corresponds to an average gas density in the casing of a maximum of $\rho_m = (\rho_s + \rho_d)/2 \approx 10$ kg/m³. Furthermore, the graph is valid for a medium frame size matching an impeller diameter of approximately 500 mm. The ordinate is the reference maximum number of impellers which can be accomodated on one shaft without being prone to subsynchronous vibrations; the abscissa is the tip-speed

Mach number of the first stage and the parameter is the first-stage flow coefficient. (Note: as an empirical statistics diagram, the graph shows prevailing tendencies rather than exact values).

The maximum (reference) number of impellers $i_{\max\,0}$ read from the graph must be corrected as follows: density twofold, deduct 1; density threefold, deduct 2; small frame size, deduct 1; large frame size, add 1; tip speed 220, add 1; tip speed 170, add 2; per additional intercooler, add 1; large side stream, deduct 1.

A denser gas requires, according to Eq. (4.8), a higher *RSR*, i.e. a shorter shaft with fewer impellers. With a smaller frame size the bearings and outer shaft seals occupy a greater portion of the bearing span than with large frame sizes. So the available length for the aero package is smaller for small frame sizes and larger for large frame sizes, requiring a modification of the maximum number of impellers. If the average impeller tip speed is lower than the graph reference of 270 m/s (for instance with gases with a high molar mass due to Mach number limitations), the maximum rotational speed $N_{\max}$ in Eq. (4.8) becomes correspondingly small, raising the value of *RSR* and thus enabling more impellers on the shaft. The effect of intercooling reduces the flow coefficient of the next impeller through a greater volume reduction for the subsequent stages, leading to a thicker and shorter past-cooler shaft portion.

So there are a lot of parameters influencing the aero-rotordynamic interaction and there are even more than described above: particular OEM specialities of arranging stages relative to each other or compressors with labyrinth shaft seals only (e.g. air, nitrogen, carbon dioxide, chlorine compressors). Also, of course, impellers could be squeezed together axially or the shaft thickness could be increased beyond the normal standards. This, however, could be more detrimental to the efficiency than tolerated by the market. Since Fig. 4.4 is based on *built* compressors, it is a satisfactory description of a complicated interdependency by means of a few simple parameters with some corrections.

The first rule of compressor design for a given specification, of course, is to arrive at the smallest possible casing diameter with the highest efficiency and the smallest number of stages, or, in aerodynamic terms, to maximize the flow coefficient and Mach number. This, however, is easier said than done.

In order to understand the complicated aero-rotordynamic interaction the following basic relationships have to be observed:

- strictly adhere to the minimum required rotor stiffness ratio,
- high flow coefficient level → increased efficiency → small casing outer diameter,
- high flow coefficient → great axial stage length and small shaft diameter → reduced stiffness,
- low flow coefficient → small axial stage length and large shaft diameter → increased stiffness,
- high stage pressure ratio → small number of stages,
- high tip-speed Mach number → low stage volume ratio → high compressibility → formidable reduction of flow coefficient for each consecutive stage → increased stiffness,

- low tip-speed Mach number → high stage volume ratio → low compressibility → very small reduction of flow coefficient for each consecutive stage → reduced stiffness.

Five examples in Figs. 4.4 and 4.5 should shed some more light on these relationships.

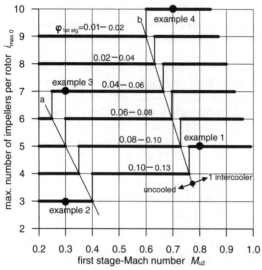

Fig. 4.4 Prevailing tendency for maximum feasible number of impellers per rotor. Reference: gas density 10 kg/m^3, impeller OD 500 mm, tip speed 270 m/s; to be corrected for actual density, frame size, tip speed, intercoolers, sidestream; a maximum Mach number for rotors each with identical impellers, b maximum Mach number for uncooled compressors.

Example 1: high flow coefficient/high Mach number. As has been shown, high values of M_{u2} are feasible for heavier gases (molar mass 30 and higher). For $\varphi = 0.12$ and $M_{u2} = 0.8$ up to five impellers can be accomodated on one shaft. The high Mach number and the intercooling lead to a rapid reduction of flow coefficients and a small stage pitch toward the compressor rear end. If more impellers were required, the flow coefficient level (and hence the efficiency) would have to be reduced, for instance to 0.05 to have eight impellers on the shaft.

Example 2: high flow coefficient/low Mach number. The compressibility is so low that practically all impellers have nearly the same flow coefficient, which increases the axial rotor length and reduces the average shaft diameter. This is the high-efficiency corner of the graph; but the diminished stiffness allows up to three impellers only (whereas for high Mach numbers, example 1, the volume reduces by some 20% from stage to stage, thus decreasing the bearing span and increasing the shaft diameter). Again, should there be a need for more than three impellers, the flow coefficient must be lowered by increasing the outer impeller diameters, resulting in a larger frame size and a lower efficiency.

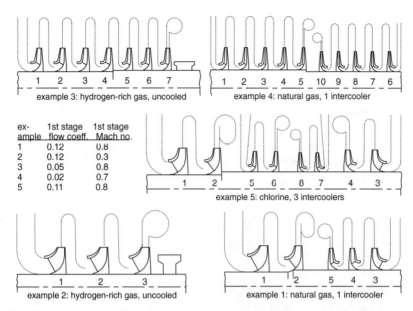

Fig. 4.5 Rotors with maximum feasible number of impellers. All the rotors shown have a stiffness ratio of $RSR = N_{c1}/N_{max} = 0.42$.

Example 3: low flow coefficient/low Mach number. Seven impellers can be placed on the shaft, which, in spite of the lower efficiency, can be of great advantage in avoiding a two-casing design with an enormous cost increase. Due to the low Mach number the flow coefficient reduces very little from stage to stage, so that groups of impellers can be made identical.

Example 4: low flow coefficient/high Mach number. The first stage flow coefficient is low and reduces still further towards the rear end of the casing (down to the minimum of approximately 0.01) so that the number of accommodatable impellers per shaft increases to as high as ten. The high Mach number requires a step-down of impeller diameters past the cooler to prevent last impeller flow coefficients below the established minimum. This is the low-efficiency corner of the graph.

Example 5: an eight-stage chlorine compressor rotor is shown in Fig. 4.5 with a high first-stage flow coefficient and Mach number and a low last-stage flow coefficient. It has the same rotor stiffness ratio as the other examples, but, due to the high molar mass, an average tip speed of only 170 m/s. The rotational speed in turn is low and thus, in conjunction with the three intercoolers, allows three more impellers on the shaft than the rotor of example 1.

The relationship between flow coefficient, stage length and shaft diameter is shown in Fig. 1.24; the stage volume ratio/Mach number dependency is displayed in Fig. 4.6, whose underlying formula is (for perfect gas behavior):

$$\frac{\dot{V}_2}{\dot{V}_1} = \left[s(\kappa-1)M_{u2}^2 + 1 \right]^{\frac{\kappa - \kappa\eta_p - 1}{\kappa - 1}} . \tag{4.9}$$

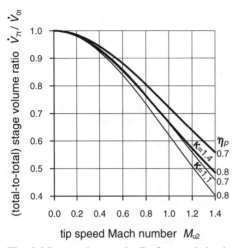

Fig. 4.6 Stage volume ratio. Perfect gas behavior; work input factor $s = 0.63$.

So the frequently asked question of how many impellers can be accommodated in one casing cannot be answered simply: it can be as low as three and as high as ten, depending on the gas and the pressure ratio. As has been shown, there is an absolute primacy of rotordynamics over aerodynamics in the form of minimum rotor stiffness requirements, preventing optimum aerodynamic (i.e. efficiency) solutions for many applications. The maximum efficiency potential can only be obtained when the number of required stages is three to five. On the other hand, the single-shaft process compressor offers the potential of up to ten impellers per casing, thus avoiding two-casing solutions in many cases.

In practice, for the specified data, a stepwise course will be followed:

- Decide on the maximum tip speed and maximum tip-speed Mach number accounting for the described constraints.
- Establish the required minimum number of stages.
- Settle the pertaining maximum flow coefficient as dictated by rotordynamics.
- Calculate the smallest impeller diameter and come up with the proper frame size.

4.2.3 Maximum Volume Flow per Frame Size

An equally frequently asked question (especially in specification forms) is: what is the maximum volume flow of a given frame size? No prospective compressor buyer wants to have a borderline case.

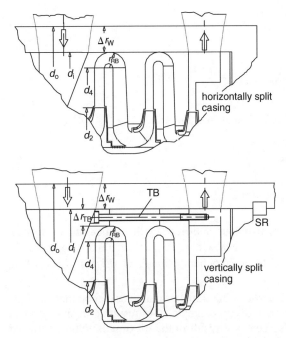

Fig. 4.7 Relationship between frame size, diffuser and impeller diameter. TB tension bolt, SR shear ring, RB return bend, w casing wall, o,i outer, inner casing.

The volume flow is:

$$\dot{V} = \varphi u_2 \frac{\pi}{4} d_2^2 . \tag{4.10}$$

A frame size can be characterized by its maximum vaneless diffuser outer diameter d_4 (Fig. 4.7). The outer diameter of the casing cylinder is

$$d_o = d_4 + 2 r_{RB} + 2\Delta r_{TB} + 2\Delta r_W , \tag{4.11}$$

where r_{RB} is the return bend radius, Δr_{TB} is the radial space required for the tension bolt in barrel casings, and Δr_W is the casing wall thickness. $\Delta r_{TB} = 0$ for horizontally split casings.

Introducing d_4 in Eq. (4.10) we obtain

$$\dot{V} = \varphi u_2 \frac{\pi/4}{(d_4/d_2)^2} d_4^2 . \tag{4.12}$$

Using the intrinsic parameter limits described above will result in the reference volume flow being allotted to 100% (Eq. (4.4)):

$$\dot{V}_{100} = \varphi_{\max} u_{2\max} \frac{\pi/4}{(d_4/d_2)_{opt}^2} d_4^2 \qquad (4.13)$$

$$= \varphi_{\max} \sqrt{C^* \left(\frac{\sigma_{100}}{R_{p0.2}}\right)_{\max} R_{p0.2\max}} \frac{\pi/4}{(d_4/d_2)_{opt}^2} d_4^2.$$

With the intrinsic parameter limits (Tables 4.1 and 4.2):

$$\dot{V}_{100} = 0.15\sqrt{170 \times 0.6 \times 950} \frac{\pi/4}{1.65^2} d_4^2 \qquad (4.14)$$

$$= 0.15 \times 311.3 \times \frac{\pi/4}{1.65^2} d_4^2 = 100\% \triangleq 1.0.$$

The diffuser ratio of 1.65 will bring about an efficiency that is near the maximum. The dimension d_4, of course, is associated with the particular frame size; e.g. for a medium frame size (pertaining to an impeller diameter of 0.5 m): $d_4 = 0.79$ m; hence we obtain $\dot{V} = 8.4$ m³/s.

Increase through oversize impeller. The impeller diameter can be increased, so that the diffuser ratio will decrease to, for example, 1.45 resulting in an efficiency trade-off of approximately 3%. This will result in the maximum volume flow per frame size being 30% greater than the reference:

$$\frac{\dot{V}_{\max}}{\dot{V}_{100}} = \left(\frac{1.65}{1.45}\right)^2 = 1.30. \qquad (4.15)$$

Decrease through lower material yield strength. If the yield strength is reduced from 950 to 590 N/mm² (applicable for gases containing H_2S causing stress corrosion cracking) the maximum volume flow is reduced by some 20%:

$$\frac{\dot{V}}{\dot{V}_{100}} = \sqrt{\frac{590}{950}} = 0.79. \qquad (4.16)$$

These volume flows are feasible only if a maximum flow coefficient of 0.15 and a stress-related tip speed of 311 m/s are feasible. This is the case only for a maximum of four impellers on the rotor (Fig. 4.4) and molar masses of between 24 and 32 maximum, keeping the tip-speed Mach numbers below 0.97 (Fig. 4.2). High numbers of stages and gases with high molar mass reduce the casing volume flow substantially.

Decrease through high number of stages. For an eight-stage compressor handling a low molar mass gas of, for example, M = 4.9 kg/kmol with a first-stage tip speed Mach number of 0.4, the maximum flow coefficient may be as low as 0.03 (Fig. 4.4) and the proportionality factor C^* (Eq. (4.2)) increases from ≈170 to ≈200, enabling a higher stress-related tip speed. This reduces the reference volume flow to as low as 22%:

$$\frac{\dot{V}}{\dot{V}_{100}} = \frac{0.03}{0.15}\sqrt{\frac{200}{170}} = 0.22 . \tag{4.17}$$

If the gas additionally contains H$_2$S, this is reduced even more :

$$\frac{\dot{V}}{\dot{V}_{100}} = \frac{0.03}{0.15}\sqrt{\frac{200}{170}}\sqrt{\frac{590}{950}} = 0.17 . \tag{4.18}$$

Decrease through heavy (i.e., higher molar mass) gases. As displayed in Fig. 4.3, the tip speed has to be lowered for heavier gases. Replacing in Eq. (4.13) the stress-related tip speed by $u_{2\max} = M_{u2\max} a_0$:

$$\dot{V} = \varphi_{\max} M_{u\,2\max} a_0 \frac{\pi/4}{(d_4/d_2)_{opt}^2} d_4^2 . \tag{4.19}$$

A typical six-stage refrigerating propane compressor with an inlet sonic velocity of 220 m/s at –40°C and a tip-speed Mach number of 0.9 has a maximum flow coefficient of 0.09. Therefore it reaches a maximum frame size volume flow of:

$$\frac{\dot{V}}{\dot{V}_{100}} = \frac{0.09}{0.15}\frac{0.9 \times 220}{311.3} = 0.38 . \tag{4.20}$$

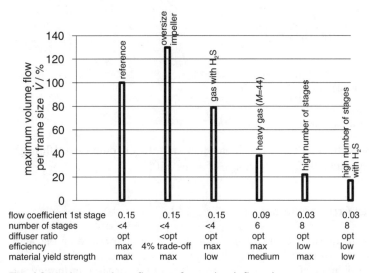

Fig. 4.8 Maximum volume flow per frame size, influencing parameters.

These results are graphically displayed in Fig. 4.8. The maximum volume flow per frame size varies considerably; it can be as low as 17% and as high as 130%, related to a reference value depending on the gas, the number of stages, the material yield strength, and other parameters. A reasonable answer can only be given in close connection with a specific application where the selected frame size ought to have a volume flow reserve of some 20% to 30% for any anticipated uprates in the future. These can be brought about by wider impellers or speed increase or other means, described in Chap. 9. For this purpose the internal space within the aero package and the impeller peripheral speed should not be used up entirely for the particular design.

4.2.4 Minimum Discharge Volume Flow of Turbocompressors

There is a general limit that distinguishes turbocompressors from displacement (e.g. reciprocating) compressors: the volume flow at the rear end of the centrifugal compressor (strictly speaking, the *suction* volume flow of the last stage). The minimum suction volume flow of the last compressor stage is given by

$$\dot{V}_{min} = \varphi_{min}(\pi/4)d_{2\,min}^{\,2}u_{2\,min} . \tag{4.21}$$

Introducing minimum values of:

- flow coefficient $\varphi_{min} \approx 0.01$,
- impeller exit width $b_{2\,min} = 3\text{–}4$ mm ≈ 3.5 mm (described above),
- established relative impeller exit width $(b_2/d_2)_{min} \approx 0.013$,
- tip speed $u_2 \approx 150$ m/s (i.e. a minimum pol. head of $y_{p\,min} \approx 10$ kJ/kg) we obtain

$$\dot{V}_{min} = \varphi_{min}(\pi/4)\left(\frac{b_{2\,min}}{(b_2/d_2)_{min}}\right)^2 u_{2\,min} , \tag{4.22}$$

$$\dot{V}_{min} = 0.01\times(\pi/4)\times\left(\frac{0.0035}{0.013}\right)^2 \times 150 \tag{4.23}$$

$$= 0.0854 \text{ m}^3/\text{s} = 307 \text{ m}^3/\text{h} \approx 300 \text{ m}^3/\text{h}.$$

This is a well-established value for single-shaft centrifugal compressors throughout the industry with a variation tolerance of some ±10%.

4.2.5 Impeller Arrangement

There remains the question of how these impellers are best arranged on the shaft. The traditional way of positioning impellers on a shaft is "one after the other" regardless of the number of sections. This so-called in-line or straight-through arrangement is the most used method if the casing contains only one section, and it used to be the exclusive method at a time when the various manufacturers had only a unidirectional line of impellers due to the dies used for blade forging or the pattern used for impeller casting. With the advent of the numerically controlled milling of impellers from a solid hub forging, the direction of rotation became arbitrary. Therefore impellers can now be arranged back-to-back without the necessity of investing in additional expensive manufacturing equipment. With this additional degree of freedom compressors can be designed more efficiently by utilizing the potential of reducing the recirculation loss through a single or multiple back-to-back impeller arrangement.

The different flow paths of the balance-piston leakage for in-line and back-to-back designs are shown in three six-stage compressor examples in Fig. 4.9. Example A has one section, example B has two back-to-back sections and example C has three sections with a twofold back-to-back wheel arrangement.

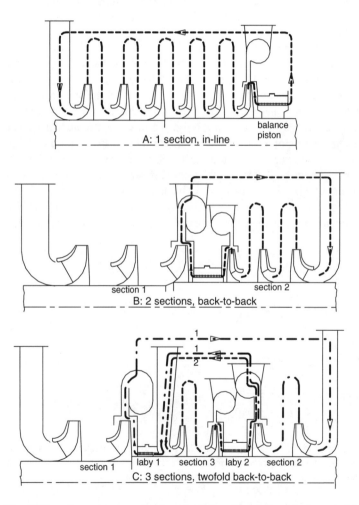

Fig. 4.9 Impeller arrangement and recirculation flow paths. 1 leakage in laby 1 recirculated through section 2; 2 leakage in laby 2 recirculated through section 3.

The function of the balance piston of an in-line compressor is to compensate the impeller axial force, which normally has the opposite direction to the incoming flow in the eye opening.

To achieve this, the inboard face of the piston is subjected to the discharge pressure of the adjacent impeller and the outboard face is connected via the recirculation line to the suction nozzle. So the leakage mass flow of the balance piston passes through all stages, consuming an extra gas power of $\Delta P = \Delta \dot{m} \Delta h$, where Δh is the total compressor Euler head. Since the piston diameter is relatively large (approximately equal to the eye diameter of the adjacent impeller) with a comparatively large clearance, and, since the labyrinth seal is subjected to (nearly) the complete pressure ratio of the machine, the recirculation in terms of power loss is relatively high (roughly between 1 and 3.5%, depending on the flow coefficient and pressure ratio, as shown in Fig. 4.10).

If the compressor has one intercooler, the two sections lend themselves to a back-to-back configuration (B in Fig. 4.9). The balance piston has been reduced to a rudimentary element in the form of a mere shaft labyrinth between stages 3 and 6, with a diameter approximately equal to the hub diameter of the adjacent impellers with a smaller clearance and a pressure ratio of roughly $(p_d/p_s)^{1/2}$. The leakage mass flow is recirculated through section 2 only. All of these circumstances lead to a compressor power loss of approximately one third compared with the in-line design (Fig. 4.10), i.e. between 0.3% and 1.2%. This represents an approximate power savings of 1% with high flow coefficients, 2% with medium and 3% with low flow coefficient compressors. The missing balance piston reduces the risk of a rub-caused damage and decreases the rotor mass, thus increasing the rotor stiffness ratio. The greatest part of the axial thrust is balanced by the impellers themselves.

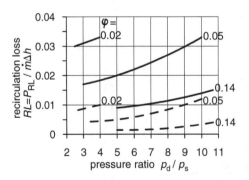

Fig. 4.10 Compressor power loss through recirculation. φ flow coefficient of first stage; *solid lines*: in-line impeller arrangement, *dashed lines*: back-to-back impeller arrangement, conventional labyrinth material (steel), clearance: 0.002 × diameter.

If the compressor has two intercoolers, the leakage flow can be minimized for the configuration shown in C of Fig. 4.9, where section 2 is back-to-back with section 1 and section 3 is back-to-back with section 2, resulting in the smallest possible pressure differences across the rudimentary "balance pistons" 1 and 2 (see also the threefold back-to-back chlorine compressor, Fig. 4.5, example 5). The leakage across labyrinth 1 passes through section 2 and the leakage through labyrinth 2 passes through section 3 only. The partial thrust not balanced by the impellers can be compensated by properly sized labyrinth diameters.

References

Fulton JW (1984) The decision to full load test a high pressure centrifugal compressor in its module prior to tow-out. In: Conf. Trans. Fluid Machinery for the Oil, Petrochemical, and Related Industries, Proc. Inst. Mech. Eng. C45/84, pp 133–138; London

5 Compressor Off-Design Operation

This chapter contains a methodical description and comparison of the five regulation modes: variable speed, suction throttling, adjustable inlet guide vanes, adjustable diffuser vanes and cooled bypassing. Two more topics are discussed. First, there is the test deducted dimensionless performance curve, showing work input factor versus exit flow coefficient, remaining (nearly) invariant by any transformations of speed, temperature and even gas composition. Second, there is the publication of test results in the expansion mode, i.e. the operation of a compressor stage beyond its choke point.

When the compressor is operated with a volume flow deviating from the design point, and if this occurs without changing the speed or the gas, the design point expands into the performance curve.

As a result of suction throttling, speed variation, adjusting inlet or diffuser vanes and bypassing further performance curves are generated which are referred to in their entirety as the compressor performance map. It is normal practice to keep suction pressure, temperature and gas data constant for the entire map, although they are not constant in reality.

By means of a suitable control system, i.e. automatic change of the adjusting parameters, the compressor can be operated at each individual point on the performance map; in other words, the compressor is regulated to control the preset discharge pressure or flow.

After the description of the five most important regulating modes, attempts will be made to answer FAQs such as: What is the best regulation for a given system resistance curve? Is there an appropriate performance curve form for variable boundary conditions? Does a characteristic curve exist which is invariant with regard to all possible changes of external conditions? Is one row of adjustable inlet guide vanes helpful for multistage machines? What causes surge and choke being the two prominent performance curve limits? Why do variable-speed performance curves of many multistage compressors differ so greatly from the fan laws well known from the text books?

5.1 Regulation Methods and Nomenclature

There are basically five modes to bring about an adaptation of the compressor behavior to the following three categories of system characteristics (in accordance with the categories in Chap. 1), as shown in Fig. 5.1.

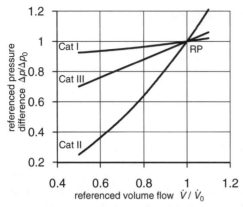

Fig. 5.1 Typical system resistance curves (approximate). Category I: basic process pressure plus pipe resistance, category II: exclusively system resistance (recycle compressors), category III: condenser curve for constant water flow (refrigerating compressors); RP rated point.

- *Category I*: flat resistance curve for almost constant discharge pressure (process pressure superimposed by pipe resistance),
- *Category II*: parabolic resistance curve to overcome the frictional resistance in apparatuses,
- *Category III*: condenser characteristic curve for constant cooling water flow in refrigerating plants.

The five regulation modes are: speed variation, suction throttling, adjusting of inlet guide vanes, adjusting of diffuser vanes and bypassing.

The actual operating point is always at the intersection of the system resistance curve and the compressor characteristic curve pertaining to the regulation mode.

The nomenclature for performance maps is displayed in Fig. 5.2 on the basis of head and efficiency vs. volume flow as per API Standard 617 (1995).

The performance curve anchor point 0 is the "rated point": it is the intersection of the highest specified head y_0 and the highest specified volume flow $\dot{V}_0$ requiring the highest speed N_0 of any specified operating point. This speed is the "one hundred percent speed". The rated point may or may not coincide with the best efficiency point BEP at the optimum volume flow $\dot{V}_{opt}$ and it may or may not be a guarantee point as per the purchasing contract.

The surge point on the unregulated performance curve ST represents the minimum stable volume flow $\dot{V}_{ST}$ at N_0; TD is the surge point at constant head brought about by variable speed, suction throttling, adjustable IGV, or bypass regulation; the pertaining volume flow is $\dot{V}_{TD}$. The surge limit line connects the surge points of the individual performance curve surge points. The maximum volume flow $\dot{V}_{CH}$ is achieved at the choke point CH where the head is zero (a condition that can be reached for a single stage only if there are one or more preceding stages).

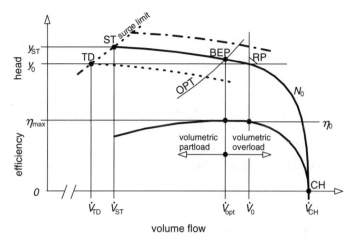

Fig. 5.2 Performance curve nomenclature (single-stage). *Solid lines*: unregulated curve, 100% speed, no suction throttling, no prerotation; *dash-dotted line*: regulated curve, overspeed or negative prerotation; *dotted line*: regulated curve, underspeed or suction throttling or positive prerotation; RP rated point (reference "0"); BEP best efficiency point; OPT line of maximum efficiency; CH choke; TD turndown; ST stability; N_0 100% speed.

The maximum efficiency line OPT connects the best efficiency points of the performance curves; operating to the right of that curve is referred to as "volumetric overload" and operating to the left of that curve is referred to as "volumetric partload".

Definitions of stability, turndown, head rise to surge, and choke are as follows.

Stability: $$ST = \frac{\dot{V}_0 - \dot{V}_{ST}}{\dot{V}_0}.\tag{5.1}$$

Turndown: $$TD = \frac{\dot{V}_0 - \dot{V}_{TD}}{\dot{V}_0}.\tag{5.2}$$

Head rise: $$HR = \frac{y_{ST} - y_0}{y_0}.\tag{5.3}$$

Choke: $$CH = \frac{\dot{V}_{CH}}{\dot{V}_0}.\tag{5.4}$$

The following comparison is based on referenced polytropic head/volume flow curves with referenced polytropic efficiencies as curve parameters, each for single compressor stages operating at a medium tip speed Mach number of between 0.6 and 0.88.

5.1.1 Variable Speed

In accordance with the fan laws for fluid-flow machines handling incompressible fluids, i.e.

$$\dot{V} \propto N \quad \text{and} \quad y_p \propto N^2, \tag{5.5}$$

the volume flow is proportional to the speed, and the head is proportional to the speed squared. This is still approximately valid for single-stage compressors with low Mach numbers (where the compressibility is negligible).

Variable speed has the following features:

- Relatively high partload efficiencies, since the compressor produces the required head only.
- Volumetric overload possible as a result of overspeed.
- Acts on all compressor stages.
- Suitable for all impeller and compressor types.
- Particularly suitable for parabolic system resistance curve and changing gas composition.
- Driver with variable speed required.

Efficiency and operational flexibility make variable speed the first choice of all-regulation modes. A typical performance map is shown in Fig. 5.3.

It should be kept in mind, however, that achieving variable speed by means of a hydraulic coupling will cost the user coupling slip losses in the case of compressor speed reduction, so that these partly offset the favorable compressor power input.

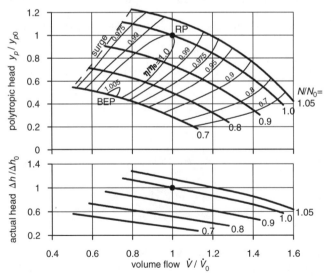

Fig. 5.3 Speed regulation; typical single-stage performance map. Test-derived results with 3D-impeller with beam-type shaft, radial inlet and return vane channel: $\varphi = 0.09$, $M_{u2} = 0.6$, $\beta_2 = 52°$, N/N_0 speed ratio, RP rated point, BEP best efficiency point.

5.1.2 Suction Throttling

If a throttle valve installed in the suction line is regarded as an integral part of the compressor, the suction-throttling performance map results, as shown in Figs. 5.4 and 5.5. The compressor suction state is defined *upstream* of the valve and each parameter curve corresponds to a constant pressure ratio across the valve (or percentage of suction pressure).

When the valve is rotated from the fully-open position towards the closed position, it produces a resistance increasing with the adjusting angle. This reduces the suction pressure downstream of the valve and subsequently the mass flow, discharge pressure, power and polytropic head (defined from upstream of the valve to discharge). The valve loss is charged to the compressor efficiency. However, in reality the volume flow, head and efficiency remain constant when defined from *downstream* of the valve to discharge.

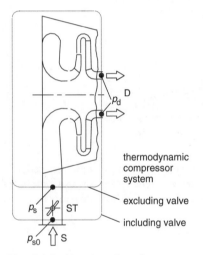

Fig. 5.4 Suction throttling. S compressor suction, D discharge into next stage, ST suction throttle (valve), p_s suction pressure downstream of valve, p_{s0} suction pressure upstream of valve (specified pressure), p_d discharge pressure.

Suction throttling has the following features:

- Low partload efficiencies because the non-required head generated by the impeller is dissipated in the valve.
- At constant head no volumetric overload possible.
- Acts on all stages of the compressor.
- Suitable for all impeller and compressor types.
- Particularly suitable for flat system resistance curve.
- For a parabolic system resistance curve there is high energy penalty.

- Not suitable for many applications with suction pressures of 1 bar or slightly above, since process licensors usually prohibit sub-atmospheric pressures for inflammable gases.
- Relatively low investment costs.

Suction throttling is the second choice in compressor regulating.

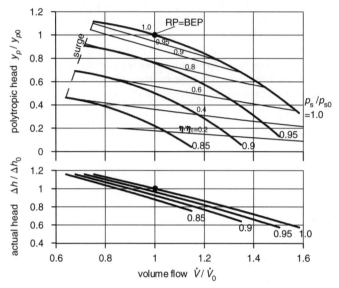

Fig. 5.5 Suction throttle regulation; typical single-stage performance map. Test derived, with 3D-impeller with beam-type shaft, radial inlet and return vane channel: $\varphi = 0.09$, $M_{u2} = 0.6$, $\beta_2 = 52°$, p_s/p_{s0} pressure ratio across valve, RP/BEP rated/best efficiency point.

5.1.3 Adjustable Inlet Guide Vanes

A row of adjustable vanes positioned ahead of the impeller produce a positive or negative prerotation (sometimes referred to as "variable geometry regulation"), see Fig. 5.6. In accordance with the Euler turbomachinery the work of compression $y_p = (c_{u2}u_2 - c_{u1}u_1)\eta_h$ is primarily influenced by the variation of the peripheral component c_{u1}; so it is really a head regulation. A certain vane angle producing a positive prewhirl does not result in an even reduction of the polytropic head along the performance curve: at volumetric overload there is a very strong effect as c_{u1} is large in relation to c_{u2}, and at partload the effect is small as c_{u1} is small in relation to c_{u2} (see also Figs. 1.21 and 1.22). So with positive prerotation there is a de facto shift of the entire curve to smaller volume flows including the surge limit because the smallest absolute flow angle at the diffuser inlet α_{3min}, essential for the incipience of surge, is only reached at a lower volume flow.

Adjustable inlet guide vanes have the following features:

- Medium partload efficiencies; compressor produces only the required head.
- Volumetric overload possible at constant head through negative prerotation.
- Affects performance of the pertaining impeller only.
- Multistage single-shaft compressors can normally only be equipped with one IGV row (in-line impellers) or two rows (back-to-back) due to lack of space.

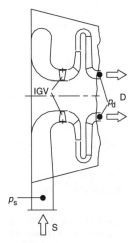

Fig. 5.6 Adjustable inlet guide vanes. S compressor suction, D discharge into next stage, IGV adjustable IGVs, p_s suction pressure, p_d discharge pressure.

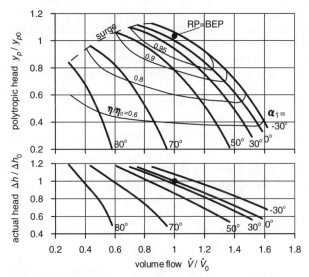

Fig. 5.7 Adjustable inlet guide vane regulation; typical single-stage performance map. Test-derived results with 3D-impeller with beam type shaft, radial inlet and return vane channel: $\varphi = 0.09$, $M_{u2} = 0.6$, $\beta_2 = 52°$, α_1 vane angle, RP rated point, BEP best efficiency point.

- Advantage of efficiency and operating range over suction throttling even with multistage machines having only one IGV row.
- Very suitable for integrally geared compressors since sufficient space is available ahead of each stage.
- Suitable for all impeller types; the effect, however, is stronger for backward curved impeller blading, since c_{u2} is smaller.
- Particularly suitable for flat system resistance curve.
- For a parabolic system resistance curve there is a smaller energy penalty than for suction throttling.
- Higher investment costs than suction throttling.

A typical performance map is shown in Fig. 5.7.

5.1.4 Adjustable Diffuser Vanes

For the sake of completeness and because it is aerodynamically enlightening, this regulating mode should be mentioned here, although it is not applied with single-shaft compressors due to an extreme lack of interstage space for the actuating mechanism. (Integrally geared compressors, though, are occasionally equipped with adjustable diffuser vanes.)

Adjustable vanes arranged in the diffuser downstream of the impeller provide along the entire performance curve a more efficient transformation of kinetic energy into static enthalpy than a vaneless diffuser. Rotating the vanes towards the closed position means decreasing the vane angle and at the same time reducing the throat between two adjacent vanes, thus reducing the entire throughflow area downstream of the impeller, as shown in Fig. 5.8. This signifies a low-loss adaptation to the continually reducing flow angle at partload. Influencing the volume flow in this way does not produce energy dissipation as with suction throttling but a nearly incidence free decrease of the through-flow area by accommodating the diffuser to the much wider impeller operating range, resulting in a stabilization of the flow in the diffuser. The stage as a whole thus obtains a comparatively large operating range at constant pressure ratio.

Adjustable diffuser vanes have the following features:
- Suitable only for integrally geared compressors.
- Relatively wide operating range at constant head due to a very favorable surge limit.
- Lowest partload efficiencies of all regulating methods.
- Volumetric overload possible through further vane opening.
- Affects performance of the pertaining impeller only.
- Suitable for all impeller types.
- Investment costs comparable with adjustable inlet guide vanes (higher than suction throttling).

As opposed to inlet guide vane regulation, the power input remains unchanged for a given mass flow at various discharge pressures, since a variable geometry diffuser cannot modify the total enthalpy change brought about in the impeller only. A typical (measured) performance map is shown in Fig. 5.9.

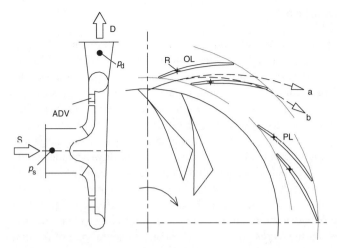

Fig. 5.8 Adjustable diffuser vanes. S suction, D discharge, ADV adjustable diffuser vanes (high solidity), p_s suction pressure, p_d discharge pressure, OL overload position, PL partload position, R axis of rotation, a streamline with ADV, b streamline without ADV.

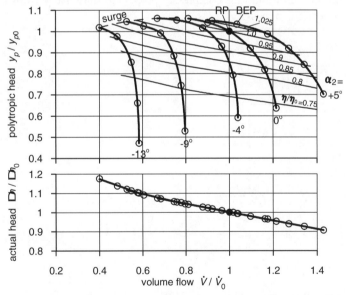

Fig. 5.9 Adjustable diffuser vane regulation; typical single-stage performance map. Test results with an S-type overhung impeller with radial inlet and volute: $\varphi = 0.07$, $M_{u2} = 0.88$, $\beta_2 = 60°$, α_2 diffuser vane inlet angle change (high-solidity diffuser vanes), RP rated point, BEP best efficiency point.

5.1.5 Bypass Regulation

Although this mode (sometimes referred to as "discharge throttling") is solely based on energy dissipation, it is quite often applied in the oil and gas industry, especially in cases where variable speed, suction throttling (sub-atmospheric suction!) are not applicable and IGVs not appropriate. Since the compressor itself does not have any regulating element whatsoever, it therefore has only one single performance curve with increasing discharge pressure towards partload. It cannot but operate at exactly the volume flow corresponding to the preset discharge pressure. Should, at the same preset pressure, the desired system flow become smaller, the compressor must, nevertheless, keep operating at the same compressor flow as before. This means that the difference between system and compressor flow has to be bypassed to the suction side, not without having been cooled, see Fig. 5.10. So there is a difference between the required compressor mass flow $\dot{m}_c$ and the process mass flow $\dot{m}$. When the bypass line is regarded as an integral part of the compressor and the bypass loss is charged to the compressor, the process relevant compressor efficiency η_c has to be modified by means of the mass flow ratio: $\eta = \dot{m} \times \eta_c / \dot{m}_c$ and the single curve expands into a map with the bypass ratio as the parameter. The farther away the (system) operating point is from the performance curve proper, the higher is the energy dissipation through recycling. In reality, however (as with suction throttling), the compressor proper has only one performance curve at $\Delta\dot{m}/\dot{m}_c = 0$ (no bypass).

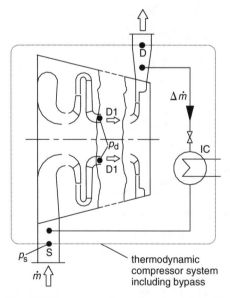

Fig. 5.10 Compressor bypassing. S suction, D1 discharge into next stage, D section or compressor discharge, IC bypass cooler, p_s suction pressure, p_d stage discharge pressure, $\dot{m}$ process mass flow, $\Delta\dot{m}$ bypass mass flow.

Bypass regulation has the following features:
- High energy loss depending on the difference between the desired and actual compressor operating point.
- Suitable for all impeller and compressor types.
- Most suitable for a constant pressure or flat system resistance curve.
- For a parabolic system resistance curve there is a very high energy penalty.

A typical performance curve is shown in Fig. 5.11.

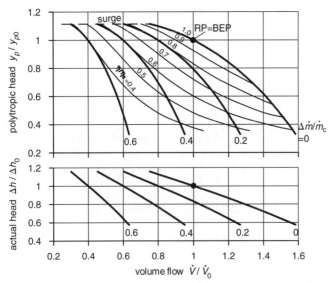

Fig. 5.11 Bypass regulation; typical single-stage performance map. Test-derived results with 3D-impeller with beam type shaft, radial inlet and return vane channel: $\varphi = 0.09$, $M_{u2} = 0.6$, $\beta_2 = 52°$, $\Delta \dot{m} / \dot{m}_c$ bypass ratio, RP rated point, BEP best efficiency point.

5.2 Customary Performance Maps for Multistage Process Compressors

In order to compare actual performance data for a multistage compressor, regulated either by speed variation or suction throttling or adjustable IGVs or bypassing, a case study has been carried out for a seven-stage acidgas compressor in gas/oil separation. This study allows us, by means of the customary map variants, to describe the peculiarities of the different methods of regulation such as variable speed mismatching, suction throttling affinity, multistage IGV effects and part-bypassing. The specification and design results are given in Tables 5.1 and 5.2 respectively.

Table 5.1 Case study: specification for acidgas compressor

Gas	vol.%	CH_4	7%	C_4H_{10}	6%	H_2S	15%
		C_2H_6	32%	C_5H_{12}	2%	CO_2	2%
		C_3H_8	33%	C_6H_{14}+	1%[1]	H_2O	2%
Molar mass	kg/kmol	37.7					
Actual suction volume flow	m^3/s	5.5					
Suction pressure/temperature	bar /°C	3.63/37					
Discharge pressure/temperature	bar/°C	35.3/167					
Pressure ratio		9.725					
Number of coolers		0					
Polytropic head	kJ/kg	174					
Number of stages		7					
Tip-speed Mach number 1st/7th stage		0.84/0.70					

[1]hexane and higher

Table 5.2 Case study: calculation results for acidgas compressor

		Regulation mode			
		Speed variation	Suction throttling	Adjustable IGVs	Bypassing
Operating point 1: Partload at constant pressure ratio					
Flow	$\dot{V}/\dot{V}_0$	0.75	0.75	0.75	0.75
Pressure ratio	p_d/p_s	9.725	9.725	9.725	9.725
Power	P_i/P_{i0}	0.78	0.83	0.80	1.0
Surge at p_d/p_s=const	$\dot{V}_s/\dot{V}_0$	0.67	0.65	0.60	–
Operating point 2: Partload at fan law parabola					
Flow	$\dot{V}/\dot{V}_0$	0.85	0.85	0.85	0.85
Pressure ratio	p_d/p_s	5.5	5.5	5.5	5.5
Power	P_i/P_{i0}	0.62	0.75	0.73	0.90

In Fig. 5.12 the performance maps for all four regulation modes for the above compressor are directly compared with each other with regard to the pressure ratio and power input versus volume flow in referenced terms (IGV regulation with *one* row of vanes upstream of the first impeller).

For comparison purposes the following two partload operating points were selected. Point 1 : 75% flow at constant (design) pressure ratio corresponding to a flat system resistance curve; point 2 : 85% flow on the fan-law parabola through the design point according to $y_p/y_{p0} = (\dot{V}/\dot{V}_0)^2$, corresponding to a parabolic system resistance curve.

As can be deduced from the above table, speed regulation is energetically the best. For point 1 power input we have: IGV regulation 2.5% higher, suction throttling 6.5% higher, bypassing 28% higher (!). For point 2 power input we have: IGV regulation 18% higher, suction throttling 21% higher, bypassing 45% higher (!). This proves that operating at constant discharge pressure does not result in essential power differences between the regulation modes, except bypassing. However, operating at a 72% head on the fan-law parabola requires paying substantial penalties for suction throttling , IGV regulation and bypassing in particular, compared to speed regulation, although the IGV surge is some 10% more favorable than the speed surge. It also becomes obvious that just one row of adjustable inlet guide vanes ahead of a seven-stage compressor still has a remarkable effect.

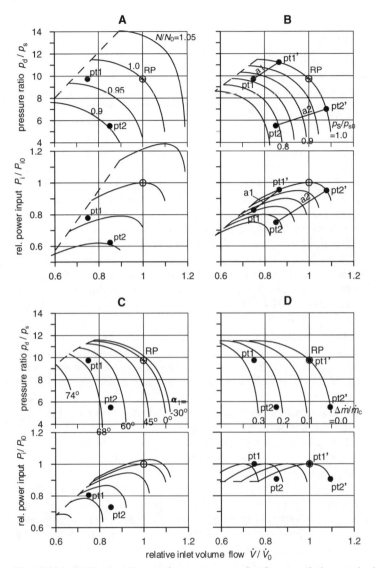

Fig. 5.12A–D Comparative performance maps for four regulation methods. Case study: 7-stage acidgas compressor: **A** speed variation, **B** suction throttling, **C** adjustable IGVs, **D** cooled bypassing, RP rated point, pt1 partload point at $p_d/p_s =$ const, pt2 partload point at fan-law parabola, pt1' and pt2' de facto compressor operating points, a1 and a2 affine lines, N/N_0 speed ratio, p_s/p_{s0} throttle ratio, α_1 IGV angle, $\Delta \dot{m}/\dot{m}_c$ bypass ratio.

The following section describes some peculiarities of the different maps with the intention of shedding some light on the inner working mechanisms of these regulation modes.

5.2.1 Variable Speed

It strikes the eye that the variable speed performance curves, Fig. 5.12A top left, are quite different in shape from each other, with a much flatter than parabolic surge line and apparent deviations from the fan-law criteria. Figure 5.13 shows additionally the fan-law parabola FLP: $y_p/y_{p0} = (\dot{V}/\dot{V}_0)^2$ with the points A through E. D is the design point at $N/N_0 = 1.0$. The actual speed line 1.05 should have passed through point E if the compressor were following the fan laws. However, the actual speed curve is above that point; the actual speed line 0.95 should have passed through point C and the actual speed line 0.9 should have passed through point B according to the fan laws. However, the actual curves are considerably below these points, representing incompressible gas behavior. The phenomenon responsible for these deviations is called "aerodynamic stage mismatching" and is defined as the nonlinear, i.e. progressive shift of the actual volume flow of a single stage relative to the foregoing stage. For instance, when stage n increases its volume flow by 5% and stage $n+1$ increases its intrinsic volume flow by 9% (e.g. because stage n operates at a lower Mach number) the mismatch ratio between the two stages is $MR = 1.09/1.05 = 1.038$. This tuning disorder or compressibility change between stages is brought about by *any* off-design operation, be it flow, speed, gas composition, suction or recooling temperature. The main cause for compressibility changes is the Mach number of stage n which, via the stage volume ratio, Eq. (4.9), determines the sizing of the throughflow area of stage $n+1$. So, for example, if M_{u2} decreases, the volume flow to stage $n+1$ increases and the pertaining existing throughflow area is undersized, resulting in volumetric overloading. So it is obvious that the mismatch effect is markedly pronounced in high-Mach-number compressors with many stages, as the above case study vividly illustrates. Mismatching (and hence fan-law deviation) is almost negligible in low-Mach-number compressors (recycle compressors handling H_2-rich gases which come close to being incompressible fluids).

Fan-law behavior is exhibited by turbomachines handling incompressible fluids only. So the speed-line sections through the fan-law points B, C and E carry the designations $(N/N_0)_{incompr}$. A vertical line from the fan-law point to the actual speed curve (line b in Fig. 5.13) represents the head deviation due to mismatching, and a horizontal line from the fan-law point to the actual speed curve (line a in Fig. 5.13) represents the flow deviation due to mismatching. Connecting all vertical intersections with actual speeds results in the "fan-law flow line" FLF, and connecting all horizontal intersections results in the "fan-law head line" FLH. The horizontal distance between FLP and FLH states how much the flow changes at constant head, and the vertical distance between FLP and FLF states how much the head changes at constant flow, all due to mismatching. The mismatch wedge formed by FLH and FLF is the visible manifestation of the aerodynamic disharmony between the stages, which grows nonlinearly with greater distance from the design point.

The compressor operating at B´ on the FLH line has the first stages operating at volumetric partload with increased head and the last stages at overload with decreased head, so that all seven stages are grouped around the best efficiency point. As a result the efficiency is much higher than at B´´and the fan-law head of B is reached. This behavior is described by sketch 1 in Fig. 5.13. So the line of the optimum composite efficiency comes close to the FLH line for the entire map.

The compressor operating at B´´on the FLF line has the first stage operating at its own best efficiency point (in case it coincides with the design point). All the subsequent stages at volumetric overload thus exhibit an efficiency and head decrease, so that the overall head decreases from B to B´´. This behavior is described by sketch 2 in Fig. 5.13.

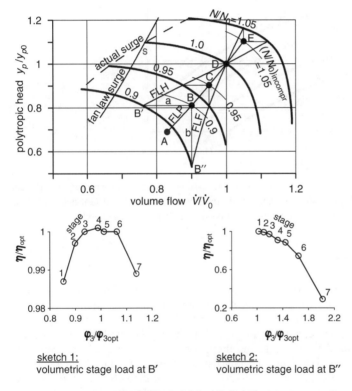

sketch 1:
volumetric stage load at B´

sketch 2:
volumetric stage load at B´´

at D: $\varphi_3/\varphi_{3opt} \approx 1.0$ for all stages

Fig. 5.13 Variable speed performance map. Case study: 7-stage acidgas compressor: $M = 38$ kg/kmol, $M_{u2} = 0.84$ (1st stage); FLP fan law parabola (incompressible gas) with 105, 95, 90% speed-curve sections (points A through E); FLH fan-law deviation at constant head (compressible gas), FLF fan law deviation at constant flow (compressible gas); fan law deviation: wedge between FLH and FLF (distances a and b).

The graph also contains the surge fan-law parabola through the surge point S. This would be the presumed surge limit line with no stage mismatching, i.e. for an incompressible gas.

So, altogether, mismatching has the effect of moving the speed map to smaller volume flows at lower-than-design speeds and to greater volume flows at greater-than-design speeds, inclusive of surge, best efficiency line and choke line.

An example of a low-Mach-number compressor with only four stages for recycle gas (molar mass 8.7) is shown in Fig. 5.14. The mismatch wedge is comparatively small, as well as the surge shift from the parabolic law, because the stage-by-stage volume reduction of 9% is a priori small due to a tip-speed Mach number of only 0.46. If the Mach number is reduced to below 0.25, there is no appreciable mismatch anymore and the variable speed performance map follows the fan-law criteria

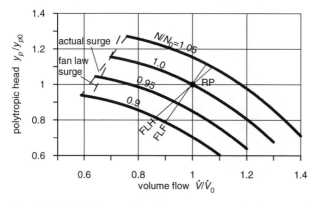

Fig. 5.14 Variable speed performance map. 4-stage recycle compressor for H_2-gas: $M = 8.7$ kg/kmol, $M_{u2} = 0.46$ (1st stage); FLH fan law deviation at constant head, FLF fan law deviation at constant flow; fan law deviation: wedge between FLH and FLF.

5.2.2 Suction Throttling

The inner aerodynamic working mechanism of suction throttling is quite different from speed regulation, although both performance maps look similar. The compressor from downstream of the suction valve to the discharge nozzle has just one single performance curve, which remains as it is during the whole process of throttling. Incorporating the valve into the thermodynamic system "compressor" and acting on the valve expands the curve into an apparent map which comes into being by simply defining the suction conditions *ahead* of the valve. Then the system gets an additional operation parameter: the pressure downstream of the valve, which varies with the valve setting. The customary throttle curve parameter is p_s/p_{s0}, i.e. pressure ratio across the valve or fraction of the specified suction pressure. It has to be kept in mind, though, that a given p_s/p_{s0} does not correspond to a constant butterfly valve angle!

In contrast to the speed variation pressure ratio, the flow and gas power input are linearly interconnected. For example, when a compressor operating at the design point is subjected to a suction pressure reduction of 10%, the mass flow and the discharge pressure and the gas power will be reduced by 10% as well, whereas the compressor itself (not accounting for the valve) will keep operating at the same actual volume flow, head and efficiency.

Any point on the unthrottled pressure ratio curve (and discharge pressure curve) is part of a straight line extending over the entire map:

$$p_d / p_{s0} = (p_d / p_{s0})_{\text{unthrottled}} \times (\dot{V} / \dot{V}_0), \qquad (5.6)$$

$$p_d = (p_d)_{\text{unthrottled}} \times (\dot{V} / \dot{V}_0), \qquad (5.7)$$

and any point on the unthrottled gas power curve is part of a straight line extending over the entire map:

$$P_i / P_{i0} = (P_i / P_{i0})_{\text{unthrottled}} \times (\dot{V} / \dot{V}_0). \qquad (5.8)$$

These lines, called affine lines, are displayed in Fig. 5.12B, top right. The affine line through point 1, having a throttle ratio of 0.867, intersects the unthrottled curve at the affine point 1´; the affine line through point 2, having a throttle ratio of 0.788, intersects the unthrottled curve at the affine point 2´. So, in order to reach point 1, the suction pressure has to be lowered to $0.867 \times 3.63 = 3.147$ bar by partly closing the valve; and, to reach point 2, the suction pressure has to be reduced to $0.788 \times 3.63 = 2.86$ bar in this example. Point 2 looks like a partload point, but, actually, it is an overload point with a referenced volume flow of 1.08. The principal advantage of suction throttling versus discharge throttling is obvious: the head is lower in any case.

The calculation of an arbitrary throttle curve (for the pressure or power chart) is as simple as:

$$p_d / p_{s0} = (p_s / p_{s0}) \times (p_d / p_{s0})_{\text{unthrottled}}, \qquad (5.9)$$

$$P_i / P_{i0} = (p_s / p_{s0}) \times (P_i / P_{i0})_{\text{unthrottled}}. \qquad (5.10)$$

So all single performance curves within the map are similar and in no way distorted as in variable speed maps. All potential operating points on one affine line have just one corresponding point on the unthrottled performance curve, i.e. one actual volume flow, one head and one efficiency. Also, there is only one surge point transformed, by the valve action, into a *straight* surge limit line, which is an affine line as well.

It is standard procedure to keep the suction conditions and gas data constant for the entire map; so a possible modification of the compressibility factor at suction and of the polytropic exponents are neglected during throttling and also the Joule–Thomson effect being characteristic with real-gas behavior resulting in a temperature decrease during throttling. Only in case of guaranteed throttling operating points do these effects have to be accounted for. The affine-line method is valid for gas power only. The mechanical losses have to be treated separately.

5.2.3 Adjustable Inlet Guide Vanes

It is astonishing that just one row of movable guide vanes ahead of a seven-stage compressor enables such a wide performance map as far as flow and head are concerned, including an appreciably better surge at constant head (turndown) than speed regulation and suction throttling. The power consumption is slightly more favorable than with suction throttling, and there is the advantage of reducing the discharge pressure without decreasing the suction pressure, which is often prohibited by the process licensor.

In order to interpret surge differences between the IGV and the other regulation modes a model compressor has been devised with two stages: the 1st stage with IGV, and the 2nd stage unregulated. The compressor surge is determined by the 2nd-stage volume flow $\dot{V}_{min\_2stg}$. The compressor surge volume flow in terms of the 1st-stage volume flow, using the 1st-stage volume ratio $\dot{V}_2/\dot{V}_1$ according to Eq. (4.9) is

$$\dot{V}_{min\_1stg} = \dot{V}_{min\_2stg} \frac{1}{(\dot{V}_2/\dot{V}_1)_{1stg}} \tag{5.11}$$

Once the adjusting vanes are utilized to reduce the flow at constant pressure ratio with an IGV angle of, for example, 74° (as in Fig. 5.12C, bottom left) the 1st-stage volume ratio increases due to lower compression. So the suction surge flow decreases, while the 2nd-stage surge flow remains unchanged.

This basic relationship does not change in principle if the 2nd stage is actually a stage group (i.e. the unregulated group).

The change in aerodynamic matching between the variable geometry of the 1st stage and the unregulated stages 2–7 brings about this surge improvement, in contrast to speed variation or suction throttling. All the surge points in Fig. 5.12C are established by the surge flow of stage no. 6.

Figure 5.15 displays additionally the fan-law parabolas through rated and unregulated surge points. In a similar way to Fig. 5.13, this shows the shift of the entire performance map to the left as a result of stage mismatching.

In Fig. 5.16 the same low-Mach-number four-stage compressor as in Fig. 5.14 is shown with IGV regulation. A comparison of the 7-stage high-Mach-number compressor and the 4-stage low-Mach-number compressor, both with IGV regulation, reveals remarkable differences:

- The stability of the 7-stage compressor is 24% versus 30% for the 4-stage one, being inherent with higher Mach number compressors.
- The turndown of the 7-stage compressor, however, is 40% versus 34% for the 4-stage one, due to aerodynamic mismatching. Using again the model compressor and Eq. (5.11) to interpret this: along the path from stability to the turndown point the 1st-stage volume ratio $\dot{V}_2/\dot{V}_1$ of the 7-stage increases by some 15% due to vane closing (less compression!). Thus the compressor turndown is improved considerably beyond the fan-law parabola, since the surge is determined by the 2nd-group surge flow $\dot{V}_{min\_2stg}$, which is constant. So $\dot{V}_2/\dot{V}_1$ is highly variable with the vane angle, i.e. in sketch 2 of Fig. 5.15, stage 1 at turndown

gains a smaller flow coefficient ratio φ_3/φ_{3opt} compared to stage 6, thus improving turndown over the fan law.

With the 4-stage compressor, however, the 1st-stage volume ratio increases negligibly from stability to turndown due to near-incompressibility, thus placing turndown just about on the fan-law parabola, since the surge is determined by the 2nd-group surge flow. Here $\dot{V}_2/\dot{V}_1$ is nearly rigid and the vane angle cannot improve the turndown over the fan law, i.e. in sketch 2 of Fig. 5.16 the flow coefficient ratio at turndown is practically identical for all stages, indicating no turndown gain.

- The IGV effect is greater with the 7-stage compressor leading to performance curves with a higher spread extending below the 70% head (great head changes per degree of angle change). Operating point 2 is feasible: as can be seen from the side sketch 1 of Fig. 5.15, the vane angle of 66° brings about an increase of the 1st-stage volume ratio from 0.71 to 0.93, so that stages 2 to 7 are forced to operate at volumetric overload, thus in turn contributing to the required head reduction.

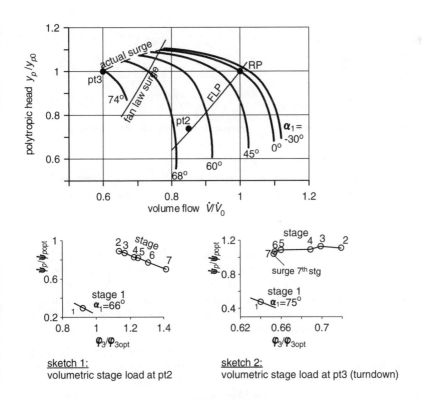

Fig. 5.15 Adjustable inlet guide vane performance map. Case study: 7-stage acidgas compressor, one row of IGVs: $M = 38$ kg/kmol, $M_{u2} = 0.84$ (1st stage), α_1 IGV angle.

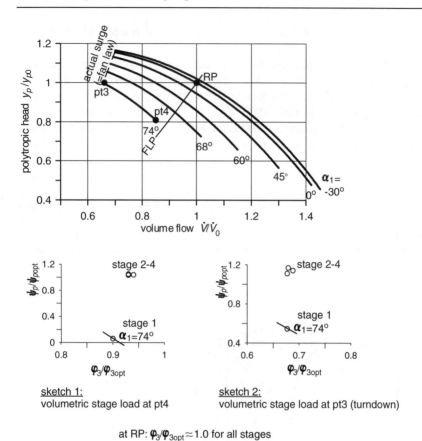

Fig. 5.16 Adjustable inlet guide vane performance map. 4-stage recycle compressor for H_2-rich gas, one row of IGVs: $M = 8.7$ kg/kmol, $M_{u2} = 0.46$ (1st stage), α_1 IGV angle.

- The 4-stage compressor performance curves are much closer together (small IGV effect and small head changes per degree of angle change). Operating point 4 at 85% flow and 72% head is hardly feasible since the 1st-stage, in spite of choking, cannot increase its volume ratio sufficiently enough, so that, as a result, stages 2 to 4 keep operating at volumetric partload, thus increasing instead of reducing head (see side sketch 1 of Fig. 5.16).

So the effectiveness of IGVs ahead of the first stage of low-Mach-number compressors is much lower than with high-Mach-number compressors, since mismatching of the latter amplifies the variable geometry outcome. The combination of just one IGV regulated stage and Mach number cause the aerodynamic mismatching to provide a high flexibility, which is often required for process compressors. On the other hand, the IGV effect with low-Mach-number compressors remains limited and does not lend itself to operation under square-law system resistance conditions, as is characteristic for these recycle compressors. Some basics on aerodynamic stage-by-stage mismatching have been published by Lüdtke (1998).

In many applications the equipment of multistage process compressors with one row of adjustable inlet guide vanes ahead of the first stage is justified because it results in greater operational flexibility and efficiency than suction throttling and energy-dissipative bypass regulation. Further details have been published by Lüdtke (1992).

5.3 Performance Curves for Variable Conditions

As stated before, it is standard practice to keep the suction conditions, the gas composition and also the recooling temperature past the intercooler(s) constant for the entire predicted performance map, although these do vary across the performance map. Since most process compressor specifications call for more than just one operating point, with variations of suction temperature and gas constituents, engineering contractors and users, in order to have one common performance map, recommend polytropic head/volume flow performance curves. This is because they believe that "a turbocompressor, at the same speed, always produces the same head at the same actual inlet volume flow regardless of the gas; this is what Euler's turbomachinery equation is all about". This, however, is only half of the truth. Whenever there is an appreciable compressibility in the flow, which is synonymous with tip-speed Mach numbers above 0.4 and more than two compressor stages, the head will change at any given volume flow whenever the operating parameters that modify the Mach number vary, i.e. the rotational speed, the suction temperature and the gas (isentropic exponent, compressibility factor, molar mass); see Eq. (1.77). Again, the name of the game is interstage aerodynamic mismatching, causing each subsequent stage to depart nonlinearly from the original state and so causing a change of the polytropic head and the efficiency. So, at constant volume flow and speed but 2% less molar mass, the high-Mach-number compressor of the above case study loses 3% head. With 5% less molar mass the head will decrease by as much as 8%, and the low-Mach-number compressor's head decreases by (only) 1% if the molar mass is reduced by 5%. So, in the case of variable boundary conditions, it could be justified to use head/flow performance maps for light gases and for a limited number of stages only (guideline: molar mass ≤ 8, tip-speed Mach number ≤ 0.45, number of stages ≤ 4); and the Mach number variations should not exceed 5%. Then the expected head errors stay below 1% because mismatching is negligible. In all other cases it is not meaningful to use polytropic head curves for variable gas composition and suction temperature.

Single-stage compressors are not exempt from these statements, since from inlet to exit there can be a considerable compressibility and hence inlet/exit aerodynamic mismatching, which will also limit the validity of assuming a constant head/volume flow curve invariant with Mach number changes.

5.4 In Search of an Invariant Performance Curve

Except for the described limited applications there is apparently no such thing as an invariant performance curve for multistage process compressors.

Also, the "reduced head and flow method",

$$y_{p\,\text{red}} = y_p (\kappa_v Z_1 RT_1)_{\text{ref}} / (\kappa_v Z_1 RT_1)$$
$$\dot{V}_{\text{red}} = \dot{V} \sqrt{(\kappa_v Z_1 RT_1)_{\text{ref}} / (\kappa_v Z_1 RT_1)}\,, \tag{5.12}$$

found in some textbooks does not yield adequate results for all process compressor applications.

So one has to concentrate on single-stage performance curves containing nothing but dimensionless coefficients to adequately predict performance under different conditions and with different gases. The procedure to be carried through (by way of an example) is as follows.

The starting point is a performance map covering a wide Mach number range (for instance 0.33 to 0.99). The measured values of a stage with a 400 mm impeller with 3D-blades, having a flow coefficient of 0.09, arranged between a plenum inlet and a return-vane passage are shown as an example in Fig. 5.17.

Measurements were taken between the suction flange (station s) and the return vane exit (station 7); see Fig. 1.20. Converting flow into flow coefficient φ, polytropic head into head coefficient ψ_p and Euler head into the work input factor s results in different ψ_p-, η_p- and s-curves for each M_{u2}, and the maximum efficiency occurs at different values of φ. So it is still an unwanted array of curves that hardly allow performance predictions with different gases.

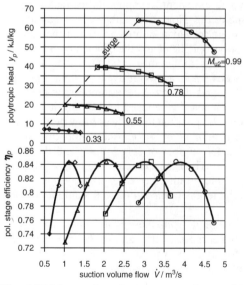

Fig. 5.17 Measured performance map for a wide Mach number range. Shrouded 3D-impeller with plenum inlet and return vane channel: $d_2 = 400$ mm, $\varphi = 0.09$.

The φ_3-Performance Curve

The picture changes drastically when the abscissa is replaced by the impeller-exit flow coefficient φ_3, which turns out to be much more meaningful for component design and analysis than the inlet flow coefficient φ, which is more an external characteristic,

$$\varphi_3 = \frac{c_{m3}}{u_2} = \frac{\dot{V}_3}{\pi d_2 b_2 u_2}, \tag{5.13}$$

defined infinitesimally downstream of the impeller exit at the through-flow area unobstructed by the blade thickness. It can be calculated by using φ from Eq. (1.66):

$$\varphi_3 = f\varphi \frac{\dot{V}_3/\dot{V}_{0t}}{4(b_2/d_2)}, \tag{5.14}$$

where the numerator is the total-to-static impeller volume ratio

$$\dot{V}_3/\dot{V}_{0t} = \left[1 + s*(\kappa_v - 1)rM_{u2}^2\right]^{\frac{\kappa_v(1-\eta_{p1})-r}{r(\kappa_v-1)}}, \tag{5.15}$$

the impeller slip is

$$s* = c_{u3}/u_2 \approx s = \frac{\Delta h}{u_2^2}, \tag{5.16}$$

the isentropic volume exponent is

$$\kappa_v \approx \frac{\kappa_{v0} + \kappa_{v7}}{2}, \tag{5.17}$$

the reaction is

$$r = \frac{\Delta h - \frac{c_3^2}{2}}{\Delta h} = 1 - \frac{\varphi_3^2 + s*^2}{2s}, \tag{5.18}$$

the polytropic total-to-total impeller efficiency is

$$\eta_{p1} = \frac{y_{p1}}{\Delta h}, \tag{5.19}$$

and the leakage factor is

$$f = \Delta \dot{m}/\dot{m}. \tag{5.20}$$

Equation (5.20) includes the leakages of the balance piston, the buffer gas, the recirculation with back-to-back impellers and the external labyrinths. For reasons of convenience and with sufficient accuracy the slip s^* can be replaced by the work input factor s.

For the first time, Sheets (1951) showed that the non-dimensional curves of single-stage compressors become very simple for a wide range of Mach numbers when plotted as a function of the impeller exit flow coefficient instead of the inlet flow coefficient. The analysis of his test data then resulted in one single head coefficient, one single efficiency curve and (nearly) one single curve for the work input factor for tip-speed Mach numbers of 0.54 to 0.90 for compressor A and as well for tip-speed Mach numbers of 0.36 to 1.01 for compressor B, although he used an array of gases with molar masses between 29 and 365 (!); see Fig. 5.18. This means that within these Mach number ranges the work input coefficient s also appeared as a single curve. Only at higher Mach numbers were the ψ- and η-curves shifted to lower values of φ_3.

His findings culminate in the statement that in the φ_3-diagram the performance of a stage is a function of the Mach number only, no matter how different are the test conditions and the gases handled.

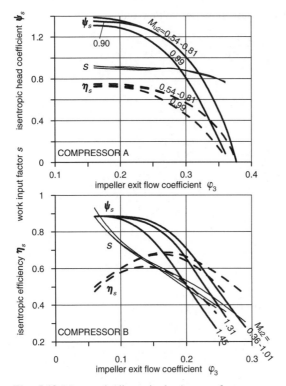

Fig. 5.18 Measured (dimensionless) φ_3-performance curves (according to Sheets (1951), redrawn). *Compressor A*: R-type overhung impeller with radial inlet and volute, test gas: air ($M = 28.96$), $d_2 = 182$ mm, $u_2 = 183$–335 m/s. *Compressor B*: impeller with backward-curved blades, beam type shaft, operating with different gases and tip speeds; test molar mass: $M = 29$-365, $d_2 = 235$ mm, $u_2 = 125$–172 m/s.

The fundamental inference of these findings is that with φ_3, s, η and M_{u2} the decisive similarity parameters were found responsible for the flow pattern and hence for the performance of a compressor stage independent of the gas and the suction temperature. It was also found that it does not matter whether the tip-speed Mach number in question is achieved with a low-molar-mass gas with high tip speed or a high-molar-mass gas with low tip speed. The combination of parameters making up the tip-speed Mach number is what counts as a flow similarity characteristic and not any single parameter thereof.

Very similar results came up on nondimensionalizing the test results in Fig. 5.17. Figure 5.19 is the so-called φ_3-graph with stage and impeller efficiency, head and work input coefficient. The findings are summarized as follows:

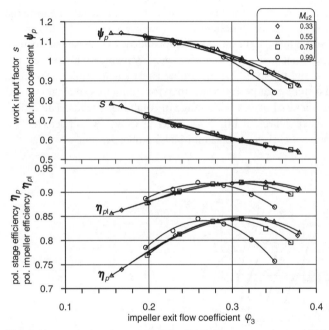

Fig. 5.19 Test-deduced φ_3-graph (Fig. 5.17). Parameter: tip-speed Mach number.

- For the complete Mach number range tested of 0.33 to 0.99 the work input factor appears as a single slightly concave line with negligible scatter.
- The polytropic head coefficient for Mach numbers 0.78 and less lies in a narrow band of less than 1% deviation; accounting for the measuring tolerance, it is practically a single line also. At higher Mach numbers the maximum flow coefficient φ_3 is reduced.
- The maximum efficiency is within a band of ±0.5% for the complete Mach number range (which is smaller than the measuring tolerance).

- For Mach numbers up to approximately 0.6 the maximum efficiency appears at a constant value of φ_3; at higher Mach numbers the maximum efficiency is gradually shifted to lower values of φ_3

So the work input, regardless of Mach number, is firmly hooked as a single function to the impeller exit flow coefficient. This confirms that the impeller exit velocity triangle, determined by φ_3 and s, is responsible for the flow and the specific actual work, whereas the impeller inlet triangle plays a secondary role following the lead of the exit, since the work input and inlet flow coefficient are not functionally associated with each other in a direct way. The head coefficient and stage efficiency for Mach numbers 0.6 and lower are connected as single functions to the impeller exit flow coefficient as well, confirming the predominant role of the exit for the stage losses in that compressibility range.

Only at higher Mach numbers do the loss sources at the impeller inlet and in the stationary ducts downstream of the impeller start to play a greater role.

Figure 5.20 sheds some more light on the interdependency of these parameters. This "total performance map" displays the efficiency, impeller exit flow coefficient, work input factor and tip-speed Mach number as parameter curves. So it can be regarded as a kind of universal performance map for a compressor stage of constant geometry that is operated over a wide range of rotational speeds. The different measured performance curves of Fig. 5.17, in terms of $\dot{V}_3/\dot{V}_{0t}$, appear as "inverse curves" and represent curves of constant Mach number M_{u2}.

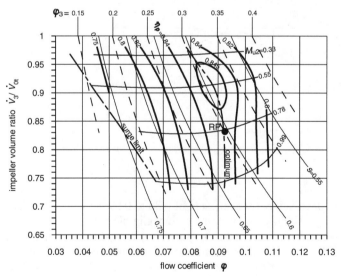

Fig. 5.20 Total performance map according to the test evaluation (Fig. 5.17): interrelationship of impeller volume ratio, efficiency, work input factor, flow coefficients, and tip-speed Mach number. RP rated point.

5.4 In Search of an Invariant Performance Curve

The diagram offers the following enlightening information on compressor behavior:

- A high efficiency level is maintained over the entire range of Mach numbers (a fact that can for instance be utilized for standardization by means of fixed geometry stages, as will be shown in Chap. 10).
- At lower-than-rated-point Mach numbers the maximum efficiency follows more or less a constant φ_3-line and hence a constant s-line. Since the impeller exit becomes too small the best efficiency point is shifted to lower flow coefficients according to the relationship:

$$\varphi \approx (\dot{V}_3/\dot{V}_{0t})_{RP}/(\dot{V}_3/\dot{V}_{0t}) \times \varphi_{RP}.$$

- At higher-than-rated-point Mach numbers the maximum efficiency begins to follow a constant φ-line. The rated point (optimum) exit flow coefficient cannot be maintained, since the losses associated with higher Mach numbers and flow coefficients φ increase. The best efficiency point is shifted to lower exit flow coefficients according to the relationship:

$$\varphi_3 \approx (\dot{V}_3/\dot{V}_{0t})/(\dot{V}_3/\dot{V}_{0t})_{RP} \times \varphi_{3RP},$$

much in the same way as with Sheet's compressor B.
- At lower-than-rated-point Mach numbers the surge limit improves considerably which is inherent with low M_{u2}.
- At higher-than-rated-point Mach numbers the surge limit deteriorates in relation to the optimum because of the higher Mach number per se and because the impeller exit becomes too wide for this operation (i.e. the work input factor increases and the diffuser inlet flow angle decreases).

These test analysis results using similarity parameters imply that for this type of process compressor stage (having relatively low pressure ratios, in contrast to centrifugal stages in turbochargers and jet engines) the suction volume flow and the work input and also the efficiency to a large extent are determined by the fluid mechanical processes culminating in the through-flow area of the impeller *exit*. The fluid mechanics at the impeller *inlet* just react to the centrifugal pull of the exit by somewhat greater or smaller incidence angles, provided, of course, that no basic mistake in sizing the inlet has been made. On the other hand, the occurrence of choking and the volumetric position of the choke point and a relatively rare form of stall are determined by the impeller-inlet flow channels.

So an invariant version of the performance map exists in the form of an s, φ_3- and η, φ_3-diagram (in short: a φ_3-diagram) for a single-stage only. This can be used for any gas, suction temperature and tip speed, provided that all the relevant parameters are in the allowable range that has been tested (or perhaps reliably proven by computational fluid dynamics methods).

It must be mentioned that a minor shortcoming of the φ_3-diagram is that the total-to-total impeller efficiency η_{pi} can only be measured with great difficulty and with an inherently great tolerance. However, it can be calculated either by conventional or CFD methods, which of course have inaccuracies as well. Errors affecting the impeller volume ratio and hence the exit flow coefficient will stretch or contract the curves horizontally in the φ_3-diagram; in the above example a simple calculation method was used, resulting in very roughly: $\eta_{pl} \approx \eta_p^{0.5}$. In view of the errors made, it is however essential to consistently use the approximation throughout the analysis and design procedure of any compressor.

Engineering contractors or users of a multistage process compressor may be disappointed that there is no simple ready-to-use approach to provide them with the kind of help they expect for their various specified multi-gas operating points. But, at least an invariant "core" performance curve for a stage has been identified and any compressor can be broken down into its φ_3-diagrams, which can then be used, on a stage-by-stage basis, by aerodynamic superimposition, to predict compressor operation with another gas.

5.5 Performance-Curve Limits

Besides the rated and the best efficiency points the surge and the choke are the most significant performance-curve points, confining the compressor operation to a limited range. The aerodynamic causes and implications of these phenomena are as follows.

5.5.1 Surge

If the flow is reduced to below the surge point, the compressor operation becomes unstable, resulting in periodic pressure and flow oscillations throughout the overall system of compressor, coolers, pipes, reservoirs and valves. The flow may change its magnitude in the forward direction alone or change its direction between forward and backward with a complete momentary halt in between. The oscillations affect the power and axial rotor thrust as well and, in the case of an uncooled compressor, will steadily increase the temperature level, unless stopped by an automatic trip. Depending on the pressure ratio, the changes may be mild and quiet or abruptly violent and very loud, jeopardizing the machine and the system. It has to be differentiated between system surge and stage stall.

System Surge

System surge is a periodic occurrence caused by the interaction of the machine and the system, which in turn may be defined as an energy reservoir (collector) on the pressure side with a sufficiently large storage capacity.

The basic performance curve extends into the 2nd quadrant with backward volume flow and positive head (sometimes referred to as the "brake curve"), as shown in Fig. 5.21. The normal curve section has a relative pressure maximum at A, a relative minimum at B, a positive slope connecting A and B, and a zero passage at C. The branch of the curve to the left of A is normally not shown, since it is forbidden or not relevant. System surge can best be explained by the extended performance curve.

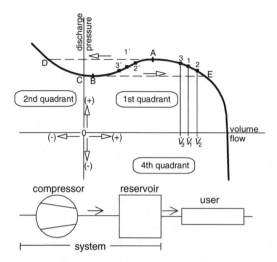

Fig. 5.21 Extended performance curve: A–B positive slope; right of A negative slope; A relative maximum; B relative minimum; C zero flow; D negative slope, negative flow.

If the compressor operating at point 1 sees a small disturbance that increases the flow to point 2, the difference $\dot{V}_2 - \dot{V}_1$ has to be stored in the reservoir, increasing its pressure, thus restoring the former equilibrium at point 1 between the compressor and collector. If the compressor sees a small disturbance from point 1 to 3, the flow difference $\dot{V}_1 - \dot{V}_3$ has to be momentarily delivered by the reservoir to the adjoining system, thus decreasing the reservoir pressure and restoring the former operating point 1. So this curve branch with its negative slope constitutes a stable operation.

If the compressor operates on the positive-sloped branch of the curve, i.e. at point 1´, and is subjected to a small perturbation from 1´ to 2´, the surplus $\dot{V}_{2'} - \dot{V}_{1'}$ must be stored in the reservoir, increasing its pressure and thus shifting the operating point further away from the original point 1´. This is the criterion of instability. The pressure will first increase until A is reached. However, since the compressor pumps more flow into the reservoir, the pressure will increase further. This can only be made possible by the compressor shifting its operating point to D in the 2nd quadrant necessarily reversing the direction of flow. So the reservoir begins to unload, decreasing the pressure until point B is reached. Further pressure reduction due to the backward flow is only feasible by the compressor instantaneously shifting its operating point to E in the 1st quadrant. At this point, however, the flow, restoring its forward direction, increases the reservoir pressure until point A is reached again and the surge cycle ADBEA is (continuously) repeated.

If the compressor undergoes a perturbation from point 1´ to 3´, the flow difference $\dot{V}_1 - \dot{V}_3$, which is momentarily not made available by the compressor, must be made up for by the collector. This inevitably reduces the pressure until B is reached, from where the compressor jumps over to E. So this perturbation also enters into the same continuous surge cycle BEADB described above.

So any forceful operation on the positive slope of the performance curve leads to a periodic surge involving the entire system. The condition for incipient system surge is

$$\frac{\partial(p_d/p_s)}{\partial \dot{V}} = 0, \qquad (5.21)$$

which is reached at point A in Fig. 5.21. The tacit assumption for this is that the collector capacity is sufficiently large that the pressure change lags behind the flow change. This means that, practically speaking, the system resistance is a quasi-constant pressure line instead of a more or less parabolic resistance curve.

If the reservoir capacity is "less than sufficiently large", the compressor can stably operate to the left of the maximum pressure point (with the horizontal tangent) which is often observed during compressor tests. Figure 5.22a shows a measured system surge point with a horizontal tangent, whereas in Fig. 5.22b another compressor was found to have its system surge point markedly to the left of the maximum pressure point due to a small collector volume (i.e. the throttle valve during test was close to the discharge nozzle). In the latter case the positive-slope branch is irrelevant for compressor operation.

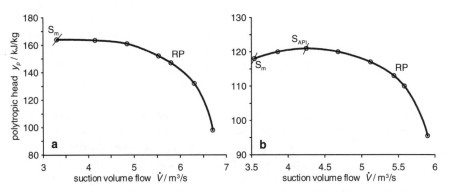

Fig. 5.22a, b System surge: measured performance curves. **a** system surge at $dp/dV = 0$; 4-stage air compressor, 1st stage: $\varphi = 0.09$, $M_{u2} = 0.81$; **b** system surge at $dp/dV > 0$; 3-stage CO_2 compressor, 1st stage: $\varphi = 0.08$, $M_{u2} = 1.0$; S_m surge point as per test, S_{API} certified surge as per API Standard 617, RP rated point.

Traupel (1966) and Eckert (1961) confirm this basic fact, saying that Eq. (5.21) is valid only for infinite collector capacities and that, in general, instability occurs where the gradient of the system resistance curve is greater than the gradient of the compressor curve.

The fact remains that system surge is not compressor intrinsic but dictated instead by the dynamic interaction of the machine with the system. Any theoretical or actual possibility of a compressor curve having a pressure maximum plus a positively sloped curve section is categorically forbidden by API Standard 617 (1995), para 2.1.4: "The head-capacity characteristic curve shall rise continuously from the rated point to predicted surge", expressing a safety philosophy. So normally, a measured curve section with positive slope (which is usually short) is declared non-existent and the pressure maximum is declared as the official surge point. This prevents the compressor from having two potential operating points for a preset discharge pressure.

So the established criterion for a conservatively predicted system surge limit is

$$\frac{\partial(p_d/p_s)}{\partial \dot{V}} \geq 0. \tag{5.22}$$

Stage Surge

It is often observed that a compressor begins to surge when the performance curve still has a substantial (negative) gradient that is far from exhibiting the horizontal pressure tangent characteristic of system surge, as shown in Fig. 5.23.

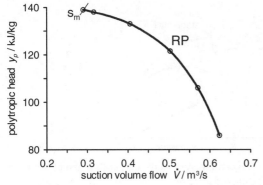

Fig. 5.23 Stage surge: measured performance curve. Typical stage surge at $dp/d\dot{V} < 0$ caused by rotating stall; 6-stage air compressor, 1st stage: $\varphi = 0.015$, $M_{u2} = 0.68$, S_m surge point as per test, RP rated point.

The cause is a phenomenon called stage surge or stage stall, characterized as rotating stall which is caused by local flow separation and reversal either in the impeller or in the stationary flow channels. The compressor operation is still stable but exhibits pressure fluctuations about a mean value. Whereas system surge is periodic and unstable, rotating stall, from the standpoint of the compressor, is a quasi-stationary and stable operating mode with forward flow. However, as far as the stalled component is concerned, it is a dynamic instability. In principle all compressor components may run into stall, like the impeller, diffuser, return vane channel, volute and especially when the component pressure gradient is positive:

$$\frac{\partial(p_d/p_s)_{component}}{\partial \dot{V}_{component}} > 0. \tag{5.23}$$

The distortion may start as a local reverse flow so that neither the stage nor the compressor may stall. However, when the flow is reduced further, the complete pertinent stage will stall, and finally the stall will spread to the entire compressor, driving it into stage surge which is as periodic and unstable as system surge. The primary cause of stage surge is flow reversal causing rotating stall, with the eventual surging of the machine being a secondary appearance. Stage surge is the surge proper of the machine and is initiated without the support of an external reservoir. The transition from stall to surge is brought about by the losses accompanying the process of stalling, which coerces the stage performance curve into either a positive slope (mild stall) or even a discontinuous modification (violent stall), as shown in Fig. 5.24. This is exactly the prerequisite for the periodic instability called surge, much in the same way as system surge, taking into account adjoining (small) reservoirs; then it is quite possible that an interaction with the external system of coolers, separators and pipes amplifies the stage instability.

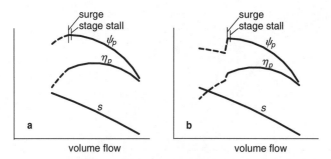

Fig. 5.24a, b Transition from stall to surge. **a** progressive or mild stall, **b** abrupt or violent stall. Head coefficient, efficiency and work input factor.

In some compressors a clear distinction can be made between the regimes of rotating stall with its pressure unsteadiness and periodic surge, which are normally close neighbors in the performance map.

In process centrifugal compressors the vaneless diffuser is by far the most common flow element to be prone to reverse flow and rotating stall and thus responsible for the onset of stage surge. This is much more so than for impeller rotating stall (formerly called "inducer stall"), which occurs only at higher relative inlet Mach numbers, which are seldom attained in process compressors (vaned high-solidity diffusers, being inherently stall-prone, are not very common in process compressors either). The kinked head curve in Fig. 5.24 is caused by the diffuser alone; the work input as the impeller characteristic remains unaffected.

The most plausible interpretation of what happens in the vaneless diffuser upon flow reduction was given by Senoo and Kinoshita (1977) as follows:

- It starts out with the absolute flow angle becoming zero at the diffuser front wall inlet, since the velocity profile at the impeller exit is highly distorted in the shroud/blade suction side corner due to a distinctive split-up into a core and wake zone, as is shown in Fig. 5.25.
- A local flow reversal – limited in its radial and circumferential extension – takes shape because the drastically reduced radial velocity component adjacent to the wall does not have sufficient kinetic energy to maintain its forward movement to overcome the radial pressure gradient generated by the conservation of angular momentum, which, of course, reaches into the boundary layer as well. Following a flow visualization photo by Osborne and Japikse (1982), Fig. 5.26 shows the flow reverse zone at the diffuser inlet.

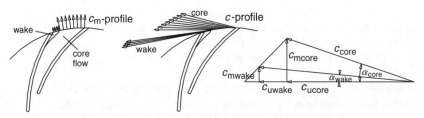

Fig. 5.25 Distorted velocity profile at reduced flow. Reduction of absolute flow angle at diffuser inlet; incipience of rotating stall.

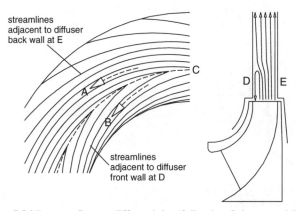

Fig. 5.26 Reverse flow at diffuser inlet (following Osborne and Japikse (1982)). A outward core flow, B inward reverse flow, C separating line, D diffuser front wall, E diffuser back wall.

- The mass averaged flow angle α_{3c}, corresponding to the zero flow angle at the wall, is the "critical inlet flow angle for reverse flow". α_{3c} is the decisive parameter for the onset of rotating stall.
- Senoo and Kinoshita (1977): "(…) it is theoretically and experimentally proven that a rotating stall takes place when the flow locally reverses".
- The phenomenon of rotating stall is described in numerous text books. Since a zone of stagnating or reversing flow blocks the through flow area locally, the radial velocity and hence the angle of the core flow increase, thus eliminating the very cause of reverse flow and shifting the separated flow zone to a neighboring location, setting in motion a rotation of the distorted regime.
- During vaneless diffuser rotating stall usually two to three reverse flow cells rotate in the forward direction (i.e. with shaft rotation) at between 5% and 22% of the rotational speed (high values for small flow coefficient stages). Impeller rotating stall cells can reach up to 82% of the rotational speed.
- When the flow is decreased further, the reverse zone extends until it reaches the diffuser exit, which is the critical condition for the compressor to run into surge. When the rotating stall has reached the diffuser exit, the losses increase rapidly, leading to a dramatic decrease of efficiency and head, with positive slope often separated by a kink from the rest of the curve (Fig. 5.24). This is a very similar condition for surge as for system surge, resulting in a similar surge cycle.
- Although the incipience of reverse flow is not yet the critical condition of surge, the critical inlet flow angle is a conservative prediction which the compressor designer is recommended to apply in order to avoid the cause "flow reversal", which in turn will a priori avoid the effect "stage surge".
- The orbiting cells in turn cause pressure fluctuations by consecutively blocking impeller blade channels during one impeller rotation, thus forcing the channels to operate consecutively at different characteristic curve points.

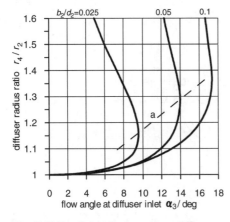

Fig. 5.27 Parallel-walled vaneless diffuser reverse-flow zone (according to Senoo and Kinoshita (1977), redrawn). *Left* of each curve is the reverse-flow zone, a beginning of rotating stall upon flow reduction (decrease of α_3).

The radial extension of the reverse-flow zone is described in Fig. 5.27. In the case of a low flow coefficient stage with $b_2/d_2 = 0.025$ having a flow angle of $\alpha_3 = 8°$ there will be RS extending from $r/r_2 = 1.05$ to 1.33; if the diffuser has a total radius ratio of 1.6, there is no backflow from the outside into the diffuser and the compressor will not start surging; it will just exhibit rotating stall. However, in the case of a high flow coefficient stage with $b_2/d_2 = 0.1$ and $\alpha_3 = 16°$ the reverse flow covers a radius ratio range of between 1.2 and 1.7, and if $r_4/r_2 = 1.6$ the compressor will start surging. With narrow diffusers the reverse flow begins near the inlet; with wide diffusers the inlet is free of any reverse flow which begins further downstream.

The critical inlet flow angle α_{3c}, according to Senoo and Kinoshita (1977), depends on the diffuser inlet Mach number M_{c3} and the referenced impeller exit width b_2/d_2 for undistorted inlet velocity profiles and parallel-wall diffusers ($b_2 = b_3$), as shown in Fig. 5.28. It can be as low as 9° and as high as 25°.

If the inlet radial velocity profile is distorted by 30%, i.e. $\Delta c_m/c_m = 0.3$, α_{3c} has to be corrected by some +4°, and if the inlet circumferential velocity profile is distorted by 30%, i.e. $\Delta c_u/c_u = 0.3$, α_{3c} has to be corrected by some +2°. In other words, for disturbed inlet conditions the diffuser stalls earlier and the compressor has a worse surge limit, i.e. a narrower operating range.

If the diffuser is pinched, i.e. $b_3 < b_2$, the flow angle will be steeper and the onset of RS will be delayed. However, according to Nishida et al. (1988), the critical flow angle α_{3c} as per Fig. 5.28 has to be modified as follows:

$$\alpha_{3c}^* = \alpha_{3c} + (17.02 - 148.4 b_2/d_2)(1 - b_3/b_2) \,. \tag{5.24}$$

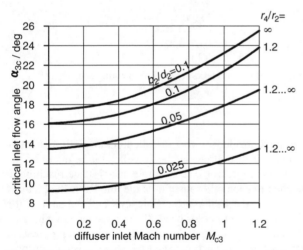

Fig. 5.28 Critical (mass averaged) flow angle at parallel-walled vaneless diffuser inlet; valid for $b_3 = b_2$ and undistorted velocity profile (according to Senoo and Kinoshita (1977), redrawn).

Vibration Excitation Caused by Rotating Stall

The circulating pressure pulsations excite shaft and blade vibrations (in the case of unshrouded impellers). These can be critical in the case of higher gas densities, where the corresponding exciting forces can be comparatively high, especially during resonance between excitation and natural frequency.

According to the findings of the FVV (1990) and Abdel-Hamid et al. (1987) the rotating-stall fundamental excitation frequency for forward cell rotation is

$$f_{ex} = \lambda f - f_{RS} = \lambda f - \lambda f(N_{RS}/N) = \lambda f[1-(N_{RS}/N)], \quad (5.25)$$

and for backward cell rotation it is

$$f_{ex} = \lambda f + f_{RS} = \lambda f + \lambda f(N_{RS}/N) = \lambda f[1+(N_{RS}/N)]. \quad (5.26)$$

The second harmonic is

$$f_{ex2} = 2\lambda f[1\pm(N_{RS}/N)], \quad (5.27)$$

where f_{ex} is the exciting frequency through rotating stall, f_{RS} is the rotating stall frequency, f is the rotational shaft frequency, N_{RS}/N is the orbiting ratio stall/shaft, and λ is the number of stall cells.

The number of cells λ and the rotation ratios are given in Table 5.3 (according to Van den Braembussche and Japikse (1996)).

Table 5.3 Cell rotation ratio

Type of rotating stall	λ	N_{RS}/N
Vaneless diffuser rotating stall	2–3	(0.05)–0.13–0.22
Abrupt impeller rotating stall	1–3	0.26–0.31
Progressive impeller rotating stall	1–3	0.67–0.82
Mild impeller rotating stall	4–5	0.14

All cells have forward rotation; N_{RS} rotational speed of cells; N operating speed

5.5.2 Choke

The maximum volume flow is controlled by the so-called choke limit characterized by attaining sonic velocity in the throat at the impeller blade inlet. At the rated point (RP) of a multistage compressor (see example of a 4-stage machine in Fig. 5.29) all stages are matched to each other insofar as the line RP is the locus of the stage design points (at or near the best stage efficiency), and the sum of the pertaining stage heads is the overall compressor head. If the 1st stage is volumetrically overloaded, an aerodynamic stage mismatch takes place due to a progressive stage-by-stage head decrease. Therefore, each consecutive stage sees a disproportional flow increase, as shown by the line OL1.

If the overload grows, mismatching will snowball with the last stage head approaching zero and sending the overall compressor head into nose dive, as shown by line OL2. This is referred to in this book as the compressor choke limit, where the last stage arrives at zero head.

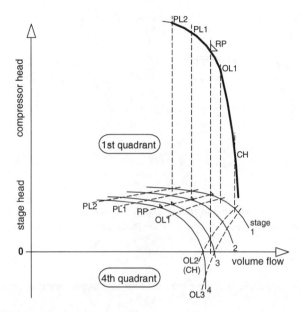

Fig. 5.29 Choke limit: 4-stage compressor performance curve and individual stage performance curves. PL partload, RP rated point, OL overload, CH choke point.

And if the foregoing stages can still generate the necessary density, the last stage can be overloaded a little more, being driven into the 4th quadrant, where it consumes rather than produces pressure ratio, as shown by the line OL3.

The choking mass flow can be calculated from the continuity equation with throat area, critical sonic velocity, density in the throat and boundary layer displacement thickness (blockage factor):

$$\dot{m}_{CH} = A a_c \rho_c (1-B). \tag{5.28}$$

The impeller inlet along the leading edge (with its length L, z blades and thickness s_b, see Fig. 5.30) is subdivided into n annuli having a pertaining diameter d_i and blade angle β_{1bi} (both functions of radius r). The throat area then becomes:

$$\Delta A = \left(\frac{d_i \pi}{z} \sin \beta_{1b} - s_b\right)\frac{L}{n} \rightarrow A = \frac{Lz}{n}\sum_{i=1}^{n}\left(\frac{d_i \pi}{z}\sin \beta_{1bi} - s_{bi}\right). \tag{5.29}$$

The sonic velocity (for perfect and real-gas behavior) is defined as:

$$a = \sqrt{\left(\frac{\partial p}{\partial \rho}\right)_s}. \tag{5.30}$$

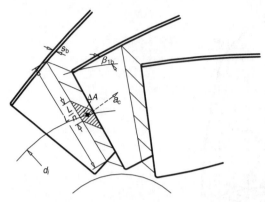

Fig. 5.30 Impeller inlet throat area. ΔA element of throat area, L length of throat, n number of elements, β_{1b} blade angle, a_c throat sonic velocity, s_b blade thickness, d_i element diameter.

From Eq. (2.6), defining an isentropic change of state (with isentropic volume exponent):

$$\frac{p}{\rho^{\kappa_v}} = const \rightarrow \ln p - \kappa_v \ln \rho = 0 \rightarrow \frac{dp}{p} - \kappa_v \frac{d\rho}{\rho} = 0. \tag{5.31}$$

As per the definition, the change of pressure with the density takes place at constant entropy:

$$\left(\frac{\partial p}{\partial \rho}\right)_s = \kappa_v \frac{p}{\rho} = \kappa_v ZRT \rightarrow a = \sqrt{\kappa_v ZRT}. \tag{5.32}$$

The critical sonic velocity in the throat is:

$$a_c = \sqrt{\kappa_{vc} Z_c R T_c}. \tag{5.33}$$

The critical temperature T_c is calculated by conserving the total enthalpy from state "0" (upstream of blades) to state "c" (throat), since there is no energy transfer. Using the (invariant) rothalpy in the rotating frame of reference we obtain

$$h_0 + \frac{w_0^2}{2} - \frac{u_0^2}{2} = h_c + \frac{w_c^2}{2} - \frac{u_c^2}{2}, \quad \text{with } w_c = a_c. \tag{5.34}$$

Using the total enthalpy for state "0":

$$h_{0t} - h_c = \int_{t_0}^{t_c} c_p(t) \, dt = \bar{c}_p (T_{0t} - T_c) = \frac{Z_c \kappa_{vc} R T_c}{2} - \frac{u_c^2}{2}. \tag{5.35}$$

Resolving for the static critical temperature to be used for the critical sonic velocity yields

$$T_c = \frac{\bar{c}_p T_{0t} + \dfrac{u_c^2}{2}}{\bar{c}_p + \dfrac{Z_c \kappa_{vc} R}{2}}.$$ (5.36)

The critical density is calculated from the basic isentropic relationship between "0" (total state) and "c" (static state) with isentropic temperature exponent:

$$\rho_c = \rho_{0t} \left(\frac{T_c}{T_{0t}} \right)^{\frac{1}{\bar{\kappa}_T - 1}}, \quad \text{with} \quad \bar{\kappa}_T = \frac{\kappa_{T0t} + \kappa_{Tc}}{2}$$ (5.37)

Equations (5.35) and (5.36) are iteratively solved. Experience has shown that the blockage factor B is approximately 0.02.

The choking mass flow, Eq. (5.28), can now be computed. This computation approach should be calibrated by actual measurements in order to correct for inaccuracies of the calculated throat area and of the blockage factor. The critical density is unquestionable in the case of a single stage; but for multistage machines choke flow calculation is only as good as the prediction of the density made available by the cumulative action of the foregoing stages.

5.5.3 Beyond Choke (Expansion Mode in the Fourth Quadrant)

If a multistage compressor is subjected to an ever-decreasing system resistance (for instance by opening the discharge valve and with no other substantial resistance in the system), the head has to adapt to the new situation by reducing its stage heads by flow increase.

However, due to aerodynamic stage mismatching, this is not achieved uniformly for all stages but progressively, so that each consecutive stage reduces its head to a greater extent than the foregoing stage. The compressor choke point is reached when the last stage's head becomes zero. The compressor head is still positive. If this is still too great to match the reduced system resistance, the compressor reacts by driving the last stage beyond its choke point, i.e. into the "fourth quadrant", as can be seen in Fig. 5.31. It shows, as an example, the measured performance curve of the last stage of an integrally geared four-stage CO_2-compressor with a 17 : 1 compressor pressure ratio at the design point.

At a compressor pressure ratio of 2.7 : 1 the last stage is already deep in the fourth quadrant: the last stage suction pressure is 10.2 bar and its discharge pressure is 2.7 bar (!). However, this stage acts by no means as a power generating turbine; on the contrary, it needs power input to overcome the excessive internal resistance caused by the abnormal operating mode. The experimental flange-to-flange data of the single stage are given in Table 5.4.

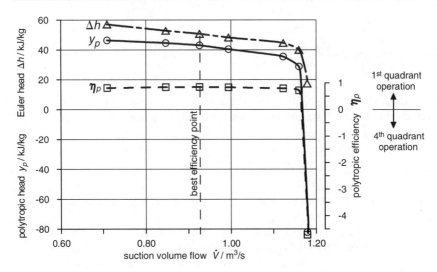

Fig. 5.31 Measured single-stage performance curve. Flange-to-flange recordings in 1st and 4th performance quadrant (positive Euler head, negative pol. head and negative efficiency).

Table 5.4 Single compressor stage, best efficiency point and beyond choke operation

	Unit	Best efficiency point	4th quadrant operation
Suction volume flow	m³/s	0.93	1.18
Isentropic head	kJ/kg	42.5	−69.0
Isentropic efficiency	–	0.838	−3.97
Polytropic head	kJ/kg	43.1	−82.1
Polytropic efficiency	–	0.849	−4.7
Euler head	kJ/kg	50.7	17.4
Gas power input	kW	793	346
Pressure ratio (total-to-total)	–	1.92:1	1: 3.75
Temperature difference (total-to-total)	K	58.8	13.4
Entropy difference (total-to-total)	J kg⁻¹K⁻¹	32.0	318.0
Velocity suction nozzle	m/s	29.6	37.5
Velocity discharge nozzle	m/s	18.3	180

How do expansion and power input with its temperature rise fit together? What strikes the eye with this 4th quadrant measuring point, besides the smaller-than-unity pressure ratio, is that the expansion head is 62% greater than the compression head and that the gas velocity in the discharge nozzle and the entropy production is some ten times as great as normal. This means for such an adiabatic system that a lot of energy (including the valuable enthalpy of the foregoing stages) is irreversibly transferred into unavailable energy, sometimes referred to as anergy. This also means that the 4th-quadrant change of state in the enthalpy–entropy diagram is predominantly horizontal, resembling superficially a throttling change of state. The 1st-quadrant change of state of the normal operating point is predominantly vertical. The analysis of the 4th-quadrant test point (which did not include pressure measurements at the impeller exit as in laboratory test setups) revealed the following course of events in the stage.

The φ_3-diagram is shown in Fig. 5.32; since the total-to-static impeller volume ratio as per Eq. (5.15) is greater than 1.0 the impeller exit flow coefficient φ_3 more than doubles in relation to the best efficiency point; and since there is reason to believe that the exit velocity triangle, Fig. 5.33, still follows the (approximate) rule of constant relative exit angle for increasing impeller exit volume flow, the work input factor plotted versus φ_3 is an approximate straight line, as in Fig. 5.19, even for such a great extension. Test data analysis is given in Table 5.5.

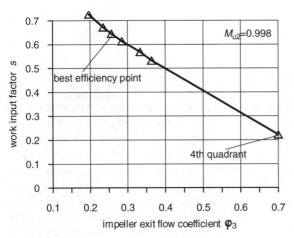

Fig. 5.32 Intrinsic performance curve: s, φ_3-diagram including 4th-quadrant test point.

Table 5.5 Test results of best efficiency point and 4th quadrant operation

	Symbol	Best efficiency point	4th quadrant operation
Flow coefficient	φ	0.049	0.0624
Work input factor	s	0.645	0.222
Impeller exit flow coefficient	φ_3	0.257	0.70
Sonic velocity in impeller throat [m/s]	a_c		269
Absolute velocity in impeller throat [m/s]	c_c		243
Absolute impeller exit velocity [m/s]	c_3	195	206
Reaction	r	0.626	-0.22
Polytropic impeller efficiency, total-to-total	η_{pi}	0.92	0.05

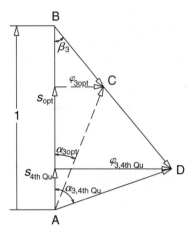

Fig 5.33 Dimensionless velocity triangle at impeller exit. ABC best efficiency point (1st quadrant), ABD expansion mode beyond choke (4th quadrant), AC and AD absolute velocities, BC and BD relative velocities.

The change of state in the 4th quadrant is shown in the enthalpy–entropy diagram, Fig. 5.34B. From the suction nozzle (total state 0t, approximately equal to the static state, since the velocity is low) the gas expands until it reaches sonic velocity in the impeller throat (critical static state c); the impeller increases the static pressure to state 3 constituting already a lower pressure than at suction; 0t–3t is the change of total enthalpy, i.e. the energy input; because the total-to-total impeller efficiency is $\eta_{si} = 0.05$, the total pressure at the impeller exit p_{3t} nearly equals the total pressure at the impeller inlet p_{0t} (at this particular measuring point only); the losses in the diffuser, volute and discharge cone are so large, due to the excessive velocity, that the static pressure decreases below the measured total discharge pressure p_{π}. Figure 5.34A, being to scale with Fig. 5.34B, shows the change of state at the best efficiency point. The test stage is shown in Fig. 5.35.

This extreme aerodynamic mismatching, with the last stage operating in the fourth quadrant, can only be accomplished by multistage compressors where the first stages provide the pressure ratio that is partly used up in the last stage again. In this way the compressor can adapt itself to an extremely small external system resistance by generating head in the front stages and consuming head in the rear stage, which acts as a quasi-throttle under energy input.

The gas expands on its path from the suction nozzle to the impeller throat, where it reaches sonic velocity. Since the impeller still produces a positive static head (see Fig. 5.34B) it necessarily generates a low static suction pressure due to the extreme low discharge pressure, thus enabling sonic velocity. The gas is recompressed in the impeller, although not attaining total suction pressure. In the diffuser and volute the gas again expands by producing massive losses and discharging with excessive flange exit velocity so that the static discharge pressure remains significantly lower than the recorded total pressure. This expansion–compression–expansion mode of operation is stable and safe as long as the thrust bearing has been sized to account for the thrust variations that are inevitable for the stage operating in the 1st or 4th quadrant. So all potential operating points, including abnormal ones, have to be carefully defined so that the machine can be designed accordingly.

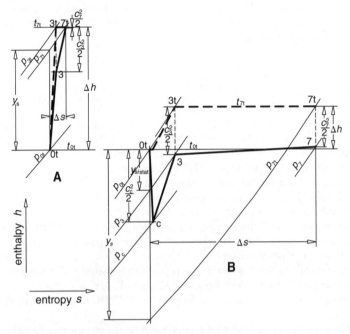

Fig. 5.34A, B Enthalpy–entropy diagram: change of state in 1st and 4th performance quadrant. **A** operation at best efficiency point (1st quadrant), **B** operation in expansion mode beyond choke (4th quadrant), 0 suction nozzle, c impeller inlet throat, 3 impeller exit, 7 discharge nozzle, *solid thick line* static change, *dashed thick line* total change, *dashed thin line* dynamic enthalpy.

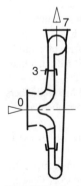

Fig. 5.35 4th-quadrant measurements: type of stage tested. Overhung shrouded impeller, axial suction: $d_2 = 293$ mm, $\varphi = 0.05$, $M_{u2} = 1.0$, $\beta_{2b} = 50°$.

Summarizing, it has been shown that normal compressor operation in the first quadrant of the flow/head graph (positive flow, positive head) can, on the one hand, at least periodically during surge, extend into the second quadrant (negative flow, positive head). On the other hand, a stage in a compressor can extend its operation stably into the fourth quadrant of the graph (positive flow, negative head) in the case of extreme volumetric overload imposed on the compressor by extremely low external system resistance.

References

Abdel-Hamid AN, Haupt U, Rautenberg M (1987) Unsteady flow characteristics in a centrifugal compressor with vaned diffuser. ASME Paper 87-GT-142, Gas Turbine Conf. Anaheim, California, 1987; The American Society of Mechanical Engineers, New York

API Standard 617 (1995) Centrifugal Compressors for Petroleum, Chemical, and Gas Service Industries. 6th edn, American Petroleum Institute, Washington, D.C.

Eckert B, Schnell E (1961) Axial und Radial-Kompressoren. 2nd edn, Springer-Verlag Berlin Göttingen Heidelberg, pp 477-478, (in German)

FVV-Forschungsvereinigung für Verbrennungskraftmaschinen (Research Association for Internal Combustion Engines) (1990) Radialverdichter-Schwingfestigkeit. Heft 456, pp 26-27, FVV, Frankfurt, Germany, (restricted, in German)

Lüdtke K (1998) Aerodynamic stage mismatching – the key to understand multistage process centrifugal compressor behavior. VDI Berichte 1425, pp 147-156. In: Tubocompressors in industrial use. VDI Conf., Hannover, Germany, 1998; VDI-Verlag Düsseldorf, Germany

Lüdtke K (1992) The influence of adjustable inlet guide vanes on the performance of multistage industrial centrifugal compressors. ASME Paper 92-GT-17, Gas Turbine and Aeroengine Congress, Cologne, Germany, 1992; The American Society of Mechanical Engineers, New York

Nishida H et al (1988) A study on the rotating stall of centrifugal compressors (1st Report: Effect of vaneless diffuser width on rotating stall). Trans. Japan Soc. Mech. Eng. 54 (499), pp 589-594, Nihon Kikaigakkai Ronbunshu (B edn)

Osborne C, Japikse D (1982) Comprehensive evaluation of a small centrifugal compressor. Creare Technical Notes, TN-347, restricted; Creare Nanover, New Hampshire

Senoo Y, Kinoshita Y (1977) Influence of inlet flow conditions and geometries of centrifugal vaneless diffusers on critical flow angle for reverse flow. Trans. ASME, J. Fluids Eng., pp 98-103; The American Society of Mechanical Engineers, New York

Sheets HE (1951) Nondimensional compressor performance for a range of Mach numbers and molecular weights. ASME Paper 51-SA-19, ASME Meeting, Toronto, Canada, 1951; The American Society of Mechanical Engineers, New York

Traupel W (1966) Thermische Turbomaschinen. 1st vol, 2nd edn, Springer-Verlag, Berlin Heidelberg New York, pp 506-507 (in German)

Van den Braembussche R, Japikse D (1996) Rotating stall in centrifugal compressors. In: Japikse D (1996) Centrifugal compressor design and performance. Chap. 5.2, Concept ETI, Wilder, Vermont

6 Aerodynamic Compressor Design: Case Study

"Design" in the industry means to size a compressor after the receipt of an order according to a philosophy with proven components. Computational fluid dynamics (CFD) is not normally applied for these routine designs on a contract basis. CFD is extensively used during the development phase of new components, which is not within the scope of this book.

The detailed compressor design (as per the specification in Table 6.1) is conducted in three steps:

1. approximate calculation of frame size, number of stages, speed and power,
2. detailed component sizing, which will lead to the final predicted performance data on the basis of the one-dimensional stream line method,
3. impeller blade design by means of the so-called quasi-3D approach.

The geometric outcome of the aero-thermodynamic design will be checked by a rotordynamic feasibility calculation. All of this is carried out in form of a case study to achieve the best possible comprehensibility and transparency.

Table 6.1 Specification of "Natural Gas Export Compressor"

	Symbol	Unit	Data
Gas composition:		mol%	
nitrogen	N_2		3.18
carbon dioxide	CO_2		1.18
methane	CH_4 (C_1)*		87.37
ethane	C_2H_6 (C_2)		5.45
propane	C_3H_8 (C_3)		1.78
iso-butane	C_4H_{10} (iC_4)		0.32
n-butane	C_4H_{10} (C_4)		0.45
iso-pentane	C_5H_{12} (iC_5)		0.11
n-pentane	C_5H_{12} (C_5)		0.09
n-hexane	C_6H_{14} (C_6)		0.07
			100.00
*Customary abbreviation in the oil and gas industry (see Table 6.9)			
Suction pressure (abs.)/temperature	p_s / t_s	bar / °C	62.7/31.2
Discharge pressure (abs)	p_d	bar	92.97
Mass flow	$\dot{m}$	kg/s	85.9
Maximum stage head	y_{pmax}	kJ/kg	40
Max. rotational speed (preferable)	N_{max}	1/min	14,000
Prime mover	asynchronous motor via gear box		
Regulation	adjustable inlet guide vanes		

6.1 Approximate Compressor Sizing

The procedure is used to check the aerodynamic compressor feasibility for a newly written specification and to get a first overview on what the envisaged compressor will look like. The outcome can also be used to send out a budget offer.

The questions to be answered by these first steps are as follows:

1. How many stages are needed for the job? This requires gas data analysis.
2. How many impellers can be accommodated on the shaft ?
3. How large is the impeller diameter? This will determine compressor frame size.
4. The required speed, power and discharge temperature will result.

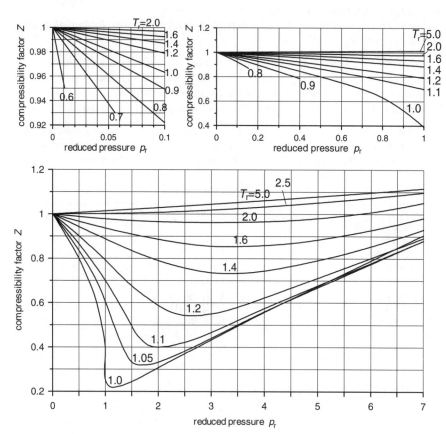

Fig. 6.1 Nelson–Obert chart: generalized compressibility factor vs. reduced pressure and reduced temperature (Nelson and Obert, (1954), redrawn). $p_r = p/p_c$ reduced pressure, $T_r = T/T_c$ reduced temperature, p_c, T_c pseudocritical values.

The compressibility factors are determined from the generalized Nelson–Obert chart (Nelson and Obert 1954) as a function of the reduced temperature and pressure, which is accurate enough for this preliminary sizing approach; see Fig. 6.1 and Table 6.2. In this "hybrid" method the isentropic exponents are replaced by the perfect-gas heat capacity ratio $c_p/c_v = Mc_p/(Mc_p\text{-}\Re)$ at atmospheric pressure (molar isobaric heat capacity Mc_p from Fig. 6.2; universal gas constant $\Re = 8.3145$ kJ kmol^{-1}K^{-1}), and the real-gas behavior is accounted for solely by the compressibility factors Z used to calculate the polytropic head, the number of stages (Tables 6.3, 6.4) and the actual volume flow (Table 6.5).

Table 6.2 Thermodynamic gas data

	Formula		Calculation result
Reduced pressure	$p_r = \dfrac{p}{\sum_{i=1}^{n}(r_i p_{ci})}$	(6.1)	suction $p_{rs} = 62.70/46.27 = 1.36$ discharge $p_{rd} = 92.97/46.27 = 2.0$
Reduced temperature	$T_r = \dfrac{T}{\sum(r_i T_{ci})}$	(6.2)	suction $T_{rs} = (31.2+273.2)/202 = 1.51$ discharge $T_{rd} = (65.0+273.2)/202 = 1.67$ (discharge temp. estimated 65 °C)
Compressibility factor	from Nelson–Obert chart Fig. 6.1		suction $Z_s = 0.89$ discharge $Z_d = 0.90$
Molar isobaric heat capacity	from Fig. 6.2		suction $Mc_{ps} = 38.13$ kJ/kmolK discharge $Mc_{pd} = 39.69$ kJ/kmolK
Isentropic exponent	$\kappa = \dfrac{c_p^0}{c_v^0} = \dfrac{Mc_p}{Mc_p - \Re} = \dfrac{\sum_{i=1}^{n}(r_i Mc_{pi})}{\sum_{i=1}^{n}(r_i Mc_{pi}) - \Re}$	(6.3)	suction $\kappa_s = 38.13/(38.13-8.3145) = 1.279$ discharge $\kappa_d = 39.69/(39.69-8.3145) = 1.265$
Molar mass	$M = \sum_{i=1}^{n} r_i M_i$	(6.4)	$M = 18.5$ kg/kmol
Gas constant	$R = \Re/M$	(6.5)	$R = 8.3145/18.5 = 0.449$ kJ kg^{-1}K^{-1}

r molar fraction, n number of gas components, isentropic exponent as per perfect gas formulae: $(n_v-1)/n_v = (n_T-1)/n_T = (\kappa-1)/\kappa\eta_p$ (6.6)

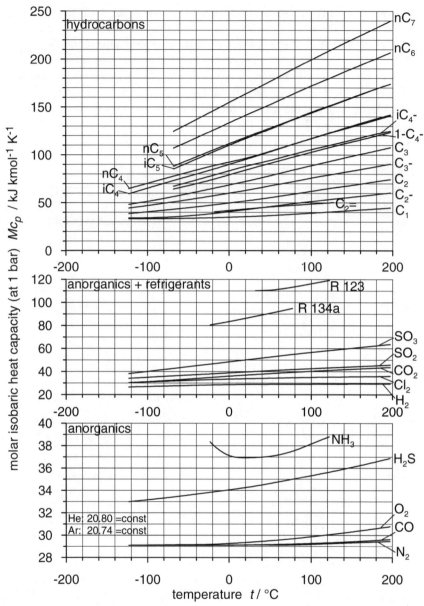

Fig. 6.2 Perfect gas molar isobaric heat capacity, for approximate calculation of isentropic exponents (= heat capacity ratios) for perfect-gas behavior (as per Eq. (6.3)).

6.1 Approximate Compressor Sizing

Table 6.3 Polytropic head

Pol. head for compressors with and without intercoolers, approximate	$y_p \approx [T_s + cT_R(1+h_c)]\dfrac{Z_s + Z_d}{2} R \dfrac{\kappa \eta_p}{\kappa - 1}\left[\pi^{\frac{\kappa-1}{\kappa\eta_p(c+1)}} - 1\right]$ (6.7)
Polytropic compressor efficiency, estimate	basic (per Fig. 1.23) 0.83 Reynolds number correction +0.015 diameter correction −0.02 blade cut-back correction +0.015 efficiency, predicted $\eta_p \approx 0.84$
Number of coolers	$c = 0$
Polytropic head	$y_p = 304.4 \times \dfrac{0.89 + 0.90}{2} \times 0.449 \times$ $\times \dfrac{1.272 \times 0.84}{0.272}\left[\left(\dfrac{92.97}{62.7}\right)^{\frac{0.272}{1.272 \times 0.84}} - 1\right] = 50.6 \text{ kJ/kg}$

T_s, T_R suction, recooling temperature, c number of intercoolers, $h_c = \Delta y_p/y_p$ relative cooler head loss, $\kappa = (\kappa_s + \kappa_d)/2$, π pressure ratio for cooled compressors; y_p sum of polytropic section heads; section pressure ratios are assumed equal.

Table 6.4 Number of stages

Nos. of stages	stage head: $y'_p = y_p/i$ (6.8)	$i = 1$: $y'_p = 50.6/1 = 50.6$ kJ/kg limit of $y_{pmax} = 40$ is exceeded, increase i ! $i = 2$: $y'_p = 50.6/2 = 25.3$ kJ/kg, this is ok!
Check tip speed	rearranging Eq. (1.73): $u_2 = \sqrt{2\psi_p y_p 10^3}$ (6.9) ψ_p as per Fig. 1.23 $u_{2\max} = \sqrt{C * \dfrac{\sigma_{100}}{R_{p0.2}} R_{p0.2}}$	$u_2 = \sqrt{2 \times 1.0 \times 25.4 \times 10^3} = 225.4$ m/s; allowable. As per Eq. (4.4): $u_{2\max} = \sqrt{170 \times 0.6 \times 830} = 291$ m/s
Check Mach number	$M_{u2} = \dfrac{u_2}{\sqrt{\kappa_s Z_s R T_s}}$	$M_{u2} = \dfrac{225}{\sqrt{1.279 \times 0.89 \times 0.449 \times 304.4 \times 10^3}} = 0.57$, allowable. As per Fig. 4.2: $M_{u2\max} = 1.1$
Nos. of stages	rotordynamic check	$i = 2$; allowable. As per Fig. 4.4: $i_{\max} = 4$ (with corrections)

Table 6.5 Volume flow, impeller diameter, rotational speed, discharge temperature

Suction volume flow	$\dot{V} = \dfrac{\dot{m} Z_s R T_s}{p_s}$	(6.10)	$\dot{V} = \dfrac{85.9 \times 0.89 \times 0.449 \times 304.4}{62.7 \times 10^2} = 1.666 \text{ m}^3/\text{s}$ dimensional check with unity braces: $\dfrac{\text{kg}}{\text{s}} \dfrac{\text{kJ}}{\text{kgK}} \dfrac{\text{K}}{\text{bar}} \left\{ \dfrac{\text{bar}}{10^5 \text{ N/m}^2} \right\} \left\{ \dfrac{10^3 \text{ Nm}}{\text{kJ}} \right\} = \dfrac{\text{m}^3}{10^2 \text{ s}}$
Flow coeff.	first try, assumed		$\varphi = 0.09$
Impeller diameter 1st try	$d_2 = \sqrt{\dfrac{4 \dot{V}}{\pi u_2 \varphi}}$	(6.11)	$d_2 = \sqrt{\dfrac{4 \times 1.666}{\pi \times 225 \times 0.09}} = 0.323 \text{ m}$
Rotational speed 1st try	$N = \dfrac{u_2}{\pi d_2}$	(6.12)	$N = \dfrac{225}{\pi \times 0.323} \dfrac{\text{m}}{\text{sm}} \left\{ \dfrac{60 \text{s}}{\text{min}} \right\} = 13{,}304 \text{ 1/min}$ (specified limit not exhausted)
Flow coeff.	corrected		$\varphi = 0.0992$
Impeller dia.	corrected		$d_2 = 0.308 \text{ m}$
Rotational speed	corrected		$N = 13{,}952 \text{ 1/min}$ (still in accordance with specification)
Discharge temperature	$T_d = T_s \left(\dfrac{p_d}{p_s} \right)^{\frac{\kappa - 1}{\kappa \eta_p}}$	(6.13)	$T_d = (273.2 + 31.2) \left(\dfrac{92.97}{62.7} \right)^{\frac{0.272}{1.272 \times 0.84}} - 273.2$ $= 63°\text{C}$

Table 6.6. Efficiency (case study, 2-stage compressor)

Volume flow factor	$a = \left[s(\kappa - 1) M_{u2}^2 + 1 \right]^{\frac{\kappa - \kappa \eta_p - 1}{\kappa - 1}}$ (6.14)[a] (see Eq. (4.9), s from Fig. 1.23)		$a = (0.61 \times 0.272 \times 0.571^2 + 1)^e$ $= 0.857$ $e = (1.272 - 1.272 \times 0.84 - 1)/0.272$
Flow coeff.	2nd stage: $\varphi_2 = a \varphi_1$		$\varphi_2 = 0.857 \times 0.0992 = 0.085$
Efficiency	1st stage: $\eta_{p1} = f(\varphi_1)$		$\eta_{p1} = 0.84$
Efficiency	2nd stage: $\eta_{p2} = f(\varphi_2)$		$\eta_{p2} = 0.84$
Compressor efficiency	$\eta_p = (\eta_{p1} + \eta_{p2})/2$	(6.15)	$\eta_p = 0.84$

[a] valid for two stages only

6.1 Approximate Compressor Sizing

The determination of the efficiency is given in Table 6.6 (study), and generally per Table 6.7

The calculation of the gas power and motor nameplate rating is shown in Table 6.8. Table 6.9 displays the customary abbreviations of hydrocarbon gases.

Table 6.7. Efficiency, head coefficient (general)

Volume flow factor (use for more than 2 stages)	$a \approx (T_R/T_s)\left(\dfrac{p_d}{p_s}\right)^{\left(\frac{\kappa-1}{(c+1)\kappa\eta_p}\right)} - 1$	(6.16)
	(for $c = 0$: $T_R/T_s = 1$)	
Flow coefficient (suction side)	φ_s	
Efficiency (suction side)	$\eta_{ps} = f(\varphi_s)$	
Head coefficient (suction side)	$\psi_{ps} = f(\varphi_s)$	
Flow coefficient discharge side	$\varphi_d = a\varphi_s$	(6.17)
Efficiency discharge side	$\eta_{pd} = f(\varphi_d)$	
Head coefficient discharge side	$\psi_{pd} = f(\varphi_d)$	
Mean flow coefficient	$\varphi_m = (\varphi_s+\varphi_d)/2$	(6.18)
Mean efficiency	$\eta_{pm} = f(\varphi_m)$	
Mean head coefficient	$\psi_{pm} = f(\varphi_m)$	
Compressor efficiency	$\eta_p = (\eta_{ps}+\eta_{pm}+\eta_{pd})/3$	(6.19)
Head coefficient	$\psi_p = (\psi_{ps}+\psi_{pm}+\psi_{pd})/3$	(6.20)

Use Fig. 1.23 to arrive at η_p and ψ_p as functions of the flow coefficient. Three-point procedure accounts for line curvature.

Table 6.8. Power input

Recirculation loss	$\Delta\dot{m}/\dot{m}$ as per Fig. 4.10		$\Delta\dot{m}/\dot{m} \approx 0.01$
Gas power	$P_i = \dot{m}[1+(\Delta\dot{m}/\dot{m})]y_p/\eta_p$	(6.21)	$P_i = 85.9\times1.01\times50.6/0.84$ $= 5{,}226$ kW
Bearing loss	$P_{BL} = R_{BL}(N/1{,}000)^2$ (R_{BL} as per Fig. 6.3)	(6.22)	$P_{BL} = 0.15\times(13{,}952/1{,}000)^2$ $= 29$ kW
Power at compressor coupling	$P_{cc} = P_i+P_{BL}$	(6.23)	$P_{cc} = 5{,}226+29 = 5{,}255$ kW
Gear loss	$P_{GL} = (1-\eta_G)\times P_M$ (η_G as per Fig. 6.3)		$P_{GL} = (1-0.985)\times 5{,}900$ $= 88$ kW
Motor nameplate rating	$P_M = (P_{cc}+P_{GL})\times 1.1$ (as per API 617, see Table 6.14)		$P_M \approx 5{,}900$ kW

164   6 Aerodynamic Compressor Design: Case Study

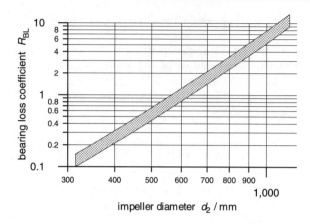

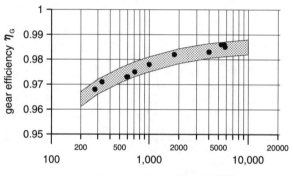

Fig. 6.3 Bearing and gear losses; rough guidelines. Tilting pad bearings (two journal plus one thrust bearing with directed oil flow); actual bearing loss: $P_{BL} = R_{BL} \times (N/1{,}000)^2$/kW, N rotational speed/rpm. Single-stage helical and double-helical gears (dots are actual applications by three manufacturers).

Table 6.9 Customary abbreviations for hydrocarbon gases in oil and gas industry

Abbreviation	Gas	Chemical formula	Abbreviation	Gas	Chemical formula
C_1	methane	CH_4	nC_4	butane	C_4H_{10}
C_2	ethane	C_2H_6	$iC_4\text{=}$	isobutylene	C_4H_8
$C_2\text{-}$	ethylene	C_2H_4	$1\text{-}C_4\text{=}$	1-butene	C_4H_8
$C_2\text{=}$	acetylene	C_2H_2	iC_5	isopentane	C_5H_{12}
C_3	propane	C_3H_8	nC_5	pentane	C_5H_{12}
$C_3\text{-}$	propylene	C_3H_6	nC_6	hexane	C_6H_{14}
iC_4	isobutane	C_4H_{10}			

6.2 Detailed Compressor Design

The detailed design is required in order to be able to send out a firm compressor quotation on the basis of an official enquiry and also to do the final design after signing of the contract. It involves sizing or selecting all compressor components. It starts out with the impeller of usually the first stage and is followed by the upstream and downstream components and then goes on to the next stage.

Sizing means "forward engineering", i.e. determining the geometry on the basis of specified operating data; selecting means "reverse engineering", i.e. determining aero-thermodynamic behavior on the basis of given geometry in the case of an existing pre-engineered impeller family. The following procedure is the sizing or design method.

Again, the design method described in this chapter is based on the experience of just one compressor manufacturer and cannot be absolutized. There are other either more scientific or more empirical methods to design a compressor leading to the final geometry of aero components.

The main design objective is to reach guaranteed head and efficiency (within the agreed tolerances) at a specified volume flow and to meet the predicted surge limit. A secondary objective may be to reach securely other desired operating points. It goes without saying that there must not be any aerodynamically caused instability zone, like rotating stall, within the complete specified performance envelope.

The detailed design is broken down into:

- impeller inlet geometry
- impeller exit geometry
- stage efficiency
- head, pressure, temperature, power
- geometry diffuser, return vane channel, volute
- axial thrust, balance piston
- lateral critical speeds.

The formulae for each item will be derived first, followed by the numerical calculation for the case study to make the procedure more transparent.

6.2.1 Impeller Inlet

Design of the inlet includes the eye diameter, blade angle and blade width and yields all relevant velocities; see Fig. 6.4.

The design procedure is iterative insofar as certain assumptions have to be made first, which will be stepwise refined. The impeller tip speed and diameter as per approximate sizing, Sect. 6.1, are used to start the detailed method; the first-stage discharge pressure and temperature are used as suction values for the second stage; if the compressor discharge pressure differs from the specification, the tip speed is varied until the desired pressure is met. All of this is carried out using the appropriate real-gas equation of state (LKP equation, Eq. (2.72), in this case study), which in turn is the basis for computing the compressibility factors, polytropic volume and temperature exponents at the stage inlets and exits, as per Chap. 2.

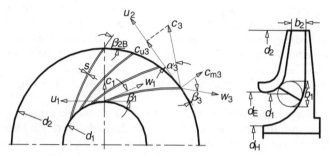

Fig. 6.4. Impeller geometry and velocities. Subscripts: 1 blade leading edge midpoint, 2 blade trailing edge, 3 infinitesimally downstream of trailing edge, E eye, H hub, B blade, b width, m meridional, u circumferential. u metal speed, c absolute velocity, w relative velocity, α absolute flow angle, β relative angle, s blade thickness.

Extension of Process Volume Flow

Since all through-flow areas of the complete flow channel have to be sized to account for the actual (static) volume flow at the exact location, the following increasing factors for the process volume flow $\dot{V}$ (defined at the inlet stagnation state) from the inlet nozzle to the first impeller eye opening have to be calculated:

- Recirculation (balance piston and optionally seal balance line):

$$f = \left(1 + \frac{\Delta \dot{m}_{BP}}{\dot{m}}\right). \tag{6.24}$$

- Expansion through heat-up by recirculation: the recirculated hot mass flow from the balance piston, and optionally from any seal balance, increases the suction temperature and hence the actual volume flow in proportion to the mixing temperature:

$$T_m = \frac{\dot{m} T_{0t} + \Delta \dot{m}_{BP} T_{BP}}{\dot{m} + \Delta \dot{m}_{BP}} = \frac{T_{0t} + (f-1) T_{BP}}{f}. \tag{6.25}$$

Heat-up expansion:

$$v_T = \frac{T_m}{T_{0t}} = \frac{1}{f} + \frac{f-1}{f} \frac{T_{BP}}{T_{0t}}, \tag{6.26}$$

where T_{BP} is the balance piston temperature and T_{0t} is the (specified) absolute stagnation inlet temperature.

- Expansion through the inlet loss between the nozzle and impeller eye (Fig. 6.5):

$$\frac{p_{0t}}{p_{0't}} = \frac{p_{0t}}{p_{0t} - \Delta p_{IL}} = \frac{1}{1-(\Delta p_{IL}/p_{0t})}. \quad (6.27)$$

The (total) pressure loss in absolute and relative terms is

$$\Delta p_{IL} = \zeta \frac{\rho_{0t}}{2} c_s^2 = \zeta \frac{c_s^2 p_{0t}}{2 Z_{0t} R T_{0t}}, \quad (6.28)$$

$$\frac{\Delta p_{IL}}{p_{0t}} = \zeta \frac{\kappa_v}{2} M_s^2. \quad (6.29)$$

Inlet loss expansion:

$$v_{IL} = \frac{\rho_{0t}}{\rho_{0't}} = \frac{1}{\left(1-\zeta(Re_s)\frac{\kappa_v}{2} M_s^2\right)}, \quad (6.30)$$

where $\zeta = 3.0\text{--}4.5$ is the loss coefficient of the plenum inlet; see Fig. 6.6. These measured values pertain to the plenum inlet test configurations with radial suction nozzle, reported by Lüdtke (1985). $M_s = c_s/a_{0t}$ is the Mach number at the inlet nozzle, $Re_s = c_s D_s/v_s$ is the Reynolds number at the inlet nozzle, κ_v is the isentropic volume exponent, v_s is the kinematic viscosity, D_s is the suction nozzle diameter and $c_s = \dot{V}_{0t} f v_T /(\pi/4) D_s^2$ is the velocity at the nozzle.

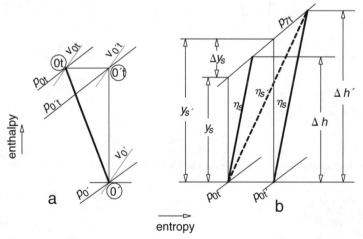

Fig. 6.5a, b Effect of plenum inlet loss: **a** volumetric expansion, **b** head increase. $p_{0t} - p_{0't}$ total pressure loss, v_0/v_{0t} volumetric expansion, Δy_s isentropic head increase.

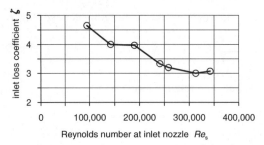

Fig. 6.6 Plenum inlet loss coefficient (radial centerline design) according to tests reported by Lüdtke (test model Fig. 6.16.); $\zeta = 2\Delta p_{IL}/\rho_{0t} c_s^2$.

- Inlet expansion from (flange) total to impeller inlet static state: the dynamic enthalpy at the impeller inlet is

$$\frac{c_{0'}^2}{2} = Z_{0t} R T_{0t} \frac{\kappa_v}{\kappa_v - 1}\left[1 - \left(\frac{p_{0'}}{p_{0't}}\right)^{\frac{\kappa_v - 1}{\kappa_v}}\right]. \quad (6.31)$$

Rearranging and using the isentropic formula, $pv^{k_v} = const$ yields the total-to-static expansion:

$$v_{IE} = \frac{v_{0'}}{v_{0't}} = \left(\frac{p_{0't}}{p_{0'}}\right)^{1/\kappa_v} = \frac{1}{\left(1 - \frac{\kappa_v - 1}{2}\frac{c_{0'}^2}{\kappa_v Z_0 R T_{0t}}\right)^{\frac{1}{\kappa_v - 1}}}. \quad (6.32)$$

Using $c_{0'} \equiv c_{m0}$ we obtain

$$v_{IE} = \frac{1}{\left[1 - \frac{\kappa_v - 1}{2}\left(\frac{c_{m0}}{u_{20}}\right)^2 M_{u2}^2\right]^{1/(\kappa_v - 1)}}. \quad (6.33)$$

- Cover-disk leakage expressed as leakage efficiency:

$$\eta_L = \frac{\dot{fm}}{\dot{fm} + \Delta \dot{m}_{CD}}. \quad (6.34)$$

So the final actual static volume flow at the impeller inlet related to the total process volume flow at the stage inlet is:

$$\frac{\dot{V}_{0'}}{\dot{V}_{0t}} = v_T v_{IL} v_{IE} \frac{f}{\eta_L}. \tag{6.35}$$

Impeller Eye Diameter

It has been reasonable practice to minimize impeller inlet losses by minimizing the relative inlet velocity.

It is evident that, on the one hand, a large eye diameter leads, with a large peripheral velocity and a small absolute velocity, to a large relative velocity but, on the other hand, a small eye diameter produces, with a small peripheral velocity and a large absolute velocity, a large relative velocity as well (Fig. 6.7).

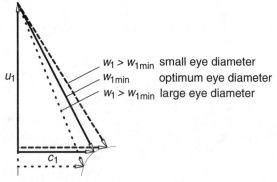

Fig. 6.7 Optimization of eye diameter (w_1-minimum criterion). Impeller inlet velocity triangle for variable eye diameter at constant hub diameter.

The minimum in between is determined as follows. Minimizing the relative velocity for zero prewhirl, $w_1^2 = c_{m1}^2 + u_1^2$ we obtain

$$\frac{c_{m1}}{u_2} = \frac{c_{m0}}{u_2}\frac{1}{\varepsilon_1} = \frac{c_E}{u_2}\frac{1}{\varepsilon_1}\frac{A_E}{A_0} = \frac{\dot{V}_{0t}\dfrac{\dot{V}_{0'}}{\dot{V}_{0t}}}{\varepsilon_1\dfrac{\pi}{4}u_2\left(d_E^2 - d_H^2\right)}\frac{A_E}{A_0}$$

$$= \frac{\dot{V}_{0t}v_T v_{IL} v_{IE}\dfrac{f}{\eta_L}\dfrac{A_E}{A_0}}{\varepsilon_1\dfrac{\pi}{4}d_2^2 u_2\left[\left(\dfrac{d_E}{d_2}\right)^2 - \left(\dfrac{d_H}{d_2}\right)^2\right]} = \frac{K\varphi}{A}, \tag{6.36}$$

with the abbreviations $K = v_T v_{IL} v_{IE} \dfrac{f}{\eta_L}\dfrac{1}{\varepsilon_1}\dfrac{A_E}{A_0}$ and the dimensionless eye area given by

$$A = \left[\left(\frac{d_E}{d_2}\right)^2 - \left(\frac{d_H}{d_2}\right)^2\right], \tag{6.37}$$

where c_{m0} is the meridional velocity ahead of the blade leading edge, c_E is the meridional velocity at the impeller eye, u_2 is the impeller tip speed for non-cut-back blades and d_2 is the impeller diameter for non-cut-back blades.

The blade blockage is given by

$$\varepsilon_1 = 1 - \frac{sz}{d_1 \pi \sin \beta_{1B}} = 1 - \frac{(s/d_{20})z}{(d_1/d_{20})\pi \sin \beta_{1B}}, \tag{6.38}$$

where s is the blade thickness, z is the number of blades, β_{1B} is the blade angle from tangent (all values at blade leading edge midpoint), $A_E = (\pi/4)(d_E^2 - d_H^2)$ is the eye area, and $A_0 = \pi d_1 b_1$ is the area ahead of the leading edge (LE).

Using the peripheral speed at the mid point of the blade leading edge we obtain the dimensionless relative inlet velocity as follows.

$$u_1 = u_2 \frac{d_1}{d_E} \frac{d_E}{d_2}, \tag{6.39}$$

$$\left(\frac{w_1}{u_2}\right)^2 = \left(\frac{K\varphi}{A}\right)^2 + \left(\frac{d_1}{d_E} \frac{d_E}{d_2}\right)^2. \tag{6.40}$$

The condition for minimizing w_1 is:

$$\frac{d\left((w_1/u_2)^2\right)}{d(d_E/d_2)} = 0 \quad \Rightarrow \quad \frac{1}{A^3} = \frac{1}{2}\left(\frac{d_1/d_E}{K\varphi}\right)^2. \tag{6.41}$$

Thus the theoretical optimum eye/tip ratio (the so-called w_1-minimum criterion) is

$$\left(\frac{d_E}{d_2}\right)_{opt} = \sqrt{\left(\frac{d_H}{d_2}\right)^2 + \left[2\left(\frac{K\varphi}{d_1/d_E}\right)^2\right]^{1/3}} \tag{6.42}$$

The hub/tip ratio d_H/d_2 is a result of rotordynamic, i.e. the shaft stiffness calculation, the range being between 0.25 and 0.45 for multistage machines (low values for high flow coefficients and high values for low flow coefficients). Rough guidelines for hub/tip ratios are given in Fig. 6.8 and are applicable for compressor rotors with up to nine impellers (as per Fig. 4.4).

In principle the hub/tip ratios could be reduced for a smaller number of impellers on the shaft. However, the influence of the hub/tip ratio on the efficiency is often overestimated and it has a limited effect only to design impellers with a varying hub/tip ratio for the same flow coefficient.

The blade inlet/eye diameter ratio d_1/d_E is approximately 0.83–0.88 for impellers with 3D-blades and approximately 1.0–1.12 for impellers with 2D-blades.

It is quite normal to make the actual eye diameter some 5% smaller than the optimum in order to reduce the cover-disk hoop stress at the impeller eye. A 5% reduction results in a mere 1% relative velocity (and hence eye Mach number) increase.

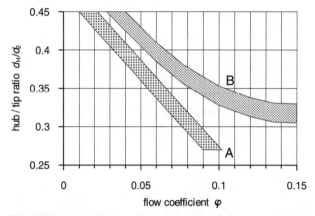

Fig. 6.8 Rough guidelines for hub/tip ratio. A low to medium flow coefficients for medium number of impellers per shaft, B medium to high flow coefficients for high number of impellers per shaft.

Blade Inlet Angle

From the inlet velocity triangle we obtain

$$\tan \beta_{1B} = \frac{c_{m1}}{u_1} = \frac{c_{m1}/u_2}{\dfrac{d_1}{d_E}\dfrac{d_E}{d_2}} = \frac{K\varphi}{\dfrac{d_1}{d_E}\dfrac{d_E}{d_2}A}. \tag{6.43}$$

Blade Inlet Width

As per continuity equation:

$$\dot{V}_{0t} v_T v_{IL} v_{IE} \frac{f}{\eta_L} = \pi d_1 b_1 c_{m0} = \pi d_1 b_1 \varepsilon_1 c_{m1}, \tag{6.44}$$

$$b_1 = \frac{\dot{V}_{0t} v_T v_{IL} v_{IE} \dfrac{f}{\eta_L \varepsilon_1}}{(\pi/4)d_2^2 u_2 4 \dfrac{d_1}{d_E}\dfrac{d_E}{d_2}\dfrac{1}{d_2}\dfrac{c_{m1}}{u_2}}. \tag{6.45}$$

Rearranging and combining with Eqs. (6.36) and (1.70) yields the inlet width in dimensionless form:

$$\frac{b_1}{d_2} = \frac{A}{4\dfrac{d_1}{d_E}\dfrac{d_E}{d_2}\dfrac{A_E}{A_0}}.\tag{6.46}$$

Cover Disk and Hub Disk Radius

In order to achieve a favorable transition between the axial and radial direction within the impeller inlet zone the cover disk and hub disk radii ought to have, in terms of impeller outer radius, the following empirical flow coefficient-dependent values.

The average cover and hub disk radius ratios (for definition see Fig. 6.9) are:
$r/r_2 = 0.13$, $R/r_2 = 0.25$ for flow coefficients $\varphi = 0.01$–0.05 (2D-impellers);
$r/r_2 = 0.23$, $R/r_2 = 0.35$ for flow coefficients $\varphi = 0.05$–0.1 (3D-impellers);
$r/r_2 = 0.35$, $R/r_2 = 0.45$ for flow coefficients $\varphi = 0.1$–0.15 (3D-impellers).

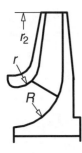

Fig. 6.9 Impeller geometry: inner cover disk and hub disk radii.

These radii can serve as guidelines for a preliminary rotor drawing and rotor-dynamic calculation in the quotation phase but would have to be checked or varied in the course of an impeller 3D- or quasi-3D- calculation to yield the final shapes of the disks and the blade.

This concludes the sizing of the impeller inlet according to the one-dimensional streamline method. The calculation in Table 6.10 pertains to the case study.

For dimensionless sizing d_{20} is the reference diameter for all other impeller dimensions and u_{20} is the reference tip speed for all velocities at the impeller inlet, since any blade cut-back at the outer impeller rim does not have an impact on the situation at the impeller inlet. The peripheral speed at the tip of cover and hub disk u_{20} is at the same time relevant for the impeller stress calculation.

As can be seen: the final results for the impeller inlet dimensions can only be arrived at iteratively.

Table 6.10 Impeller inlet, numerical calculation (after iterating for specified discharge pressure)

	Eq.	Symbol	Unit	Stage 1	Stage 2
Process mass flow		$\dot{m}$	kg/s	85.9	85.9
Suct. press. total at flange (abs.)		p_{0t}	bar	62.7	76.82
Suction temp. total at flange		T_{0t}	°C	31.2	48.2
Disch. press. tot. at flange (abs.)		p_{7t}	bar	76.82	92.97
Discharge temp. total at flange		T_{7t}	°C	48.2	64.7
Molar mass		M	kg/kmol	18.5	18.5
Gas constant		R	kJ/kgK	0.4494	0.4494
Compressibility factor suction	(2.73)	Z_{0t}		0.883	0.889
Compress. factor discharge	(2.73)	Z_{7t}		0.889	0.897
Isentropic vol. exponent suction	(2.60)	κ_{v0t}		1.339	1.358
Isentropic vol. exponent disch.	(2.60)	κ_{v7t}		1.358	1.381
Isentropic temp. exponent suct.	(2.61)	κ_{T0t}		1.319	1.312
Isentr. temp. exponent disch.	(2.61)	κ_{T7t}		1.312	1.303
Density suction $\rho_{0t}=(p_{0t}\times 10^2)/(Z_{0t}RT_{0t})$		ρ_{0t}	kg/m³	51.93	59.86
Density discharge $\rho_{7t}=(p_{7t}\times 10^2)/(Z_{7t}RT_{7t})$		ρ_{7t}	kg/m³	59.86	68.26
Sonic velocity $a=\sqrt{(\partial p/\partial \rho)_s}=\sqrt{\kappa_v ZRT\times 10^3}$		a_{0t}	m/s	402.1	417.5
		a_{7t}	m/s	417.5	433.7
Dynamic viscosity, estimate (VDI-Wärmeatlas (1994))		μ_{0t}	Ns/m²	1.3×10^{-5}	1.4×10^{-5}
Process suction volume flow $\dot{V}_{0t}=\dot{m}Z_{0t}RT_{0t}/(p_{0t}\times 10^2)$		$\dot{V}_{0t}$	m³/s	1.654	1.435
Outer impeller diameters:					
disk diameter (=ref. dia.)		d_{20}	m	0.318	0.316
blade dia. (cut-back 4.4%)		d_2	m	0.304	0.302
Iteration of tip speed to obtain specified discharge pressure		u_2	m/s	222.38	220.92
Tip speed Mach number at d_2		M_{u2}		0.553	0.529
Impeller tip speed at d_{20}		u_{20}	m/s	232.62	231.16
Flow coeff. referenced to d_{20}		φ		0.0895	0.0792
Recirculation	(6.24)	f		1.01	1.01
Heat-up expansion	(6.26)	v_T		1.001	1.0
Inlet loss expansion	(6.30)	v_{IL}		1.004	1.0
Total-to-static expansion	(6.33)	v_{IE}		1.018	1.016
Shroud leakage efficiency	(6.34)	η_L		0.99	0.99
Static/process volume flow	(6.35)	$\dot{V}_0/\dot{V}_{0t}$		1.044	1.037
Rot. speed $N=(u_2\times 60)/(\pi d_2)$		N	1/min	13,971	13,971
Suction/discharge nozzle dia.		D_s/D_d	m	0.35	0.3
Velocity at suct./disch. flange		c_s/c_d	m/s	17.4	18.0
Suct./disch. nozzle Mach no.		M_s/M_d		0.043	0.041
Blade inlet/eye diameter ratio		d_1/d_E		0.831	0.847
Rel. blade thickness/no. blades		$s/d_{20}/z$		0.01/18	0.01/18
Blade blockage	(6.38)	ε_1		0.82	0.82
Eye area/area ahead LE		A_E/A_0		0.99	0.98
Hub/tip ratio (requ. by RSR)		d_H/d_{20}		0.33	0.359

Table 6.10 (cont.)

Abbreviation	Eq.	Symbol	Unit	Stage 1	Stage 2
Abbreviation		K		1.260	1.24
Optimum eye diameter ratio	(6.42)	$(d_E/d_{20})_{opt}$		0.665	0.655
Min. relative velocity ratio	(6.40)	$(w_1/u_{20})_{min}$		0.6481	0.644
Act. eye diameter ratio (− 5%)		d_E/d_{20}		0.629	0.619
Diameter ratio		d_1/d_{20}		0.523	0.524
Dimensionless eye area	(6.37)	A		0.2867	0.2543
Acceleration nozzle/ahead LE $(D_s/d_{20})^2 v_{IL} v_{IE}(A_E/A_0)/(A\eta_L)$		c_{m0}/c_s		4.32	
Actual relative velocity ratio	(6.40)	w_1/u_{20}		0.6542	0.651
Actual absolute velocity ratio	(6.36)	c_{m1}/u_{20}		0.393	0.386
Relative impeller inlet width	(6.46)	b_1/d_{20}		0.139	0.124
Hub diameter		d_H	m	0.105	0.113
Eye diameter		d_E	m	0.2	0.195
Blade inlet diameter		d_1	m	0.166	0.166
Blade angle, mid-LE	(6.43)	β_{1B}	deg	37.0	36.4
Impeller inlet width		b_1	mm	44.2	39.2
Rel. velocity mid-LE		w_1	m/s	152.2	150.5
Abs. velocity mid-LE		c_{m1}	m/s	91.4	89.2
Abs. velocity ahead of mid-LE		c_{m0}	m/s	74.9	73.1

6.2.2 Impeller Exit

The basis for calculating the impeller exit width b_2 is the continuity equation (whereby the static exit volume flow is split up into the volume flow proper $\dot{V}_3$, the secondary factors recirculation loss f and the shroud leakage efficiency η_L):

$$\dot{V}_3 \frac{f}{\eta_L} = \pi d_2 b_2 c_{m3} \Rightarrow b_2 = \frac{\dot{V}_3(f/\eta_L)}{\pi d_2 c_{m3}}. \tag{6.47}$$

This leads to the relative impeller exit width:

$$\frac{b_2}{d_2} = \frac{\dot{V}_{0t} \dfrac{\dot{V}_3}{\dot{V}_{0t}} \dfrac{f}{\eta_L}}{u_2 \dfrac{\pi}{4} d_2^2 \dfrac{c_{m3}}{u_2} 4} = \varphi \frac{\dot{V}_3/\dot{V}_{0t}}{4\varphi_3} \frac{f}{\eta_L} \tag{6.48}$$

The total-to-static impeller volume ratio $\dot{V}_3/\dot{V}_{0t}$ multiplied with the flow coefficient constitutes the static volume flow at the impeller exit, and the exit flow coefficient φ_3 determines the gas velocity at the through flow area $\pi d_2 b_2$ (infinitesimally outside the impeller flow channel).

The impeller exit, covering the blade angle β_{2B} and the relative exit width b_2/d_2, ought to be designed to achieve the highest possible polytropic head coefficient at the highest possible efficiency, the widest possible operating range and an acceptable head rise to surge. Of course, since one cannot have all at the same time, this results in a well-balanced compromise between these different goals.

Blade Exit Angle

For the sake of these goals an exit angle of 50° (from tangent) is a good choice for high and medium flow coefficients, as shown in Fig. 6.10. Higher angles would yield curves that are too flat; lower angles would result in head coefficients that are too low. However, β_{2B} ought to be reduced for flow coefficients below 0.06 in order to maximize b_2/d_2, which has a substantial impact on the efficiency of narrow impellers.

Exit Flow Coefficient

Furthermore, it has been shown from well-proven experience that the impeller exit flow coefficient φ_3 for this type of shrouded impellers has to be a function of the flow coefficient φ (Fig. 6.10) rather than being a constant (for instance 0.3) for the complete impeller line. This prevents high flow coefficient impellers from becoming axially too wide, which is detrimental to efficiency and operating range. This is because high secondary flows tend to accumulate low energy fluid in the shroud/suction side corner of the impeller exit and form a large wake zone. Low values of φ_3 for narrow impellers, on the other hand, increase the width, assuring higher efficiencies and work input factors without compromising on the operating range. This is because narrow stages have a low critical diffuser inlet flow angle α_3 which triggers rotating stall.

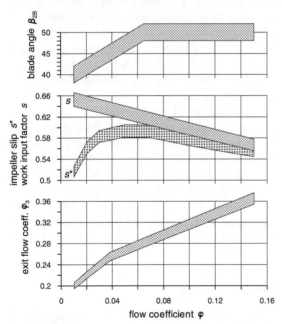

Fig. 6.10 Impeller exit flow coefficient, impeller slip factor, work input factor and blade exit angle as functions of the inlet flow coefficient (empirically determined values for the best efficiency point). s includes work of shroud leakage and disk friction.

Number of Blades

Setting the number of impeller blades is a compromise between aerodynamics and manufacturing: adding one more blade would increase the polytropic head coefficient by 1.3% but would reduce the throat, i.e. the smallest clear span between two adjacent blades in the inlet region, making blade milling more difficult and costlier. So a blade number of $z = 20$ for 2D-impellers has proved satisfactory and $z = 18$ for 3D-impellers is very feasible for the smaller throat due to the upstream extension of the blade leading edge. The even number is very practical in case of any damage to a single blade due to foreign object impact. In the latter case the damaged part of the blade can be removed, as well as the corresponding part of the opposite blade, enabling temporary operation with the damaged rotor to be quickly resumed without running into major unbalance problems.

Exit Flow Angle

Figure 6.11 sheds some more light on the exit flow coefficient/slip factor relationship in the context of the absolute and relative flow angles α_3 and β_3 respectively, i.e. it shows the shape of the exit velocity triangles for the impeller line and the enormous difference of the triangles between low and high flow coefficient impellers.

The (dimensionless) velocity triangles in the diagram are oriented as in Fig. 1.22. The bands s and s^* constitute the loci of the velocity triangle tips (of the impeller line as in Fig. 3.4) where the absolute and relative velocities intersect. A high flow coefficient impeller (on the right-hand side of the band) has an absolute flow angle of $\alpha_3 = 33°$, a relative flow angle of $\beta_3 = 40°$, an exit flow coefficient of $\varphi_3 = 0.36$, and a work input factor of $s = 0.58$, i.e. a steep triangle. A low flow coefficient impeller (on the left-hand side of the band) has an absolute flow angle of $\alpha_3 = 17°$, a relative flow angle of $\beta_3 = 30°$, an exit flow coefficient of $\varphi_3 = 0.2$ and a work input factor of $s = 0.65$, i.e. a flat triangle. The pertaining flow coefficient φ-range of the 2D- and 3D-impeller line is shown at the bottom of Fig. 6.11. The true triangles are represented by the band s^* (the difference between the practical and true triangles described in Sect. 1.7). S-type impellers (as per Fig. 3.3) have work input factors of around 0.7, and R-type impellers ($\beta_{2B} = 90°$) have a traditional work input factor of 0.875 at a relative exit flow angle of 72°.

As a rule of thumb, the relative flow angles are $\beta_2 \approx 0.75 \times \beta_{2B}$ for 3D-impellers and $\beta_2 \approx (0.63–0.74) \times \beta_{2B}$ for 2D-impellers, reflecting the deviation of the flow angle from the blade angle due to the influence of the relative vortex within the blade channel (β_{2B} is the blade exit angle).

Impeller Slip Factor and Work Input Factor

The impeller slip factor $s^* = c_{u3}/u_2$, which is the dimensionless thermodynamic work, can be empirically calculated from the blade angle, blade number and exit flow coefficient at the best efficiency point according to Wiesner (1966), who based his investigations on the classical slip-factor method of Busemann (1928):

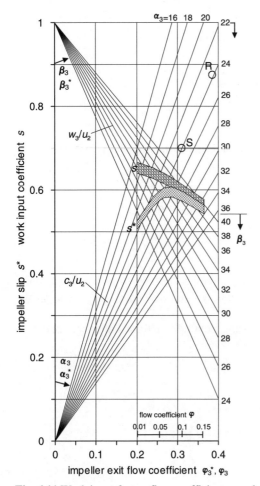

Fig. 6.11 Work input factor, flow coefficients vs. shape of impeller exit velocity triangle for the best efficiency point. $s = f(\varphi_3)$ practical velocity triangle, $s^* = f(\varphi_3^*)$ true velocity triangle, α_3 absolute flow angle in deg, β_3 relative flow angle in deg, φ inlet flow coefficient, R R-type impeller with inducer ($\beta_{2B} = 90°$), S S-type impeller ($\beta_{2B} = 60°$).

$$s^* = 1 - \frac{\sqrt{\sin \beta_{2B}}}{z^{0.7}} - \frac{\varphi_3}{\tan \beta_{2B}} + \Delta s^*, \tag{6.49}$$

where Δs^* is an experimentally determined correction.

The work input factor $s = \Delta h/u_2^2$ is the dimensionless thermodynamic work plus disk friction plus shroud leakage work; since it can easily be measured on the test bed, the original equation for s^* is rewritten for s and, as a side effect, the (different) correction Δs then becomes more consistent:

$$s = 1 - \frac{\sqrt{\sin \beta_{2B}}}{z^{0.7}} - \frac{\varphi_3}{\tan \beta_{2B}} + \Delta s, \qquad (6.50)$$

where $\Delta s = -0.017 \pm 0.005$; φ_3 is defined for the best stage efficiency point.

The slip/work input factor relationship is

$$\frac{s^*}{s} = \eta_L - \frac{C_M}{2\pi} \frac{\eta_L}{f} \frac{1}{s\varphi(\dot{V}_3 / \dot{V}_{0t})}. \qquad (6.51)$$

The frictional torque on the hub and shroud disks of the impeller is

$$M_F = C_M \frac{\rho_3}{2} \omega^2 r_2^5. \qquad (6.52)$$

The torque coefficient according to Daily and Nece (1960) is

$$C_M = \frac{0.102(a/r_2)^{0.1}}{Re^{0.2}} = \frac{0.102(a/r_2)^{0.1}}{(u_2 r_2 / v_3)^{0.2}}, \qquad (6.53)$$

where a is the average gap of disks/diaphragm; $a/r_2 = 0.02$ is an acceptable ratio; v_3 is the kinematic viscosity at the impeller exit and ω is the angular impeller velocity.

From the derivation of Eq. (6.51), using Eq. (1.48) and assuming zero prewhirl, the gas power is given by

$$\dot{m}s u_2^2 = \dot{m}s^* u_2^2 \frac{1}{\eta_L \eta_F}. \qquad (6.54)$$

The disk friction efficiency (using Eq. (1.47)) is

$$\eta_F = \frac{P_i - M_F \omega}{P_i}; \quad \frac{s^*}{s} = \eta_L \eta_F = \eta_L \left(1 - \frac{M_F \omega}{P_i}\right), \qquad (6.55)$$

$$\frac{M_F \omega}{P_i} = \frac{C_M \rho_3 \omega^3 r_2^5}{2 \dot{m}s u_2^2} = \frac{C_M \rho_3 u_2 r_2^2}{2 \dot{m}s}. \qquad (6.56)$$

From the continuity equation $\dot{m}/\eta_L = \rho_3 c_{m3} \pi 2 r_2 b_2$ and Eq. (6.48) we obtain

$$\frac{s^*}{s} = \eta_L - \frac{C_M}{2\pi} \frac{\eta_L}{f} \frac{1}{s\varphi(\dot{V}_3 / \dot{V}_{0t})}. \qquad (6.57)$$

The approximate slip/work input factor ratio is shown in Fig. 6.12: it is not surprising that for low-flow-coefficient impellers the shroud leakage and disk friction can add as much as 15% to 22% to the thermodynamic work. It should be kept in mind, though, that some of the influencing parameters depend on the specific application like recirculation loss from the balance piston and shroud leakage.

Figure 6.11 also contains the work input coefficient s with the abscissa being the exit flow coefficient without accounting for shroud leakage; both values are easily calculable from test bed measurements, resulting in fictitious velocity triangles used as reference basis for compressor design. It has to be kept in mind that the relevant absolute flow angles are smaller and the relative flow angles are greater than the actual ones.

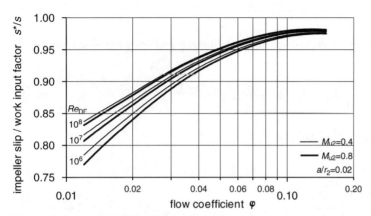

Fig. 6.12. Ratio of impeller slip/work input factor (approximate). $Re_{DF} = u_2 r_2/v_3$ disk friction Reynolds number, M_{u2} tip speed Mach number, r_2 impeller outer radius, a average gap between impeller disks and diaphragm, $f = f(\varphi) = 1.04–1.005$ recirculation loss, $\eta_L = f(\varphi) = 0.95 – 0.995$ shroud leakage.

Impeller Volume Ratio

The volume change within the impeller from inlet total to exit static is primarily influenced by the Mach number, and secondarily by the isentropic exponent, efficiency, exit flow coefficient and work input factor. This should come out in the equation which is derived as follows:

$$\frac{\dot{V}_3}{\dot{V}_{0t}} = \frac{T_{3s}}{T_{0t}} \frac{p_{0t}}{p_{3s}} \frac{Z_{3s}}{Z_{0t}} = \frac{\pi_{Is}^{e}}{\pi_{Is}} Z_r = \frac{1}{\pi_{Is}^{(1-e)}} Z_r \qquad (6.58)$$

where
 subscript 3s static state impeller exit,
 subscript Is impeller static,
 π_{Is} impeller static pressure ratio (exit static over inlet total),
 Z_r compressibility factor ratio (exit static over inlet total),
 $e = (\kappa-1)/(\kappa\eta_{pIs})$ abbreviation,
 η_{pIs} polytropic static impeller efficiency (inlet total to exit static).

The following approximations are used: $s^* \approx s$; $\varphi_3^* \approx \varphi_3$; perfect gas formulae are applied; Z_r and $\kappa \approx \kappa_v$ are used to correct for real gas behavior.
 The total-to-total work (Euler-head) is:

$$\Delta h = \Delta h_{Is} + \frac{c_3^2}{2} = s u_2^2, \qquad (6.59)$$

where Δh_{Is} is the impeller enthalpy difference, inlet total to exit static.

The reaction is readily calculable once φ_3 and s are known:

$$r = \frac{\Delta h - \frac{c_3^2}{2}}{\Delta h} = 1 - \frac{c_3^2}{2su_2^2} = 1 - \frac{\varphi_3^2 + s^2}{2s}. \tag{6.60}$$

Using the basic head/pressure ratio relationship and tip-speed Mach number we obtain

$$\Delta h_{Is} = RT_{0t}\frac{\kappa}{\kappa-1}\left(\pi_{Is}{}^e - 1\right) = \Delta h - \frac{c_3^2}{2}, \tag{6.61}$$

$$RT_{0t}\frac{\kappa}{\kappa-1}\left(\pi_{Is}{}^e - 1\right) = r\Delta h = rsu_2^2, \tag{6.62}$$

$$\pi_{Is} = \left[1 + s(\kappa-1)r\frac{u_2^2}{\kappa RT_{0t}}\right]^{\frac{1}{e}}, \tag{6.63}$$

$$\pi_{Is} = \left[1 + s(\kappa-1)rM_{u2}^2\right]^{\frac{1}{e}}, \tag{6.64}$$

$$\frac{\dot{V}_3}{\dot{V}_{0t}} = \frac{Z_r}{\left[1 + s(\kappa-1)rM_{u2}^2\right]^{\left(\frac{1}{e}-1\right)}}. \tag{6.65}$$

The total-to-static impeller efficiency can be replaced by the total-to-total impeller efficiency when considering the following basic facts. The kinetic impeller exit energy subtracted from the Euler head Δh equals the total-to-static impeller enthalpy difference, and the kinetic impeller exit energy subtracted from the polytropic total-to-total impeller head y_{pI} equals the polytropic total-to-static impeller head:

$$\Delta h - \frac{c_3^2}{2} = \Delta h_{Is}; \quad y_{pI} - \frac{c_3^2}{2} = y_{pIs}. \tag{6.66}$$

Combining both equations with $y_{pI} = \Delta h\eta_{pI}$ and Eq. (6.60) we obtain

$$\frac{y_{pIs}}{\Delta h_{Is}} = \eta_{pIs} = \frac{y_{pI} - \frac{c_3^2}{2}}{\Delta h - \frac{c_3^2}{2}} = \frac{\Delta h\eta_{pI} - \frac{c_3^2}{2}}{\Delta h - \frac{c_3^2}{2}}$$

$$= \frac{\eta_{pI} - \frac{c_3^2}{2\Delta h}}{1 - \frac{c_3^2}{2\Delta h}} = \frac{\eta_{pI} + r - 1}{r}. \tag{6.67}$$

The exponent in Eq. (6.65) then becomes

$$\frac{1}{e} - 1 = \frac{\kappa}{\kappa-1} \frac{\eta_{pI} + r - 1}{r} - 1 = \frac{r + \kappa\eta_{pI} - \kappa}{r(\kappa-1)}. \tag{6.68}$$

With Eq. (6.68) the final formula for the impeller volume ratio becomes

$$\frac{\dot{V}_3}{\dot{V}_{0t}} = \frac{Z_3/Z_{0t}}{\left[1 + s(\kappa-1)rM_{u2}^2\right]^{\frac{r+\kappa\eta_{pI}-\kappa}{r(\kappa-1)}}} \tag{6.69}$$

where, $\kappa = \dfrac{\kappa_{v0t} + \kappa_{v3}}{2}$ \hfill (6.70)

The total-to total impeller efficiency is still the only unknown parameter. It is extremely difficult to measure exactly the impeller efficiency, even under laboratory conditions, due to the highly distorted velocity profiles in the radial and peripheral directions at the impeller exit. Since the measuring uncertainties are relatively great, it has become common practice to calculate approximately the total impeller efficiency by using the measured stage efficiency and combining it with the theoretical reaction and the diffuser efficiency.

Definitions

Reaction:

$$r = \frac{\Delta h - \dfrac{c_3^2}{2}}{\Delta h}. \tag{6.71}$$

Polytropic stage efficiency, total-to-total:

$$\eta_p = \frac{y_p}{\Delta h}. \tag{6.72}$$

Polytropic impeller efficiency, total-to-total:

$$\eta_{pI} = \frac{y_{pI}}{\Delta h}. \tag{6.73}$$

Polytropic impeller efficiency, inlet total-to-exit static:

$$\eta_{pIs} \approx \frac{y_{pIs}}{\Delta h - \dfrac{c_3^2}{2}} = \frac{y_{pI} - \dfrac{c_3^2}{2}}{\Delta h - \dfrac{c_3^2}{2}}. \tag{6.74}$$

Polytropic diffuser efficiency, inlet static-to-exit total:

$$\eta_D \approx \frac{y_p - y_{p\,\text{Is}}}{\dfrac{c_3^2}{2}}. \tag{6.75}$$

"Diffuser" efficiency covers the losses in the diffuser, crossover bend and return vane channel (optionally volute plus conical exit diffuser).

The following interrelationships exist between the different efficiencies (see Fig. 1.19, and replace isentropic by polytropic heads).

$$r = 1 - \frac{\varphi_3^2 + s^2}{2s}. \tag{6.76}$$

$$\eta_p = \eta_{p\text{I}} - (1-r)(1-\eta_D) = r\eta_{p\,\text{Is}} + (1-r)\eta_D. \tag{6.77}$$

$$\eta_{p\text{I}} = \eta_p + (1-r)(1-\eta_D). \tag{6.78}$$

$$\eta_{p\,\text{Is}} = \frac{\eta_{p\text{I}} + r - 1}{r}. \tag{6.79}$$

Equation (6.78) allows a fairly good estimate of the total-to-total impeller efficiency which is needed to size the impeller exit: the static impeller efficiency and diffuser efficiency can both be assumed equal to the stage efficiency (i.e. the compression line in the enthalpy/entropy diagram is a single straight line); then the total impeller efficiency is already satisfactory. Normally the diffuser efficiency can be assumed 2 to 4 points lower than the static impeller efficiency, resulting in a kinked line in the h, s-diagram. Both results for the total-to-total impeller efficiency do not differ very much and the values for the impeller volume ratio show even smaller differences.

A not too inaccurate but highly empirical rule of thumb is:

$$\eta_{p\text{I}} \approx \sqrt{\eta_p}. \tag{6.80}$$

Applying loss models for all the stator elements would not yield more reliable results for the total impeller efficiency and it appears doubtful whether the high expenditure would be justified.

All assumptions or theoretical models, how inaccurate they may be, have to be applied consistently though.

The impeller volume ratio is a handy "genuine" design parameter which is readily available; the rather arbitrary and artificial impeller deceleration ratio w_2/w_1, suggested in many traditional text books, becomes superfluous.

The great influence of the Mach number can clearly be seen in Figs. 6.13 and 6.14, whereas all other parameters play a secondary role only.

By now all parameters for sizing the impeller exit are at hand or obtainable; some have to be assumed first and can be refined later during the course of the calculation.

Table 6.11 contains the numerical calculation of the impeller exit for the case study.

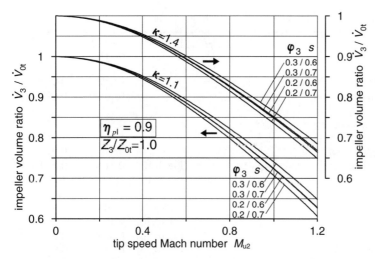

Fig. 6.13 Impeller volume ratio vs. tip-speed Mach number. κ average isentropic volume exponent across impeller, $\eta_{pl} = 0.9$ polytropic impeller efficiency (total-to-static), Z compressibility factor, φ_3 impeller exit flow coefficient, s work input factor.

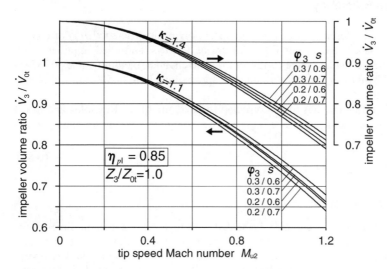

Fig. 6.14 Impeller volume ratio vs. tip speed Mach number. κ average isentropic volume exponent across impeller, $\eta_{pl} = 0.85$ polytropic impeller efficiency (total-to-static), Z compressibility factor, φ_3 impeller exit flow coefficient, s work input factor.

Table 6.11 Impeller exit width, numerical calculation

	Symbol	Unit	Reference	Stage 1	Stage 2
Disk diameter	d_{20}	m		0.318	0.316
Blade outer diameter	d_2	m		0.304	0.302
Flow coefficient referenced to d_{20}	$\varphi(d_{20})$			0.0895	0.0792
Flow coeff. referenced to d_2: $\varphi = (d_{20}/d_2)^3 \varphi(d_{20})$	φ			0.1024	0.0907
Tip speed Mach number at d_2	M_{u2}			0.553	0.529
Recirculation (from bal. piston)	f			1.01	1.01
Shroud leakage efficiency	η_L			0.99	0.99
Exit flow coefficient	φ_3		Fig. 6.10	0.314	0.310
Blade exit angle (from tangent)	β_{2B}	deg	Fig. 6.10	50	50
Number of blades	z			18	18
Work input factor	s		Eq. (6.50)	0.604	0.607
Isentropic exponent	κ		Eq. (6.70)	1.35	1.37
Reaction	r		Eq. (6.76)	0.616	0.617
Pol. total impeller efficiency, est.	η_{pI}		Eq. (6.78)	0.92	0.92
Compressibility factor inlet	Z_{0t}			0.883	0.889
Compressib. factor impeller exit	Z_3			0.886	0.893
Impeller volume ratio	$\dot{V}_3 / \dot{V}_{0t}$		Eq. (6.69)	0.915	0.923
Relative impeller exit width	b_2/d_2		Eq. (6.48)	0.0761	0.0688
Impeller exit width	b_2	mm		23.1	20.8

6.2.3 Stage Efficiency

The underlying reference values for the basic polytropic efficiency η_{pRef} of a complete compressor stage as per Fig. 1.23 are as follows: shrouded impellers with backward curved blades, stages with return vane channel, medium values of Mach and Reynolds number, impeller diameter (approximately 400–500 mm), and surface roughness.

The basic efficiency has to be corrected for size, inlet loss, Reynolds and Mach number, blade trimming, diffuser ratio, surface roughness, volute loss and geometry variants:

$$\eta_p = \eta_{p\,\text{Ref}} + \sum \Delta \eta_{pi} \qquad (6.81)$$

Size Effect Correction

Compressor components can only be scaled with regard to their main dimensions; relative surface roughness, labyrinth clearances and blade thicknesses to a certain extent cannot be scaled one hundred percent because that would jeopardize operational safety. In addition, impellers with smaller outer diameters have per se a smaller Reynolds number (for a given gas and density level), which diminishes the efficiency. These influences plus manufacturing tolerances decreasing with size are covered by the term "size effect", which is actually an imperfect scaling impact. The approximate size correction $\Delta \eta_{d2}$ is +2% for $d_2 = 800$ mm and –5% to –2% for $d_2 = 200$ mm depending on the flow coefficient.

Inlet Loss Correction

Since suction-nozzle sizing by velocity or Mach number have both turned out to be unsatisfactory, it is suggested that we should determine the nozzle diameter by limiting the inlet losses from the flange to the impeller eye.

This total pressure loss increases the head of the adjoining impeller, which is synonymous with a lower stage efficiency. The loss is caused by the directional change of the flow, circumventing the shaft, acceleration and highly turbulent wall friction.

The related isentropic stage head increase is (see Fig. 6.5b):

$$y_r = \frac{y'_s - y_s}{y_s}, \tag{6.82}$$

where y_s is the isentropic head without inlet loss and y'_s is the increased isentropic head (accounting for inlet loss). With Eq. (2.48) we obtain

$$y_r = \frac{\left(\dfrac{p_{7t}}{p_{0t} - \Delta p_{in}}\right)^{\frac{\kappa_v - 1}{\kappa_v}} - 1}{\left(\dfrac{p_{7t}}{p_{0t}}\right)^{\frac{\kappa_v - 1}{\kappa_v}} - 1} - 1, \tag{6.83}$$

where p_{0t} is the total pressure at the suction flange and p_{7t} is the total pressure at the stage discharge.

The pressure loss of the plenum inlet is

$$\Delta p_{in} = \zeta \frac{\rho_{0t}}{2} c_s^2 = \zeta \frac{c_s^2 p_{0t}}{2 Z_{0t} R T_{0t}}, \tag{6.84}$$

$$\frac{\Delta p_{in}}{p_{0t}} = \zeta \frac{\kappa_{v0}}{2} M_s^2, \tag{6.85}$$

where $\zeta = f(Re)$ is the loss coefficient as per Fig. 6.6, κ_{v0} is the isentropic exponent at the inlet and M_s is the Mach number at the flange.

With Eq. (6.83) we obtain

$$y_r = \frac{\left(\dfrac{p_{7t}/p_{0t}}{1 - \zeta \dfrac{\kappa_v}{2} M_s^2}\right)^{\frac{\kappa_v - 1}{\kappa_v}} - 1}{(p_{7t}/p_{0t})^{\frac{\kappa_v - 1}{\kappa_v}} - 1} - 1, \tag{6.86}$$

where p_{7t}/p_{0t} is the pressure ratio of a single stage downstream of the inlet nozzle.

The stage head increase has to be translated into an efficiency decrease $\Delta\eta_{in}$. From Fig. 6.5b it is evident that

$$\frac{y'_s}{\eta_s} = \Delta h'; \quad \frac{y_s}{\eta_s} = \Delta h; \quad \frac{y_s}{\eta'_s} = \Delta h'; \quad \eta'_s = \eta_s - \Delta\eta. \tag{6.87}$$

Using Eq. (6.86), the related isentropic head increase also becomes the related Euler head increase:

$$y_r = \frac{\Delta h' \eta_s}{\Delta h \eta_s} - 1 = \frac{\Delta h'}{\Delta h} - 1 = \frac{y_s \eta_s}{\eta'_s y_s} - 1 = \frac{\eta_s}{\eta_s - \Delta\eta_{in}} - 1 = \frac{1}{1 - (\Delta\eta_{in}/\eta_s)} - 1, \tag{6.88}$$

$$\Delta\eta_{in} = \frac{y_r}{y_r + 1} \eta_s. \tag{6.89}$$

The isentropic efficiency can be replaced by the polytropic efficiency, thus yielding the final formula for the efficiency degradation due to the plenum inlet loss:

$$\Delta\eta_{in} = \frac{y_r}{1 + y_r} \eta_p, \tag{6.90}$$

where η_p is the efficiency without any inlet loss.

The value of y_r is shown in Fig. 6.15 as a function of the pressure ratio of the single stage downstream of the plenum inlet with parameters nozzle Mach number and isentropic exponent.

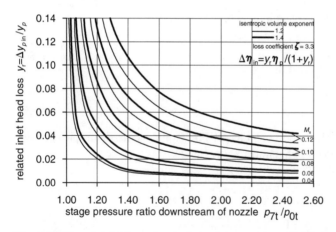

Fig. 6.15. Plenum inlet loss of a single compressor stage. M_s Mach number at suction nozzle, η_p efficiency of stage downstream of nozzle without inlet loss.

The conclusions of this nozzle sizing method are:

- It is thought to be logical to size suction nozzles by setting a ceiling on the inlet loss, e.g. $\Delta\eta_{in} \leq 0.01$ for normal cases and $\Delta\eta_{in} \leq 0.02$ for extreme cases.

- This leads to low inlet nozzle Mach numbers with hydrogen-rich gases (having low stage pressure ratios) and high inlet-nozzle Mach numbers with refrigerants (having high stage pressure ratios).
- Thus the inlet nozzle gas velocities are brought to a consistent level of around 15 to 25 m/s for all gases.

The model used for measuring the pressure loss (see Fig. 6.6) and the velocity distributions are shown in Fig. 6.16, as reported by Lüdtke (1985).

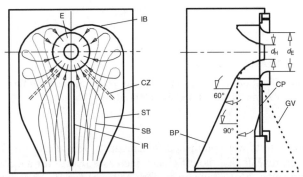

Fig. 6.16. Plenum inlet test model for measuring pressure loss according to Lüdtke (1985, redrawn). E impeller eye (annulus), IB intake baffle, IR intake rib, BP back plate, CP cover plate (Plexiglas), GV geometry variant, d_H hub diameter, d_E eye diameter, ST top stream lines near CP (*solid lines*), SB bottom streamlines near BP (*dotted lines, left*), CZ converging zone, streamlines deducted from oil film traces on BP and CP.

The streamlines at the top of the model (drawn as solid lines) are laced towards the impeller eye. The bottom streamlines (dotted lines) circumvent the hub body, creep up the side walls to the top and, after partly flowing back, reach the impeller eye annulus. This "counterflow" generates two converging flow regimes clearly indicated by oil film traces on the cover plate. The geometry variant, which has a meridional back plate inlet angle of 90° instead of 60° (required by the changeover from cast to fabricated casings), was not found to be detrimental.

Blade Cutback Correction

Blade cutback, also known as blade trimming, involves reducing the blade exit diameter without changing the outer diameter of hub and cover disk, as shown in Fig. 6.18a. The extending disks form a rotating annular diffuser, thus producing lower wall friction losses than in a comparable stationary diffuser, because the relative velocity w is smaller and more radially directed than the absolute velocity c. Moreover, a rotating diffuser is less prone to incipience of rotating stall by re-accelerating the stagnating boundary layer. The efficiency, however, can only be increased up to a maximum amount of blade trimming. In the case of exaggerated trimming, the blade length along the shroud is unduly reduced, not being able to provide the necessary diffusion anymore, with the result of a diminishing efficiency.

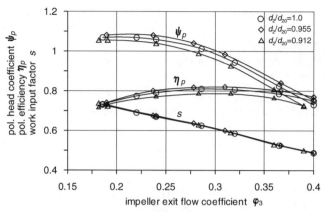

Fig. 6.17 Influence of blade cutback on single-stage compressor performance according to Lindner (1983, redrawn). $d_{20} = 400$ mm, $d_4/d_{20} = 1.45$, $\varphi = 0.093$, $M_{u2} = 0.77$.

Systematic blade cutback tests were published by Lindner (1983). Figure 6.17, taken from the latter paper, reveals that the efficiency increases by 1% for a 4.5% cutback and decreases by some 2% (compared to zero cutback) for an 8.8% cutback. The work input factor is not appreciably influenced by these cutbacks. The impeller tested had a flow coefficient of $\varphi = 0.093$ with 3D-blades and a related untrimmed-blade exit width of $b_{20}/d_{20} = 0.0665$.

Blade cutback is applied to increase efficiency, adjust side stream pressure(s), and increase shaft stiffness due to an increased hub/blade tip ratio.

Diffuser Diameter Ratio Correction

The "diaphragm package" downstream of the impeller – normally consisting of a vaneless diffuser, a crossover bend and a return vane channel – serves as a means (a) to convert the high kinetic energy at the impeller exit into static pressure and (b) to deswirl the flow before it enters the next impeller. As odd as it may sound, the longer the flow path in the diffuser/return vane unit, the higher is the total-to-total stage efficiency (suction s to discharge d in Fig. 6.18), as long as diffuser diameter ratios of around 1.8 are not exceeded. Lindner reported on test results which were obtained with the same impeller described under "Blade cutback", gaining approximately 2.5% efficiency when changing the diffuser ratio from 1.45 to 1.65 at tip-speed Mach numbers of 0.55 to 0.93. The configuration was as in Fig. 6.18b, and the test findings are given in Figs. 6.19 and 6.20.

Since the total-to-total efficiency of impeller + diffuser (stations s to 4, Fig. 6.18) must go down for an extended diffuser, the reason for the improvement can only be the return vane channel. At considerably reduced inlet kinetic energy, the channel responds with lower head losses $\Delta h_{5d} = \zeta c_5^2/2$; the loss coefficient ζ is reduced additionally, as the deswirling is distributed over a longer flow path.

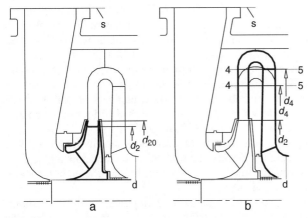

Fig. 6.18a, b Stage tests: **a** impeller blade cut-back, **b** variable diaphragm radius ratio. Measurement between suction flange (plane s) and return channel exit (plane d); d_2 = 400/382/365 mm, d_{20} = 400 mm, φ = 0.093, d_4/d_{20} = 1.65/1.45.

Figure 6.19 also shows that the low diffuser ratio yields a lower head rise to surge with the horizontal tangent on the pressure/volume curve being closer to the rated point. Thus, this contributes to a lower flexibility of the compressor to cope with changes of the molar mass, which is often the case in the oil and gas industry. High incidence losses at the return-vane leading edges during partload are presumably responsible for this inadequate curve slope. Therefore, long diffusers are definitively the better solution when discussing efficiencies plus operational flexibility.

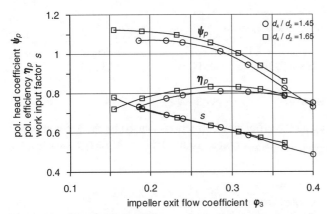

Fig. 6.19 Influence of the diffuser ratio on the compressor performance (including return-vane channel radius ratio, Fig. 6.18) according to measurements by Lindner (1983, redrawn). d_2 = 400 mm, φ = 0.093, M_{u2} = 0.77, stage with plenum inlet and return vane channel.

The growing efficiency with increasing diffuser ratio was confirmed by shop test results of two similar three-stage contract compressor sections with flow coefficients of between 0.09 and 0.04 and Mach numbers of 1.0 to 0.8 The increase of the average diffuser ratio from 1.52 to 1.67 resulted in an efficiency increase of 2.3%.

Low flow-coefficient stages can also be improved in the efficiency by applying longer vaneless diffusers. The very similar three- and four-stage contract compressor sections tested (flow coefficients of about 0.012, Mach numbers of 0.85 to 0.55), differ in their average diffuser ratios of 1.65 versus 1.48. This brings about an efficiency difference of 2.4% in favor of the long diffuser section.

Both contract-compressor measurement results were published by Lüdtke (1999).

As can be derived from these test results, the findings of the development tests by Lindner are basically confirmed: some 2–3% reduction in the efficiency for stages with some 12% smaller diffuser radius ratio.

Figure 6.20 is a graphical display of these (strictly empirical) findings. It should be mentioned, though, that theoretical calculations came up with somewhat smaller efficiency differences for the same diffuser ratio differences.

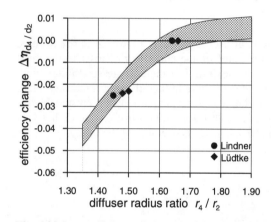

Fig. 6.20 Stage efficiency: influence of the diffuser radius ratio according to measurements by Lindner (1983) and Lüdtke (1999). Extrapolated for $r_4/r_2 < 1.45$ and > 1.65.

Mach Number Correction

From accumulated test data the efficiency corrections caused by the Mach number were deduced. Guidelines for $\Delta\eta_{Mu2}$ range from –1% for $M_{u2} = 0.9$ to –3% for $M_{u2} = 1.2$.

Reynolds Number Correction

Since the Reynolds number represents the ratio of mass forces over viscous forces, the losses in a flow channel will be lower when the Reynolds number increases, e.g. in the case of a higher pressure level. The "International Compressed Air and

Allied Machinery Committee" (Strub et al., ICAAMC (1987)) published Reynolds number correction formulae for the performance of centrifugal compressors, stating how, for instance, the efficiency is influenced by the Reynolds number variation between tests (serving as reference) and specified conditions (for which a reliable prediction is required), carried out with the same compressor.

Almost all of the (more or less standardized or repeatedly applied) compressor stages used for an actual compressor design have been tested, either in systematic development test programs or in numerous acceptance tests of contract machines. So it is justified to use the ICAAMC efficiency correction formula for the design of new compressors as well:

$$\frac{1-\eta_{p-act}}{1-\eta_{p-ref}} = \frac{0.3 + 0.7 \frac{\lambda_{act}}{\lambda_{cr-act}}}{0.3 + 0.7 \frac{\lambda_{ref}}{\lambda_{cr-ref}}}. \tag{6.91}$$

Rearrangement yields the Reynolds number correction:

$$\Delta\eta_{Re} = \eta_{p-act} - \eta_{p-ref} = 1 - (1-\eta_{p-ref}) \frac{0.3 + 0.7 \frac{\lambda_{act}}{\lambda_{cr-act}}}{0.3 + 0.7 \frac{\lambda_{ref}}{\lambda_{cr-ref}}} - \eta_{p-ref}. \tag{6.92}$$

The friction factors λ are calculated with the Colebrook–White formula, Eq. (1.91); for actual and for reference conditions these are:

$$\lambda_{act} = \left[1.74 - 2\log_{10}\left(\left(2\frac{Ra}{b_2}\right)_{act} + \frac{18.7}{Re_{act}\sqrt{\lambda_{act}}}\right)\right]^{-2}. \tag{6.93}$$

$$\lambda_{ref} = \left[1.74 - 2\log_{10}\left(\left(2\frac{Ra}{b_2}\right)_{ref} + \frac{18.7}{Re_{ref}\sqrt{\lambda_{ref}}}\right)\right]^{-2}. \tag{6.94}$$

For infinite (critical) Reynolds number, the actual conditions are

$$\lambda_{cr-act} = \left[1.74 - 2\log_{10}\left(2\frac{Ra}{b_2}\right)_{act}\right]^{-2}. \tag{6.95}$$

For infinite (critical) Reynolds number, the reference conditions are

$$\lambda_{cr-ref} = \left[1.74 - 2\log_{10}\left(2\frac{Ra}{b_2}\right)_{ref}\right]^{-2} \tag{6.96}$$

The arithmetical average roughness from the centerline of the peaks is

$$Ra = (1/l) \int_0^l |y| \, dx \qquad (6.97)$$

The detailed stage efficiency calculation is shown in Table 6.12.

Table 6.12 Stage efficiency, numerical calculation

	Symbol	Reference	Stage 1	Stage 2
Basic efficiency (= reference)	$\eta_{p\text{-ref}}$	Fig. 1.23	0.835	0.835
Size correction	$\Delta\eta_{d2}$		−0.003	−0.003
Stage pressure ratio	p_{7t}/p_{0t}		1.225	1.21
Kinematic viscosity nozzle [m²/s]	v_{0t}		2.48×10⁻⁷	
Reynolds number nozzle	Re_{in}		2.43×10⁷	
Inlet loss coefficient	ζ	Fig. 6.6	3.0	
Related isentropic head increase	y_r	Eq. (6.86)	0.019	
Inlet loss correction	$\Delta\eta_{in}$	Eq. (6.90)	−0.015	
Blade cutback correction	$\Delta\eta_{cb}$	Fig. 6.17	+0.01	+0.01
Diffuser ratio	d_4/d_2		1.8	1.67
Diffuser ratio correction	$\Delta\eta_{d4/d2}$	Fig. 6.20⁽¹⁾	0	0
Reynolds number (atmospheric air), reference	Re_{ref}	Eq. (1.88)	250,000	250,000
Related average roughness ($Ra = 5\mu m$), reference	$(Ra/b_2)_{ref}$		2.2×10⁻⁴	2.4×10⁻⁴
Friction factor, reference	λ_{ref}	Eq. (6.94)	0.0168	0.0169
Friction factor, critical, reference	$\lambda_{cr\text{-}ref}$	Eq. (6.96)	0.014	0.014
Reynolds number, actual: $Re_{act} = u_2 b_2 \rho_{0t}/\mu_{0t}$	Re_{act}	Eq. (1.88)	2.06×10⁷	1.96×10⁷
Related average roughness ($Ra = 4.3\mu m$), actual	$(Ra/b_2)_{act}$		1.9×10⁻⁴	2.1×10⁻⁴
Friction factor, actual	λ_{act}	Eq. (6.93)	0.0136	0.0139
Friction factor, critical, actual	$\lambda_{cr\text{-}act}$	Eq. (6.95)	0.0136	0.0139
Reynolds number correction	$\Delta\eta_{Re}$	Eq. (6.92)	+0.020	+0.021
Mach number correction	$\Delta\eta_{Mu2}$		0	0
Volute shape correction	$\Delta\eta_v$	est., volute 10, Fig. 3.14	–	−0.012
Corrected pol. stage efficiency	η_p	Eq. (6.81)	0.847	0.851

⁽¹⁾ values > 0 not used for case study to be on safe side.

6.2.4 Head, Pressure, Temperature, Power

The impeller static polytropic head (inlet total, exit static; using Eqs. (6.60) and (6.79)) with unity braces, $1\,N = 1\,kg\,m/s^2$ and $1\,kJ = 1{,}000\,Nm$, is given by

$$y_{p\,Is} = \eta_{p\,Is}\left(\Delta h - \frac{c_3^2}{2}\right) = \frac{\eta_{p\,I} + r - 1}{r} r \Delta h,$$

$$y_{p\,Is} = (\eta_{p\,I} + r - 1) s u_2^2 \frac{m^2}{s^2}\left\{\frac{N}{kg\,m\,s^{-2}}\right\}\left\{\frac{kJ}{1{,}000\,Nm}\right\} \quad (6.98)$$

$$= (\eta_{p\,I} + r - 1) s u_2^2 \frac{1}{1{,}000}.$$

The impeller static head allows us to calculate the impeller exit pressure and temperature required for determining the shroud leakage and the axial thrust for setting the impeller interference fit.

The basic head/pressure ratio relationship (e.g. formally Eq. (2.51)) yields the impeller exit static pressure, using Eq. (2.63):

$$p_3 = p_{0t}\left(\frac{y_{p\,Is}}{Z_{0t} R T_{0t}} \frac{n_v - 1}{n_v} + 1\right)^{\frac{n_v}{n_v - 1}}, \quad (6.99)$$

where $\quad \dfrac{n_v - 1}{n_v} = \dfrac{\kappa_v - 1}{\kappa_v} + \dfrac{1 - \eta_p}{\eta_p} \dfrac{\kappa_T - 1}{\kappa_T}, \quad (6.100)$

and $\quad n_v = \dfrac{n_{v3} + n_{v0}}{2} \approx \dfrac{n_{v7} + n_{v0}}{2}. \quad (6.101)$

The static impeller exit temperature, using Eqs. (2.14) and (2.64), is given by

$$t_3 = T_{0t}\left(\frac{p_3}{p_{0t}}\right)^{\frac{n_T - 1}{n_T}} - 273.15, \quad (6.102)$$

where $\quad m = \dfrac{n_T - 1}{n_T} = \dfrac{ZR}{c_p}\left(\dfrac{1}{\eta_p} - 1\right) + \dfrac{\kappa_T - 1}{\kappa_T}, \quad (6.103)$

and $\quad n_T = \dfrac{n_{T3} + n_{T0}}{2} \approx \dfrac{n_{T7} + n_{T0}}{2}, \quad (6.104)$

where p_3 and T_3 are approximate insofar as η_{pI} is not exact and $s^* \approx s$ and $\varphi_3^* \approx \varphi_3$.

The stage polytropic (total-to-total) head with unity braces, $1\,N = 1\,kg\,m/s^2$ and $1\,kJ = 1{,}000\,Nm$ is given by

$$y_p = \psi_p \frac{u_2^2}{2} = \eta_p s u_2^2 \frac{m^2}{s^2} \left\{ \frac{N}{kg\,m\,s^{-2}} \right\} \left\{ \frac{kJ}{1{,}000\,Nm} \right\} = \eta_p s u_2^2 \frac{1}{1{,}000}. \quad (6.105)$$

The stage discharge pressure, from Eq. (2.51), is given by

$$p_{7t} = p_{0t} \left(\frac{y_p}{Z_{0t} R T_{0t}} \frac{n_v - 1}{n_v} + 1 \right)^{\frac{n_v}{n_v - 1}}, \quad (6.106)$$

where Z_{0t} is taken from Eq. (2.73), n_v from Eqs. (2.63) and κ_v and κ_T from Eqs. (2.75)–(2.78) and Eqs. (2.83)–(2.85).

The stage discharge temperature, from Eq. (2.57), is given by

$$t_{7t} = T_{0t} \left(\frac{p_{7t}}{p_{0t}} \right)^m - 273.15, \quad (6.107)$$

where m is taken from Eq. (2.64), c_p from Eqs. (2.81)–(2.90); n_v and m are taken as the arithmetical average values between the stage suction and the discharge.

Table 6.13 Motor nameplate rating

		Formulae, references
Compressor gas power[1]	P_i	Eq. (6.110)
Process variation factor	f_{proc}	1.0–1.05; for uncertain process cond.
Derating factor	f_{de}	1.0–1.02; for wear, particle deposits[2]
Uprating factor	f_{up}	1.0–1.15; for anticipated uprates[3]
Bearing loss	P_{BL}	Fig. 6.3, Eq. (6.22)
Shaft seal loss	P_{SL}	2–5 kW for dry gas seals
Gear loss	P_{GL}	$(0.015-0.035) \times P_M$, Fig. 6.3
Power at motor coupling	P_{MC}	$(P_i \times f_{proc} \times f_{de} \times f_{up}) + P_{BL} + P_{SL} + P_{GL}$
Power tolerance for 100% flow, head[4]	f_{tol}	1.04; API 617, Para 2.1.3 and 4.3.6.1.2
Power tol. for 100% flow, excess head[5]	f_{tol}	1.07; API 617, Para 4.3.6.1.4
Motor nameplate rating	P_M	$P_{MC} \times f_{tol} \times 1.1$[6]; API 617, Para 3.1.5

[1] Maximum of all specified operating points.
[2] Accounting for efficiency degradation through erosion, flow channel obstruction, roughness increase, gap widening; note: pertaining head decrease to be compensated by design head reserve.
[3] In conjunction with space reserves within compressor.
[4] In case excess discharge pressure can be adjusted to 100% through suction throttling, closing IGVs, or speed reduction.
[5] API 617 allows maximally 105% head at 100% flow and a pertaining 107% power (in case the discharge pressure cannot be adjusted to 100% by means of regulation).
[6] The 10% power margin may be considered adequate to account for process variation and wear, so $f_{proc} = f_{de} = 1.0$. Minimum motor rating as per API 617 Para 3.1.5: $P_M = 1.04 \times 1.1 \times P_{MC} = 1.144 \times P_{MC}$ or $P_M = 1.07 \times 1.1 \times P_{MC} = 1.177 \times P_{MC}$. In the case the contract calls for zero power tolerance, the factor f_{tol} equals 1.0, and any uncertainty must then be included in the gas power P_i.

6.2 Detailed Compressor Design

The compressor Euler head is given by

$$\Delta h = \sum_{n=1}^{i}(y_p/\eta_p)_n, \qquad (6.108)$$

where i is the number of stages.

The compressor efficiency is given by

$$\eta_p = \frac{y_p}{\Delta h} = \frac{\sum_{n=1}^{i}(y_p)_n}{\Delta h}. \qquad (6.109)$$

The gas power is given by

$$P_i = \dot{fm}\Delta h \qquad (6.110)$$

The motor power rating, carried out in accordance with API Standard 617, is shown in Table 6.13.

The case study calculation for head, pressure, temperature and power is shown in Table 6.14.

Table 6.14 Numerical calculation of head, pressure, temperature and power

	Symbol	Unit	Reference	Stage 1	Stage 2
Impeller polytropic static head	y_{pIs}	kJ/kg	Eq. (6.98)	16.01	15.91
Polytropic volume exponent suction	n_{v0t}		Eq.(6.100)	1.422	1.439
Pol. volume exponent discharge	n_{v7t}		Eq.(6.100)	1.442	1.463
Polytropic volume exponent average	n_v		Eq. (6.101)	1.432	1.451
Static pressure, impeller exit	p_3	bar	Eq. (6.99)	71.40	86.76
Pol. temperature exponent suction	n_{T0t}		Eq. (6.103)	1.370	1.359
Pol. temperature exponent discharge	n_{T7t}		Eq. (6.103)	1.361	1.350
Pol. temperature exponent average	n_T		Eq. (6.104)	1.366	1.355
Static temperature, impeller exit	t_3	°C	Eq. (6.102)	42.0	58.6
Static density, impeller exit	ρ_3	kg/m³		56.73	64.9
Stage polytropic total head	y_p	kJ/kg	Eq. (6.105)	25.30	25.21
Stage discharge pressure (total)	p_{7t}	bar	Eq. (6.106)	76.82	92.97
Stage discharge temperature (total)	t_{7t}	°C	Eq. (6.107)	48.2	64.7
Stage discharge density (total)	ρ_{7t}	kg/m³		59.8	68.3
Compressor Euler head	Δh	kJ/kg	Eq. (6.108)		59.49
Compressor polytropic head	y_p	kJ/kg	Eq. (6.109)		50.51
Polytropic compressor efficiency	η_p	-	Eq. (6.109)		0.849
Gas power	P_i	kW	Eq. (6.110)		5,161
Bearing loss	P_{BL}	kW	Eq. (6.22) Fig. 6.3		30
Seal loss	P_{SL}	kW			4
Gear loss	P_{GL}	kW	Fig. 6.3		90
Power at motor coupling	P_{MC}	kW			5,285
Required motor nameplate rating $P_M=1.144\times 5{,}285$	P_M	kW	Table 6.13		6,046 ≈6,000

6.2.5 Diffuser Aerodynamics

It is essential to calculate the change of state in the annular diffuser in order to arrive at the inlet conditions at the return vane channel and the volute. The calculation has to account for the friction of the flow along the diffuser walls.

The change of the angular momentum within an infinitesimal element of a stream tube is balanced by the frictional moment acting on the side walls of this element:

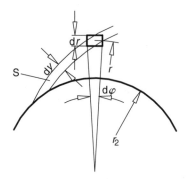

Fig. 6.21 Flow element in vaneless diffuser. S is a stream tube of width dy and depth b.

$$\dot{m}\,d(rc_u) = M_f \tag{6.111}$$

The frictional moment equals the wall shear stress in the tangential direction multiplied by the two-sided wall area times the radius; see Fig. 6.21:

$$M_f = -2\tau_u r^2 \, dr \, d\varphi . \tag{6.112}$$

The continuity equation of the stream tube at the impeller exit is

$$\dot{m} = b_2 r_2 \, d\varphi c_{m3} \rho_3 . \tag{6.113}$$

The wall shear stress in the direction of the flow and in the tangential direction with friction coefficient c_f is given by

$$\tau = c_f \frac{\rho}{2} c^2 \Rightarrow \tau_u = c_f \frac{\rho}{2} c^2 \cos\alpha ,$$

$$\tau_u = c_f \frac{\rho}{2} c_u c = c_f \frac{\rho}{2} c_u \sqrt{c_u^2 + c_m^2} . \tag{6.114}$$

Hence the change of the angular momentum becomes

$$d(rc_u) = -\frac{c_f \rho c_u \sqrt{c_u^2 + c_m^2}\, r^2}{b_2 r_2 c_{m3} \rho_3} dr , \tag{6.115}$$

$$\int_{r_2}^{r} d(rc_u) = rc_u - r_2 c_{u3} = -\int \frac{c_f \rho c_u \sqrt{c_u^2 + c_m^2}\, r^2}{\rho_3 c_{m3} b_2 r_2}\, dr\,. \tag{6.116}$$

This formula was first published by Traupel (1962).

The angular momentum of the viscous flow at an arbitrary location of the diffuser can now be calculated on the basis of the angular momentum at the impeller exit, with F being the frictional term:

$$rc_u = r_2 c_{u3} - F = r_2 c_{u3} - \frac{1}{\rho_3 c_{m3} b_2 r_2} \int_{r_2}^{r} c_f \rho c_u \sqrt{c_u^2 + c_m^2}\, r^2\, dr\,. \tag{6.117}$$

For a no-loss flow, i.e. inviscid flow ($c_f = 0$) the formula assumes the well-known form:

$$rc_u = r_2 c_{u3} = const. \tag{6.118}$$

Equation (6.117) remains valid when converted to non-dimensional form:

$$R'' V_u'' = R' V_u' - \frac{1}{B_2 \sin \alpha_3} \int_{R'}^{R''} c_f \sigma V_u \sqrt{V_u^2 + V_m^2}\, R^2\, dR\,, \tag{6.119}$$

where the following abbreviations are used:

$$V_u = \frac{c_u}{c_3}\,; \quad V_m = \frac{c_m}{c_3}\,; \quad R = \frac{r}{r_2}\,; \quad B_2 = \frac{b_2}{r_2}\,; \quad \sigma = \frac{\rho}{\rho_3}\,; \quad \sin \alpha_3 = \frac{c_{m3}}{c_3}\,. \tag{6.120}$$

This is the formula first published by Traupel (1962); the stepwise procedure follows Traupel's philosophy.

The calculation is carried out in finite radius intervals where R' is the beginning and R'' is the end of the interval, with $R'' = R' + \Delta R$. The starting point is at the impeller exit, where $R' = r_2/r_2 = 1$ and where all the relevant values are known.

The friction factor c_f is defined by the wall shear-stress formula:

$$\tau = c_f \frac{\rho}{2} c^2\,. \tag{6.121}$$

On the other hand, the shear force acting on a fluid element, as shown in Fig. 6.22, is balanced by the pressure difference across that element:

$$\tau d\pi L = \Delta p \pi \frac{d^2}{4}\,. \tag{6.122}$$

Using the equivalent pipe friction factor λ, the pressure difference becomes:

$$\Delta p = 4 c_f \frac{L}{d} \frac{\rho}{2} c^2 = \lambda \frac{L}{d_h} \frac{\rho}{2} c^2\,. \tag{6.123}$$

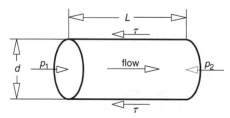

Fig. 6.22 Fluid element under pressure difference and shear stress.

The vaneless diffuser hydraulic diameter d_h, defined by four times the flow area divided by the wetted perimeter (Fig. 6.21), is given by

$$d_h = \frac{4A}{P} = \frac{4\,dy\,b}{2\,dy} = 2b.\tag{6.124}$$

The sand roughness k_s in the original Colebrook–White formula is replaced by $\approx 2 \times Ra$, where Ra is the arithmetic average roughness of a machined surface.

Thus the diffuser friction factor is defined as the equivalent pipe friction factor:

$$\lambda = \left[1.74 - 2\log_{10}\left(2\frac{k_s}{d_h} + \frac{18.7}{Re_D\sqrt{\lambda}}\right)\right]^{-2}$$

$$\Rightarrow \lambda_D = \left[1.74 - 2\log_{10}\left(2\frac{Ra}{b_2} + \frac{18.7}{Re_D\sqrt{\lambda}}\right)\right]^{-2}.\tag{6.125}$$

Diffuser Reynolds number (at the inlet):

$$Re_D = \frac{c_3 b_2}{\nu_3} = \frac{c_3 b_2 \rho_3}{\mu_3},\tag{6.126}$$

where ν is the kinematic viscosity and μ is the dynamic viscosity.

The static density-ratio curve along the diffuser σ (playing a role, especially at higher Mach numbers) can also be determined in a stepwise iterative manner. However, since this is a lengthy and tedious procedure which does not justify the effort, a shorter pragmatic method is used here.

Since the static density at the impeller exit and the total density at the stage exit are known already, the somewhat concave ρ-curve versus r/r_2 is approximated by a straight line, as shown in Fig. 6.23 (so that at the diffuser midpoint the assumed density is slightly lower and at the exit slightly higher than in reality):

$$\sigma \approx 1 + \frac{(\rho_{7t}/\rho_3)-1}{(r_4/r_2)-1}\left(\frac{r}{r_2}-1\right).\tag{6.127}$$

The case study 1st stage diffuser calculation is displayed in Table 6.15.

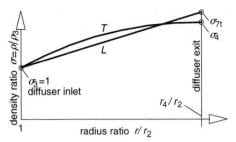

Fig. 6.23 Course of density in the diffuser: T true density, L linearized density.

Table 6.15 Diffuser 1st stage, numerical calculation

	Symbol	Reference	Stage 1
Reynolds number, disk friction	Re_{DF}	Eq. (6.53)	1.43×10^8
Gap disks/diaphragm	a/r_2		0.02
Torque coefficient	C_M	Eq. (6.53)	0.0016
Shroud leakage efficiency	η_L	Eq. (6.34)	0.99
Slip/work input ratio	s^*/s	Eq. (6.57)	0.986
Impeller slip	s^*		0.596
Actual exit flow coefficient $\varphi_3^* = \varphi_3/\eta_L$	φ_3^*		0.317
Actual absolute flow angle $\alpha_3^* = \mathrm{atan}(\varphi_3^*/s^*)$ [deg]	α_3^*		28.0
Abs. velocity inlet $c_3 = (\varphi_3^{*2}+s^{*2})^{0.5}u_2$ [m/s]	c_3		150.1
Diffuser Reynolds number	Re_D		1.45×10^7
Related average Roughness ($Ra = 6.3$ μm)	Ra/b_2		2.7×10^{-4}
Equivalent pipe friction coefficient	λ	Eq. (6.125)	0.0147
Friction factor diffuser walls $c_f = \lambda/4$	c_f	Eq. (6.123)	0.0037
Dimensionless radius, inlet	R'		1.0
Dim.less circumf. velocity inlet $V_u' = V_{u3} = \cos\alpha_3^*$	V_u'	Eq. (6.120)	0.883
Dim.less radial velocity inlet $V_m' = V_{m3} = \sin\alpha_3^*$	V_m'	Eq. (6.120)	0.469
Dimensionless density, inlet	σ	Eq. (6.120)	1.0
Dimensionless diffuser width, inlet	B_2	Eq. (6.120)	0.152

For discrete radial steps, Eq. (6.119) is transformed; the difference of the angular momentum equals the frictional term, $-F$:

$$\Delta(RV_u) = R''V_u'' - R'V_u'$$

$$= -\frac{1}{B\sin\alpha_3^*}\left(c_f\sigma V_u\sqrt{V_u^2 + V_m^2}\right)R^2\Delta R = -F . \qquad (6.128)$$

where c_c, σ, V_u, V_m and R are average values across the element.

The circumferential velocity V_u'' at the end of the interval is calculated from V_u'. V_m'' at the end of each interval is calculated from the beginning of the first interval, $V_m' = V_{m3} = \sin\alpha_3$:

$$R''V_u'' = R'V_u' - F. \tag{6.129}$$

$$V_m'' = \frac{V_{m\_1stInterval}'}{R''\sigma''} = \frac{\sin\alpha_3^*}{R''\sigma''}. \tag{6.130}$$

Equation (6.129) mainly accounts for the friction, and Eq.(6.130) accounts for the compressibility. The final conditions at element n constitute the initial conditions at element $n+1$:

$$V_{u\_n+1}' = V_{u\_n}''; \quad V_{m\_n+1}' = V_{m\_n}''. \tag{6.131}$$

We also have:

$$R^2\Delta R = \frac{1}{4}(R''+R')(R''^2 - R'^2). \tag{6.132}$$

For the first iteration step we can assume that

$$V_u\sqrt{V_u^2 + V_m^2} \approx V_u'^2, \tag{6.133}$$

since V_m is small with respect to V_u. The iteration converges rapidly.

The stepwise viscous/compressible calculation is summarized in Table 6.16.

Table 6.16 Viscous/compressible diffuser calculation, stepwise

Interval	R''	V_u''	V_m''	$R''V_u''$	σ''	F	α	α_{incomp}	$\dfrac{R''V_u''}{R_2 V_{u3}}$
1	1.104	0.795	0.422	0.8782	1.007	0.0048	27.9	28.1	0.9946
2	1.219	0.716	0.379	0.8730	1.015	0.0052	27.9	28.2	0.9887
3	1.346	0.644	0.341	0.867	1.023	0.0057	27.8	28.4	0.9822
4	1.485	0.580	0.306	0.861	1.033	0.0063	27.8	28.6	0.9751
5	1.640	0.521	0.274	0.854	1.043	0.0068	27.7	28.8	0.9674
6	1.810	0.468	0.246	0.847	1.055	0.0075	27.7	29.0	0.9590

α calculated flow angle in degrees from the tangent (viscous/compressible flow); all other parameters are dimensionless; α_{incomp} flow angle for incompressible flow.

At the diffuser exit (at the end of the last interval) the ratio in the last column is

$$\frac{(R''V_u'')_{last}}{R_2 V_{u3}} = \frac{R_4 V_{u4}}{R_2 V_{u3}} = \frac{V_{u4}}{V_{u3}\dfrac{R_2}{R_4}} = \frac{V_{u4}}{V_{u4id}} = \frac{c_{u4}}{c_{u4id}}. \tag{6.134}$$

The calculated flow angle at the diffuser exit is 27.7°; so the inlet angles of the return vanes have to be fixed accordingly.

In the inviscid/incompressible case the angle would stay constant at 28° (i.e. constituting a logarithmic spiral stream line) and $(R''V_u'')/(R_2 V_{u3})$ would remain 1.0 along the entire diffuser. Under the influence of friction alone (viscous/incompressible) the angle would increase in the diffuser by 1° (next to last column). Friction plus compressibility would bring about a slight decrease of the flow angle. The friction is low due to the relatively small wall roughness Ra and the relatively great diffuser width; so the reduction of the angular momentum (last column) is only 4% for this long diffuser. In diffusers of low flow coefficient stages and lower Mach numbers the angle increases and the angular momentum decreases considerably.

6.2.6 Return Vane Channel

The deswirling action in the return vane channel has to achieve an even velocity profile in the eye of the downstream impeller with as little losses as possible. Therefore:

– The velocity level in the multi-channel duct has to be as low as possible.
– The flow angle reduction has to be evenly distributed along the path length.

Since the velocities at the channel inlet and exit are given by the diffuser exit and the inlet to the next impeller, respectively, it is suitable to decelerate the velocity first and then accelerate towards the exit. This will result in a lower mean velocity level and hence in lower losses. Therefore, the deflection of the flow per unit of streamline length has to be smaller in the first channel half and can be increased in the second half, where the velocity is lower. Two models are discussed in the literature: linear increase of the vane angle and linear decrease of the angular momentum. The design result for the case study is shown in Fig. 6.24 (note that the streamline starts at the throat, location 6´, in Fig. 3.12, therefore, the mean angle starts at 37° instead of 27°!). The reduction of the angular momentum rc_u is not quite linear but slightly concave and the gradient of the mean vane angle along the streamline in the middle of the channel increases in the second half.

Good overall results for the stage are achieved with this geometry, made up of a double arc contour, each on the pressure and suction side.

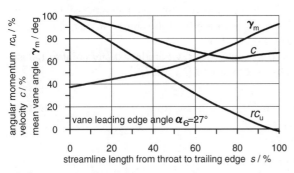

Fig. 6.24 Case study: return vane channel parameters. Double-arc vanes, γ_m mean vane angle across channel width.

The vane thickness is dictated by the bolt diameter (see Fig. 3.12), and the relative thickness has to be between 7% and 10%. This results in a relatively large leading edge radius, increasing the incidence loss but making the return vane insensitive to changes of the flow angle as they occur throughout the entire performance envelope.

As stated before, the best remedy for keeping the return vane losses low is to have a long preceding vaneless diffuser, which assures an a priori low inlet velocity.

6.2.7 Volute and Conical Exit Diffuser

Any through-flow area at an arbitrary volute azimuth as per the continuity equation (see Fig. 1.20) is

$$A = \frac{\dot{V}_\vartheta}{c_u} = \frac{\vartheta}{360} \frac{\dot{V}_V}{c_{u4id} \dfrac{c_{u4}}{c_{u4id}} \dfrac{r_4}{r}} = \frac{\vartheta}{360} \frac{\dot{V}_V}{c_{u3} \dfrac{c_{u4}}{c_{u4id}} \dfrac{r_2}{r_4} \dfrac{r_4}{r}}$$

$$= \frac{\vartheta}{360} \frac{\dot{V}_V}{s*u_2 \dfrac{r_2}{r} \dfrac{c_{u4}}{c_{u4id}}}. \tag{6.135}$$

The scroll volume flow (balance piston recirculation and shroud leakage flow do not pass through volute; therefore f and η_L do not appear in the equation; difference static/total as well as compression in conical exit diffuser neglected) is

$$\dot{V}_V \approx \dot{V}_{0t} \frac{T_{7t}}{T_{0t}} \frac{p_{0t}}{p_{7t}} \frac{Z_{7t}}{Z_{0t}}, \tag{6.136}$$

where ϑ is the azimuth (wrap angle), r is the CG radius, $\dot{V}_\vartheta$ is the partial volume flow from 0° to ϑ and c_u is the circumferential velocity at r. c_{u4} is the actual circumferential velocity at the diffuser exit calculated according to the procedures described in Sect. 6.2.5 and c_{u4id} is the ideal (i.e. inviscid/incompressible) circumferential velocity at the diffuser exit resulting from the conservation of the angular momentum; c_{u4}/c_{u4id} is given by Eq. (6.134).

The scroll area in relation to its center-of-gravity radius at any point of the azimuth can now be calculated by making use of Eq. (6.57):

$$\frac{A}{r} = \frac{\vartheta}{360} \frac{2\dot{V}_{7t}}{d_2 u_2 s} \frac{AF}{s* \dfrac{c_{u4}}{c_{u4id}}} = k\vartheta, \tag{6.137}$$

where AF is the area factor (explained below).

The total volute wrap angle, due to lack of space, does usually not cover 360°. Instead, the curved section of the scroll is cut off at $\vartheta_5 = 300$–340°, the so-called "cut-off" then being 60° to 20°. From ϑ_5 the scroll is made up of a straight section from location 5 to 6 (see Fig. 1.20), with increasing area to receive the residual incoming flow, much in the same way as in the curved section. The final value $A_5/r5_2$ of the curved section is

$$\frac{A_5}{r_5} = \frac{\vartheta_5}{360} \frac{2V_{7t}}{d_2 u_2 s} \frac{AF}{\frac{s^*}{s} \frac{c_{u4}}{c_{u4id}}} . \tag{6.138}$$

So the given parameters allow us to calculate the flow area A in relation to its center-of-gravity radius r only. The volute wrapping mode (external, central or internal) is decisive for r as a function of ϑ. The external volute (r increasing) therefore will be larger with lower losses than the internal volute (r decreasing), which will result in a more compact design. In most applications the compact design is preferred and the slightly higher volute losses can in most cases be compensated by other means of efficiency improvements.

The area factor AF normally is 1.0. However it can assume smaller values, i.e. 0.7–1.0, which will shift the entire stage characteristic curve to smaller volume flows with a higher head rise to surge, which may be beneficial for discharge pressure control, as shown in Fig. 6.25. The higher head rise is brought about by an increased part load efficiency because the optimum point of the volute itself shifts to a lower volume flow. The effect is quite similar to diffuser pinching.

Figure 6.25 shows two measured single-stage performance curves where the original scroll was replaced by one with $AF = 0.707$, i.e. an approximately 30% smaller through-flow area along the entire azimuth.

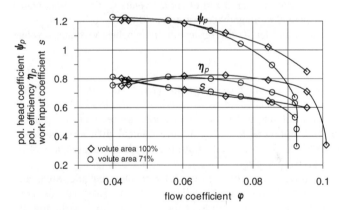

Fig. 6.25 Influence of the volute size on the compressor performance. Tests conducted on a single-stage compressor with axial suction, overhung semi-open impeller (S-type), one-sided off-set volutes with circular cross-sections, $d_2 = 355$ mm, $M_{u2} = 0.96$.

The maximum efficiency dropped by less than 1%, the optimum volume flow decreased by approximately 14%, the choke point decreased by 16%, the head decreased by 1.8% at the rated point (note that this must be accounted for during compressor design, e.g. the increase of the impeller diameter), the head rise to surge increased from 6 to 10%, and the surge point decreased by 3% only (because surge is triggered in the diffuser, not in the scroll).

The straightened-out duct between location 5 and 6, as shown in Fig. 6.26, is a transitional zone where the vortex flow changes over to pipe flow. It is a collector rather than a diffuser due to the influx along line 5–6 at the inner rim.

The area ratio of the transitional zone is

$$\frac{A_6}{A_5} \geq \frac{360}{\vartheta_5}. \tag{6.139}$$

The ($\geq$)-sign accounts for the vortex character of the flow, which may still be partially present in the duct. The transitional duct is supposed to form a harmonic entity with the exit diffuser.

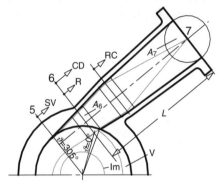

Fig. 6.26. Case study: conical exit diffuser. 5 end of curved volute section, 6 final volute cross-section, 7 discharge flange, SV straight volute section, CD conical diffuser, R rectangular, RC rectangular/circular transition, L diffuser length, Im impeller, V volute, ϑ volute azimuth, A area.

Each volute is followed by a conical exit pipe diffuser, at location 6 to 7. It has been good practice to design it according to Runstadler et al. (1975). Figure 6.27 shows the relationship between length, area ratio and pressure recovery for conical diffusers. The line of first appreciable stall for pure conical diffusers, as determined by McDonald and Fox (1965), should not be exceeded because the stall is locally unstable and might trigger surge.

Since most of the exit diffusers in compressors have rectangular inlet areas and circular exit areas, the stall tends to occur at smaller area ratios. However, the considerable inlet swirl, on the other hand, stabilizes the boundary layer, delaying the onset of stall and thus offsetting the adverse geometry effect. Pressure recovery factors of $c_p = (p_{2stat}-p_{1stat})/(p_{1tot}-p_{1stat}) \geq 0.45$ are acceptable, preferably ≥ 0.5; this means that area ratios of $A_7/A_6 \geq 3$ and length/diameter ratios of $L/D_1 \geq 4.5$ are to be preferred. In the case of a rectangular inlet the equivalent inlet diameter is $D_1 = \sqrt{4A_6/\pi}$. A 30% higher inlet blockage leads to approx. 7% lower values of c_p.

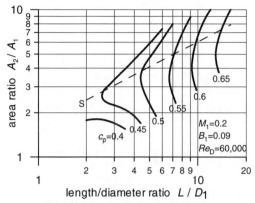

Fig. 6.27 Pressure recovery factor for conical diffusers (frustum of cone geometry) according to Runstadler et al. (1975, redrawn). S line of first appreciable stall (according to McDonald and Fox (1965)), M_1 inlet Mach number, B_1 boundary layer blockage at inlet, Re_D inlet Reynolds number, L diffuser length, D_1 equivalent inlet diameter.

Numerical Calculation of Volute and Exit Diffuser at 2nd Stage

For the case study, for space reasons, an internal volute with constant outer radius R_o was selected; inner radius R_i and center-of-gravity radius $r = (R_o + R_i)/2$ decrease from $\vartheta = 5°$ (location of scroll tongue) to $\vartheta_s = 305°$ (cut-off point of this particular scroll).

The area/radius ratio, from Eq. (6.137), in numerical terms is

$$\frac{A}{r} = k\vartheta = \frac{2 \times 1.258 \times 1000}{0.302 \times 220.9 \times 0.607 \times 0.985 \times 0.955} \times \vartheta \quad (6.140)$$

$$= 0.1835 \times \vartheta / \text{mm}.$$

The case-study volute-sizing results are shown in Table 6.17.

A law for the cross-sectional shape of the through-flow area has to be assumed to arrive at the actual area versus the azimuth. Up to an axial width limit of $b_{max} = 105.9$ mm all through-flow areas are assumed to be of quadratic shape. Once that limit is reached, the area increases radially inward, forming rectangular cross-sections. In the transitional duct from station 5 to 6 the cross-section remains rectangular and from station 6 to 7, in the exit diffuser, a transition from rectangular to circular takes place.

For the quadratic section of the volute the area is $A = b^2$, and the width is

$$b = \sqrt{\left(\frac{k\vartheta}{4}\right)^2 + k\vartheta R_o} - \frac{k\vartheta}{4}. \quad (6.141)$$

Table 6.17 Volute sizing

	Symbol	Unit	Reference	Stage 2
Volute				
Stage exit volume flow	$\dot{V}_{7t}$	m³/s	Eq. (6.136)	1.258
Reynolds number for disk friction	Re_{DF}		Eq. (6.53)	1.58×10⁸
Gap hub/cover disks/diaphragm	a/r_2			0.02
Torque coefficient	C_M		Eq.(6.53)	0.00158
Shroud leakage efficiency	η_L		Eq. (6.34)	0.99
Slip/work input ratio	s^*/s		Eq. (6.57)	0.985
Circumferential velocity ratio, scroll inlet	c_{u4}/c_{u4id}		Eq. (6.134)	0.955
Volute outer radius (constant vs. azimuth)	R_o	mm	case inner diameter	330
Final azimuth of curved section of volute	ϑ_5	deg	per design	305
Area/radius ratio, at $\vartheta_5 = 305°$	A_5/r_5	mm	Eq. (6.138)	56.0
Max. vaneless diffuser ratio (at tongue)	$(d_4/d_2)_{max}$			2.13
Minimum vaneless diffuser ratio (at ϑ_5)	$(d_4/d_2)_{min}$			1.73
Maximum axial width of volute	b_{max}	mm	per design	105.9
Azimuth quadratic/rectangular transition	$\vartheta_{5'}$	deg		220.5
Area/radius ratio factor	k	mm/°	Eq. (6.140)	0.1835
Exit diffuser				
Inlet dimensions	$b \times a$	mm		105.9×188
Equivalent inlet diameter	D_1	mm		159.2
Length ratio	L/D_1			5.14
Exit diameter (stage discharge)	D_d	mm		300
Area ratio	A_7/A_6			3.55
Pressure recovery coefficient	c_p		Fig. 6.27	≈ 0.51

For the rectangular section the area is $A = ab_{max}$ and the long (radial) side of the rectangular cross-section is:

$$a = \frac{\vartheta R_o}{\frac{b_{max}}{k} + \frac{\vartheta}{2}} \tag{6.142}$$

The center-of-gravity (CG) radius then becomes: $r = A/(A/r)$.

All volute parameters as functions of the azimuth ϑ are shown in Fig. 6.28: A/r, of course, is a linear function; the decreasing r is characteristic for internal scrolls, reducing the area A and hence reaccelerating the flow from the tongue onwards (in central scrolls r is constant, and in external scrolls r is increases with ϑ).

The described volute, as shown in Fig. 6.29, differs from the traditional textbook versions in a number of points:
- The quadratic/rectangular cross-section allows us to control the axial volute width to achieve the required shaft stiffness ratio and allows milling out of a solid block with small surface roughness.
- The long vaneless diffuser with an effective mean ratio of 1.9 places the tongue far away from the impeller exit, thus reducing peripheral pressure fluctuations at off-design operation and benefiting the efficiency.

- The long vaneless diffuser reduces the through-flow velocity level in the volute, thus decreasing friction.
- The swirl velocity, responsible for the bulk of the losses, does not play a significant role due to the diminished radial velocity in the long diffuser.

The high efficiency, the adaptability in the axial and radial direction to various geometric requirements and the options for milling and welding (by avoiding any circular cross-sections with rough surfaces due to casting) are good reasons to replace typical circular or oval textbook volutes by this novel diffuser/scroll "diaphragm package".

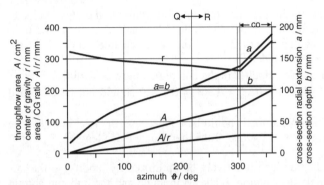

Fig. 6.28 Case study, volute parameters. Q quadratic cross-section, R rectangular cross-section, co volute cut-off, a radial side of rectangular cross-section, b axial side of rectangular cross-section.

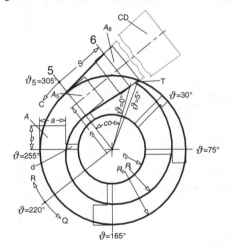

Fig. 6.29 Case study: axial view of volute. ϑ azimuth, T volute tongue, $A = ab$ throughflow area, A_5 final area curved section, r_5 center of gravity, A_6 final area of volute, co volute cut-off, a radial extension, b depth, R_o outer volute radius (constant), R_i inner volute radius (variable), r_2 impeller radius, Q quadratic section ($a = b$), R rectangular section (b = const), C curved section, S straight section (transition to conical diffuser), d annular diffuser, CD conical diffuser, 5 end of curved volute section, 6 final volute cross-section.

6.3 Impeller Blade Design

The quasi-three-dimensional technique (the so-called Q3D-method), originally developed by Wu (1952), is used here to design the impeller blades in the meridional view (R/Z-plane) and in the axial view (R/Θ-plane). The finite element calculations are carried out on the hub-to-shroud, the so-called $S1$ surface, and on the blade-to-blade, the so-called $S2$ surface. Dallenbach (1961) formulated some of the empirical criteria for favorable velocity distributions which are still valid today, see Fig. 6.30:

- At the most critical streamline, i.e. the impeller shroud, the flow velocity should decrease more rapidly in the blade inlet area, then level out in the middle and stay constant in the exit area of the blade channel; thus excessive boundary layer growth is avoided.
- Reverse flow (i.e. negative velocity) in the hub region must be avoided.
- The maximum blade loading $BL_{max} = \Delta w/w_m = 0.7{-}1.0$ should not be exceeded along the blade.
- The point of maximum blade load should be neither near the inlet nor near the exit, but rather at around 60% to 70% of the meridional blade length.

The blade-loading diagram can iteratively be optimized by modifying the initially selected blade angle distribution, blade wrap angle and the hub and shroud meridional contours. It suffices to choose just two blade angle lines at the hub and shroud, because the blade elements for process compressors are usually straight lines, required for economical flank milling out of a solid piece of material.

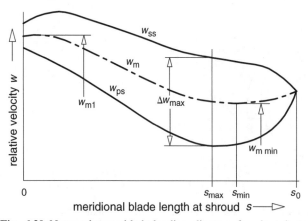

Fig. 6.30 Nomenclature: blade-loading diagram for shroud streamline. $BL_{max} = \Delta w_{max}/w_m$ maximum blade loading, $MDR = w_{m1}/w_{m\,min}$ maximum diffusion ratio, w_{ss} blade suction side, w_m mean between blades, w_{ps} blade pressure side, Δw velocity difference suction/pressure side, s_{max} location of maximum blade load, s_{min} location of minimum mean velocity.

It should be kept in mind, though, that the Q3D-method is inviscid, i.e. isentropic, which means that it is not an absolute but a qualitative and relative assessment approach requiring "calibration" in form of experimental results. Mishina and Nishida (1983), who systematically tested various impellers with different blade loading, deceleration ratios, and flow coefficients, found that:
- Deceleration along the shroud is considered most important for impeller performance.
- When the deceleration is excessive, the meridional exit velocity profile becomes distorted, resulting in an efficiency decrease both of the impeller and vaneless diffuser and early rotating stall.
- The magnitude and location of the maximum blade loading have a minor impact on the impeller efficiency only.

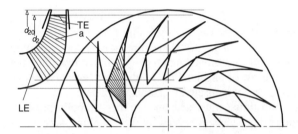

Fig. 6.31 Case study: 3D-impeller with twisted backward curved blades (1st stage), blade cut-back at outer diameter. LE blade leading edge, TE blade trailing edge, d_{20} disk diameter, d_2 trailing edge diameter, a straight-line blade elements for 5-axis NC flank milling.

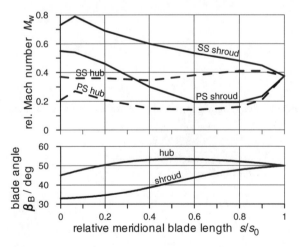

Fig. 6.32 Impeller blade angles and calculated Mach numbers (for $M_{u2} = 0.83$!) along hub and shroud meridional streamlines, SS blade suction side, PS blade pressure side, $M_w = w/a_0$ relative Mach number.

The results of the iterative Q3D-calculations for the case-study 1st impeller are displayed in Fig. 6.31 (hub-to-shroud and blade-to-blade geometry) and Fig. 6.32 (blade angles and blade loading diagram). Note that the impeller blade loading shown on the graph is valid for a design tip-speed Mach number of $M_{u2} = 0.83$ (being part of the impeller modular design concept). So the actual velocity level for the case study impeller is lower and more "straightened out".

The design Mach number loading is in agreement with the criteria described above: smooth deceleration along the shroud suction side and at the inlet rapidly decelerating along the shroud mean streamline and slightly reaccelerating towards the exit.

6.4 Axial Rotor Thrust

Axial thrust is basically generated by a pressure imbalance across the impeller. Hub thrust, acting in the opposite direction of the incoming flow in the eye, prevails over shroud and eye thrust, acting in the same direction of the incoming flow. Thus the impeller is subjected to a great force trying to move it against the incoming flow. The cumulative thrust of all impellers on the shaft, especially with impellers arranged in-line, has to be compensated by a balance piston.

Calculation of shroud and hub thrust must account for the parabolic pressure distribution in the impeller/diaphragm gaps governed by the radial pressure equation; see Fig. 6.33.

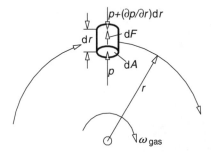

Fig. 6.33. Flow element in impeller/diaphragm gap, balance of forces. F force, A area, p pressure, ω_{gas} gas angular velocity.

From the equilibrium of forces we obtain

$$p\,dA + dA\,dr\rho r(\omega_{gas})^2 = p\,dA + \frac{\partial p}{\partial r}dr\,dA, \qquad (6.143)$$

which leads to

$$\frac{dp}{dr} = \rho r(\omega_{gas})^2 = \rho r C^2 \omega^2, \qquad (6.144)$$

where ω_{gas} is the average angular gas velocity in the gap, ω is the angular velocity of the disk, $C = \omega_{gas}/\omega$ is the core rotation factor, A is the area and ρ is the gas density.

Using the known impeller exit pressure p_3 and density ρ_3 and integrating, we obtain

$$p_3 - p(r) = \rho_3 \omega^2 r_2^2 \frac{C^2}{2}\left[1-\left(\frac{r}{r_2}\right)^2\right]. \qquad (6.145)$$

By defining a dimensionless pressure distribution coefficient

$$1-c_p(r) = 1 - \frac{p_3 - p(r)}{\rho_3 \omega^2 r_2^2} = 1 - \frac{C^2}{2}\left[1-\left(\frac{r}{r_2}\right)^2\right], \qquad (6.146)$$

the measurement of the pressure distribution in the impeller/diaphragm gaps can be non-dimensionalized in order to be used generally.

Thus, the pressure distribution coefficient and core rotation factor are directly linked via the radius ratio, i.e. once the pressure distribution in an impeller/diaphragm gap is experimentally determined (which is not too difficult to achieve), the core rotation factor can be calculated.

Figure 6.34 shows static pressure measurements in the shroud and hub gap. What strikes the eye is that, on the one hand, the shroud pressure is lower than the hub pressure at the impeller exit. The pressure difference of 5% at the radius ratio of 1.0 is attributable to the difference of the meridional streamline curvatures between hub and shroud that are typical for a flow coefficient of around 0.09 with a relative exit width of $b_2/d_2 = 0.066$.

Measurements on an impeller with $\varphi = 0.025$ and $b_2/d_2 = 0.025$ showed a much smaller difference of the impeller exit pressures at the hub and shroud.

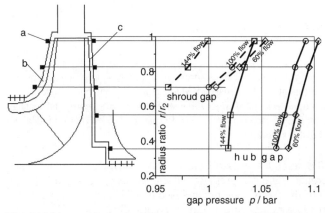

Fig. 6.34 Measured pressure in hub and shroud gap. Suction pressure 0.954 bar, flow coefficient 0.09, tip-speed Mach number 0.54, $b_2/d_2 = 0.066$, a static pressure tap, b shroud gap (centripetal leakage flow), c hub gap (centrifugal leakage flow).

On the other hand, the pressure along the shroud gap decreases more rapidly towards the seal than along the hub gap, which can be attributed to a higher core rotation factor in the shroud gap. This increases the hub thrust considerably compared with the shroud thrust for the same annulus and has to be accounted for thrust-bearing selection. The measurements on this medium flow coefficient impeller ($\varphi = 0.09$) were evaluated in detail; the results are presented in Fig. 6.35 and can be interpreted as follows.

The dimensionless pressure distribution coefficient $1-c_p$ and the core rotation factor C (both as per Eq. (6.146)) are plotted versus the radius ratio r/r_2. The centrifugal (radial outward) leakage flow in the hub gap creates a moderate pressure difference ($1-c_p = 1.0$ to 0.93 at $r/r_2 = 0.36$ radius ratio, Fig. 6.35 right) because the centrifugal direction is the natural direction the flow is inclined to take. And the fact that the leakage influx from the shaft seal is (almost) whirlfree is the reason why the core rotation factor is practically constant along the characteristic curve of the stage (at $C \approx 0.4$ the core of the gas rotates at an average rate of 40% of the impeller, regardless of the stage operating point, Fig. 6.35 left).

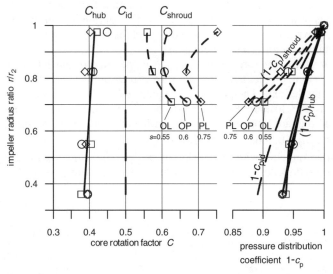

Fig. 6.35 Impeller/diaphragm gap: test-deduced pressure distribution coefficient and core rotation factor. Flow coefficient 0.093; tip-speed Mach number 0.55; $b_2/d_2 = 0.066$; *solid lines* hub gap with centrifugal leakage; *dashed lines* shroud gap with centripetal leakage; OP optimum (100% flow); OL overload (144% flow); PL partload (60% flow); id idealized (freely rotating enclosed disk), s work input factor.

The situation on the shroud side is different. The centripetal (radial inward) leakage flow must create a greater pressure difference because its direction is against the natural Δp of the freely rotating enclosed disk. This leads to a pressure distribution $1-c_p$ from 1.0 at radius ratio 1.0 to ≈ 0.88 at radius ratio 0.71. The pertaining core rotation factor responsible for this comparably high pressure difference is between 0.55 and 0.75, i.e. the gas in the gap rotates with 55% to 75% of the impeller angular velocity, being some 40% to 90% (!) higher than in the hub gap. The driving force for this high core rotation is the strong vortex c_{u3} of the leakage flow, branched off from the impeller exit, which is high at partload and lower at overload, thus explaining the fanning out of the core rotation and hence of the pressure distribution for different operating points of the stage. Therefore, the actual curve parameter in Fig. 6.35 for the centripetal gap flow is the work input factor, which is $s = 0.55$ for overload (OL), $s = 0.6$ for optimum (OP) and $s = 0.75$ for partload (PL) (these numbers allow interpolation and limited extrapolation). So $1-c_p$ has to be lower at surge than at choke and C has to be higher at surge than at choke.

Figure 6.35 also contains the pressure distribution curve $1-c_{pid}$ for an idealized core rotation of $C_{id} = 0.5$ often mentioned in the text books. So the actual hub core rotation with centrifugal leakage is some 20% lower and the actual shroud core rotation with centripetal leakage is some 30% higher than this hypothetical value.

It should be mentioned that, in the case of an adjoining balance piston, the hub-gap pressure distribution follows a similar pattern to the shroud-gap pressure distribution, because the hub gap leakage flow changes direction from centrifugal to centripetal. Equation (6.146) with Fig. 6.35 allows us to calculate the gap pressure at any radius with the unity braces: $1\text{N} = 1 \text{ kg m s}^{-2}$ and $10^5 \text{N/m}^2 = 1\text{bar}$:

$$p = p_3 - c_p \rho_3 \omega^2 r_2^2 \frac{\text{kg}}{\text{m}^3} \frac{1}{\text{s}^2} \text{m}^2 \left\{ \frac{\text{N}}{\text{kg m s}^{-2}} \right\} \left\{ \frac{\text{bar}}{10^5 \text{ N m}^{-2}} \right\}$$

$$= p_3 - c_p 10^{-5} \rho_3 \omega^2 r_2^2 \text{ / bar}, \qquad (6.147)$$

where c_p is a function of r and can be read from Fig. 6.35.

Or, which may be more convenient, using the core rotation factor (Eq. (6.145)), we obtain the expression:

$$p = p_3 - 10^{-5} \rho_3 \omega^2 r_2^2 \frac{C^2}{2}\left[1 - \left(\frac{r}{r_2}\right)^2\right] \text{ / bar}. \qquad (6.148)$$

C is approximately 0.4 for hub gaps with centrifugal leakage flow and approximately 0.6 to 0.7 at the radius ratio of the seal for shroud gaps with centripetal leakage flow. In final compressor stages with an adjoining balance piston the flow in the hub gap is centripetal, so the higher core rotation factors apply.

It turned out that the measured impeller exit pressure at the hub, p_{3hub} agreed well with the calculated mean (i.e. one-dimensional) impeller exit pressure p_3; so Eqs. (6.147) and (6.148) can be applied directly to calculate the pressure distribution in the hub gap. However, p_3 has to be corrected to arrive at the impeller exit pressure at the shroud. The guide line is: $p_{3shroud} \approx 0.95 \times p_3$ for impellers with $b_2/d_2 = 0.066$; $p_{3shroud} \approx p_{3hub} \approx p_3$ for impellers with $b_2/d_2 = 0.025$.

The resulting pressures at the cover disk and shaft seal can also be used to determine the shroud and shaft leakage flows. For thrust purposes it seems justified to linearize the flat pressure curves, yielding just one average pressure for the shroud and the hub side, each in the form:

$$p_S = (p_{3S} + p_{D1})/2 \quad \text{and} \quad p_H = (p_{3H} + p_{S2})/2, \tag{6.149}$$

with p_{D1} and p_{S2} calculated at the proper radii using Eqs. (6.147) and (6.148).

The axial thrust is broken down into the momentum force in the impeller eye, the static eye force, the shroud force, the hub force, and the balance piston force, as shown in Fig. 6.36a, b. Now the five thrust quantities can be calculated as follows.

(Dynamic) momentum thrust at impeller eye:

$$F_M = \dot{m} c_E, \tag{6.150}$$

where c_E is the velocity at the impeller eye, accounting for the extension of the process volume flow as per Sect. 6.2.1.

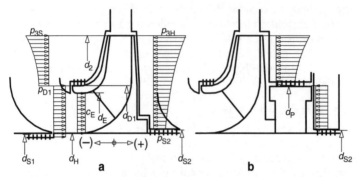

Fig. 6.36a, b Thrust breakdown: pressure distribution at shroud, hub, impeller inlet and balance piston. **a** initial or middle stage, **b** stage with adjoining balance piston, (−) (+) definition of thrust direction.

(Static) inlet thrust (with static suction pressure p_1):

$$F_I = \frac{\pi}{4}\left(d_{D1}^2 - d_{S1}^2\right)p_I. \tag{6.151}$$

Shroud thrust:

$$F_S = \frac{\pi}{4}\left(d_2^2 - d_{D1}^2\right)p_S. \tag{6.152}$$

Hub thrust:

$$F_H = \frac{\pi}{4}\left(d_2^2 - d_{S2}^2\right)p_H. \tag{6.153}$$

If the impeller has an adjacent balance piston, d_{S2} has to be replaced by d_P.

The balance piston thrust at the outer face (for in-line impeller arrangement only) is given by

$$F_P = \frac{\pi}{4}\left(d_P^2 - d_{S2}^2\right)p_P, \tag{6.154}$$

where $p_\text{P} = p_\text{0t} + \Delta p_\text{recirc} \approx p_\text{0t}$ (piston outer face is connected to the suction line, hence the pressure loss). The total gas thrust of the stage is

$$F = F_\text{M} + F_\text{I} + F_\text{S} - F_\text{H} - F_\text{P} \tag{6.155}$$

Summing up the stage thrusts to arrive at the compressor gas thrust (which is necessary to select the thrust bearing) has to account for the proper signs of each single force; see Fig. 6.37.

As can be seen in the diagram, the selection of a relatively small balance-piston diameter of 170 mm results in a negative compressor thrust throughout the entire performance map. In other words it pushes the rotor towards the thrust bearing, but this does not have any disadvantage whatsoever. On the contrary, the selected piston diameter assures a comparatively small balance piston loss of 0.8%. In the case of a piston diameter of 240 mm, resulting in a positive thrust direction, the piston leakage loss would be 1.9%, i.e. increasing the power input by 60 kW! This increase is due to the increased labyrinth through-flow area (diameter plus clearance) and the increased pressure upstream of the labyrinth, which can be derived from Fig. 6.35.

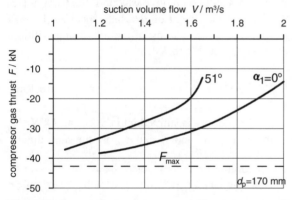

Fig. 6.37 Case study: compressor gas thrust map. α_1 adjustable inlet guide vane angle, F_max maximum thrust bearing load (per manufacturer, direction of thrust: (+) pulling thrust (away from bearing), (–) pushing thrust (towards bearing), d_p balance piston diameter.

Lempart (1992) published a compilation of thrust curves versus volume flow for various in-line and back-to-back arranged compressors, where he showed how the compressor thrust and the slope of the thrust curve can be influenced by the balance-piston diameter.

Zimmermann et al. (1983) have presented the most comprehensive investigation on disk friction so far as part of a research project with the "Research Association Internal Combustion Engines" jointly sponsored by the German and Swiss turbomachinery manufacturers. It covers pressure distribution, core rotation, frictional torque, axial thrust for centrifugal and centripetal leakage flows with and without swirl, with wide and narrow gaps for a total of 34 disk shapes as they are used for centrifugal compressors, steam and gas turbines. The pressure distributions in Figs. 6.34 and 6.35 agree satisfactorily with the findings of Zimmermann.

It has been shown that the basically parabolic pressure distributions with rotating faces get distorted by the amount and direction of the leakage flows. Centripetal influx exaggerates and centrifugal influx levels out the radial pressure difference, and the degree of leakage flow swirl influences to a large extent the core rotation factor. So in the absence of accurate experimental pressure distributions for the type of impellers used, the above thrust formulae are approximations, requiring a sufficient capacity reserve for the selected bearing. Therefore, it appears reasonable that API Standard 617 postulates that the calculated gas plus coupling thrust under the most adverse conditions must not be more than 50% of the ultimate bearing load rating.

Coupling Thrust

Since flexible-element couplings wield a force on the bearing when it is axially deflected, the thrust must be calculated on the basis of the maximum allowable deflection.

With toothed gear-type couplings as well, the shaft exerts a force on the bearing, e.g. by thermal expansion, and the teeth do not give way until the friction factor is used up. With a "safety" friction coefficient of $\mu = 0.25$, specified by API 617, the coupling lock-up thrust on the thrust bearing of the nth casing within a train is described by the numerical equation (Fig. 6.38):

$$F_C = \pm \frac{9.55 \times 10^6 \times \mu}{N} \left(\frac{P_{in}}{d_{in}} + \frac{P_{out}}{d_{out}} \right) / N, \qquad (6.156)$$

where N is the speed / min^{-1}, P_{in} is the power transmitted into the casing / kW, P_{out} is the power transmitted out of the casing/kW and d_{in} and d_{out} are the shaft diameters under the couplings/mm.

The direction of the gear type coupling thrust is indeterminate. So it has to be added to as well as subtracted from the gas thrust to come up with the complete thrust envelope.

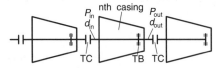

Fig. 6.38 Lock-up thrust of toothed type couplings (example: three-casing compressor train). *P* mechanical power, *d* shaft diameter under coupling, TC coupling, TB thrust bearing.

6.5 Leakage Flow through Labyrinth Seals

The inevitable parasitic flows through the labyrinth seals of cover disk, shaft and balance piston can be satisfactorily calculated with the following numerical formula (from BORSIG Pocket Book (1994)):

$$\Delta \dot{m} = \frac{\zeta \varepsilon A p_1}{\sqrt{10^5 Z_1 R(273+t_1)}} \text{ /kg/s} \tag{6.157}$$

where

$\zeta \approx 0.7$ coefficient of flow for interlocking seals; radial step/clearance ratio > 2
$\zeta \approx 1.2$ coefficient of flow for see-through seals; pitch/clearance ratio ≈ 8
$\varepsilon = f(n, p_2/p_1)$ seal factor; Fig. 6.39
$A = D\pi s/\text{mm}^2$ gap area; D/mm seal diameter
s/mm clearance (rule of thumb: $s \approx 0.002 \times D$)
p_1/p_2/bar pressure upstream/downstream of seal
Z_1 compressibility upstream of seal
R/kJ kg^{-1}K^{-1} gas constant
t_1/°C temperature upstream of seal
n number of sealing tips.

A rule of thumb is: $n \approx 5$ for cover-disk interlocking seals; $n \approx 7$ for interstage shaft see-through seals; $n \approx 20$ to 30 for balance-piston interlocking seals on in-line compressors. p_1 and p_2 can be calculated using Eqs. (6.147) and (6.148).

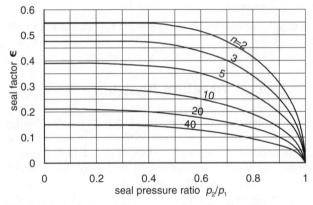

Fig. 6.39 Labyrinth seal factor: n number of sealing tips.

With the sizing of the balance piston the aero-thermodynamic compressor design is brought to a close. The selection of the shaft end seals (dry gas seals), journal and thrust bearings (tilting pad type), and coupling (elastic diaphragm type), as per specification, enable the final rotor design to be achieved.

6.6 Final Compressor Design

The compressor (sketch shown in Fig. 6.40) consists of a radially split (so-called barrel) casing of 990 mm outer diameter, which is a cylindrical forged shell with two forged bolted-on end covers. The barrel design is mandatory because the discharge pressure of 93 bar exceeds the feasibility limit for axially split casings, which is approximately 70 bar. Most barrel compressors nowadays have end covers held in place by shear rings. However, with smaller casings such as the one in the case study, bolted end covers are preferred since they offer more space to accommodate the various seal supply lines. The bearing housings in turn are bolted to the end covers. The axially split diaphragms, held together by tension bolts, form the so-called inner casing, which, being wrapped around the rotor, is inserted axially into the outer casing. The suction pressure of 63 bar is sealed off by two dry gas seals with additional inert-gas labyrinth seals.

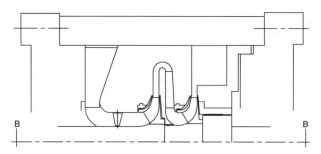

Fig. 6.40 Case study: sketch of aero components. B location of journal bearing, bearing diameter 85 mm, bearing span 1,038 mm.

6.7 Rotordynamics: Lateral Vibration Analysis

The lateral rotordynamic calculation, on the basis of the "Transfer Matrix Method", accounts for the mass distribution, flexural stiffness distribution, mass moments of inertia of the impellers, non-isotropic stiffness and damping coefficients of the journal bearings, optionally cross-coupling forces by the labyrinth seals, and three distributions of the residual unbalance. It enables the computation of the rigid and elastic-support natural frequencies, vibration amplitudes at the bearings and unbalance positions, bearing forces and the logarithmic decrements. The calculation of the bearing stiffness and damping coefficients takes into account the variable peripheral oil viscosity. The method requires that the rotor be subdivided into appropriate shaft segments with constant diameter. The calculation yields the following results.

6.7.1 Lateral Critical Speeds

The actual natural frequencies on elastic supports are determined by the unbalance response analysis. For that purpose the unbalance U (residual after balancing) is distributed on the shaft in such a way to bring out the 1st, 2nd and 3rd rotor mode shape. For the 1st mode U is placed at rotor midspan. For the 2nd mode $U/2$ is placed near bearing 1 and $U/2$ with 180° out of phase near bearing 2. For the 3rd mode a modified unbalance $U3$ is placed at the inboard shaft end (at the coupling hub).

The residual unbalances according to API 617 (1995) can be expressed in the form of numerical equations:

$$U = J(6{,}350\, m_{\text{rotor}} / N_{\min})\ /\text{g mm}, \tag{6.158}$$

$$U3 = J(6{,}350\, m_{\text{overhg}} / N_{\max}). \tag{6.159}$$

where m_{rotor} is the rotor mass in kg, m_{overhg} is the mass of the shaft portion outboard of the bearing, N is the rotational speed in rpm and J is a multiplier.

The maximum allowable major semi-axis of the vibration ellipse (zero-to-peak) at the two probe locations near the bearings is

$$A_{\text{API}} = 12.7 \sqrt{\frac{12{,}000}{N}}\ /\mu\text{m}. \tag{6.160}$$

The multiplier J has to be not less than 2.0 to exactly reach A_{API}. The higher is J the better, and the less sensitive is the rotor.

The results of the calculation under the influence of oil-film elasticity and damping are the unbalance response curves in the form: vibration amplitude versus rotational speed for the three rotor modes; see Fig. 6.41. The relative curve maxima are natural frequencies that may or may not be critical speeds.

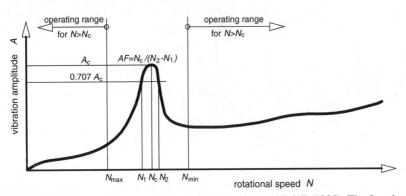

Fig. 6.41 Unbalance response analysis: definitions (from API 617 (1995), Fig. 8, redrawn).

If the amplification factor AF (for its definition see Fig. 6.41), being the criterion for a critical natural frequency, is less than 2.5, the amplitude maximum is not considered a critical speed and the compressor may be operated exactly at that natural frequency. If the amplification factor is greater than 2.5, the natural frequency is a critical speed requiring certain separation margins between the operating and resonance speed. The required distances between operating and critical speeds are described in Table 6.18.

The distance ratio states how far apart from the operating speed the critical speed has to be, corresponding to the API definition of the separation margin: for $N_c > N_{\max}$: $SM = 26-6/(AF-3)$ and for $N_c < N_{\min}$: $SM = 16+6/(AF-3)$.

Table 6.18 Separation margins according to API Standard 617 (1995)

Amplification factor	Speed distance ratio for $N_c > N_{max}$ (critical above operating speed)	Speed distance ratio for $N_c < N_{min}$ (critical below operating speed)
$AF < 2.5$	none	none
$2.5 < AF < 3.55$	$N_c/N_{max} \geq 1.15$	$N_c/N_{min} \leq 0.95$ (6.161)
$AF > 3.55$	$N_c/N_{max} \geq 1+[26-6/(AF-3)]/100$	$N_c/N_{min} \leq 1-[16+6/(AF-3)]/100$ (6.162)

6.7.2 Stability

The rotor stiffness ratio *RSR* according to Eq. (4.8) requires knowledge of the first rigid-support critical speed which, related to the maximum continuous speed, determines whether the rotor is stable, i.e. resistant to subsynchronous vibrations. The physical background was described by Nielsen and Van den Braembussche (1997) as follows:

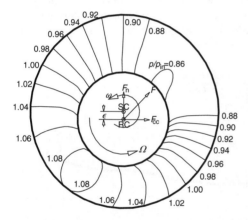

Fig. 6.42 Calculated static pressure field for a whirling rotor in an eccentric smooth axial seal; after Baskharone (1994, redrawn and supplemented). p gap pressure, p_{in} seal inlet pressure, ω whirl rotational speed, Ω rotor operational speed, ε eccentricity (deflection), RC rotor center, SC stator center, F pressure field force on rotor, F_n normal force component, F_{cc} cross-coupling force (perpendicular to rotor deflection).

Every shaft rotating in a (labyrinth) seal will exhibit an eccentricity ε due to bending and oil-film elasticity. The shaft rotating with angular velocity Ω pumps the gas by shear force towards the minimum clearance, where it meets a raised resistance leading to a pressure build-up upstream of the minimum clearance. Thus, considerable tangential pressure differences are created in the non-concentric cavity, as shown in Fig. 6.42, with a pressure side at approximately 7 o'clock and a suction side at 1 o'clock with a vertically downward deflected shaft. It is evident that the pressure distribution causes not only a restoring force in the direction of the deflection $F_n = K\varepsilon$ but also a tangential restoring force. The latter is also called the cross coupling force $F_{cc} = k\varepsilon$ and is always perpendicular to the deflection. K is the spring coefficient of the "direct stiffness" and k is the spring coefficient of the "cross-coupling stiffness". F_{cc} causes the deflected rotor to whirl around the fixed stator center on the rotor orbit at the angular velocity ω, whereby, of course, the minimum clearance moves around the circumference at velocity ω as well, although in the beginning it was located vertically below the shaft.

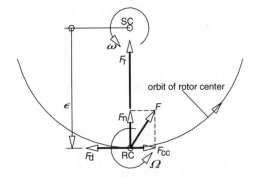

Fig. 6.43 Forces on whirling rotor; after Miskovish and Brennen (1992, redrawn and modified). ω whirl rotational speed, Ω rotor operational speed, ε eccentricity (deflection), RC rotor center, SC stator center, F_r restoring force due to deflection, F pressure field force on rotor, F_n normal force component, F_{cc} cross-coupling force (perpendicular to rotor deflection), F_d damping force.

Figure 6.43 displays all relevant forces acting on a whirling rotor. If, irrespective of the reason, a shaft is radially deflected by the value of ε from its equilibrium, it is first of all subjected to a restoring force oppositely proportional to the deflection $F_r = K^*\varepsilon$ (normal direction). The damping force $F_{rd} = -D\omega\varepsilon$ is perpendicular to the deflection and oppositely proportional to the circumferential velocity $\omega\varepsilon$ of the orbiting rotor. The normal pressure field force $F_n = K\varepsilon$ amplifies the restoring force F_r, whereas the cross-coupling force $F_{cc} = k\varepsilon$ weakens the damping force F_d, because it acts in the direction of the whirl rotational speed.

The objective of these rotordynamic stability calculations is to limit the destabilizing cross-coupling force, which can be considerably increased if the incoming gas is not vortex-free

According to Nielsen and Van den Braembussche the behavior of rotor vibrations excited by cross-coupling forces is governed by the parameters:

– Core rotation factor: $C = c_u/u$
– Whirl frequency ratio: $W = k/(D\Omega)$
– Rotor stiffness ratio: $RSR = N_{\text{c1-rig}}/N = \Omega_{\text{c1-rig}}/\Omega$
  where

c_u average gas circumferential velocity in seal gap
u rotor circumferential velocity
k cross coupling spring constant
D damping coefficient
$N_{\text{cl-rig}}$ 1st rigid-support critical speed
$\Omega_{\text{cl-rig}}$ 1st rigid-support critical angular velocity.

If the whirl frequency ratio, which is nearly proportional to the core rotation factor, is greater than the rotor stiffness ratio, the cross-coupling force in the seal will dominate over the damping force, which acts in the opposite direction. Therefore the rotor will destabilize with $k/D > \Omega_{\text{cl-rig}}$ leading to unstable behavior that increases the vibration amplitude. According to Ehrich (1990) the rotor becomes unstable, if the cross-coupling force F_{cc} is greater than the damping force F_d. Since, according to Benckert and Wachter (1978), k increases with the gas density, $\Omega_{\text{cl-rig}}$ has to be increased for denser gases, i.e. the shaft has to be shortened or made thicker. This is the reasoning behind Fulton's empirical criterion taking care of the stiffness/density interrelationship:

$$RSR \geq \frac{1}{3.24 - 0.36\ln\left(\dfrac{\rho_s + \rho_d}{2}\right)}. \tag{6.163}$$

If shaft stiffening is not possible, the whirl frequency ratio has to be reduced by deswirling the flow (i.e. decreasing the core rotation factor) before it enters the seal. This can be accomplished by swirl brakes. There is more on this in Nielsen and Van den Braembussche (1997).

6.7.3 Logarithmic Decrement

According to the definition, the logarithmic decrement (or in short: "log dec") of a rotor is the natural logarithm of the ratio of two consecutive maxima of the decaying vibration amplitude after the rotor has been exposed to a sudden perturbation (see Fig. 6.44):

$$\ln q = \ln\left|\frac{A(t_n)}{A(t_{n+1})}\right|. \tag{6.164}$$

The calculations for the "basic log dec", i.e. without accounting for cross-coupling effects, is normally carried out as long as the average gas density across the compressor does not exceed approximately 80 kg/m³. The minimum requirement is

$$\ln q_{\text{basic}} \geq 0.3. \tag{6.165}$$

For higher densities the cross-coupling effects of all labyrinth seals have to be part of the complete lateral vibration analysis; in this case the minimum requirement for the rotor "actual log dec" is

$$\ln q_{\text{act}} \geq 0.1. \tag{6.166}$$

6.7 Rotordynamics: Lateral Vibration Analysis

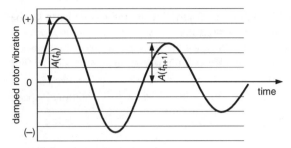

Fig. 6.44 Definition of logarithmic decrement: damped rotor vibration after sudden perturbation.

Table 6.19 Case study: numerical results of lateral vibration analysis

	Symbol	Unit	1st mode	2nd mode	3rd mode
Rotational speed	N	1/min	13,971	13,971	13,971
Rotor mass	m_{rotor}	kg	151.5	151.5	–
Overhung mass, inbord	m_{overhg}	kg	–	–	16.6
Unbalance	U	gmm	68.9	–	–
	$U/2$	gmm	–	±34.4	–
	$U3$	gmm	–	–	7.6
Critical speed, elastic, calculated	N_c	1/min	8,871	(1)	19,996
Amplification factor, calculated	AF		2.8	(1)	1.6
Separation margin, required	SM_{requ}	%	5.0	(1)	0
Separation margin, calculated	SM_{calc}	%	36.5	(1)	43
Peak amplitude, allowable	A_{API}	μm	11.8	11.8	11.8
Peak amplitude, calculated	A_{calc}	μm	0.937	(1)	0.43
Multiplier for unbalance to reach A_{API}: $J = A_{API}/A_{calc}$	J		12.56	(1)	27
Average density compressor	ρ_m	kg/m³	60.1		
Rotor stiffness ratio, required	RSR_{requ}		≥0.566		
Critical speed, rigid, calculated	$N_{c\text{-}rig}$	1/min	9,259	31,246	43,945
Rotor stiffness ratio, calculated	RSR_{calc}		0.663		
Basic log. decrement, calculated	$\ln q$		1.17		

Amplitudes relate to coupling side location of vibration probe.
[1] no appreciable peak.

The detailed results of the lateral vibration analysis for the case study are given in Table 6.19. It reveals that the rotor stays within all limits:

- To reach the allowable peak amplitude the rotor needs 12 and 27 times the API standard residual unbalance, i.e., much more than the required minimum multiplier of $J = 2$; so it is extremely insensitive to enlarged unbalances (e.g. particle deposition).
- There is only one appreciable elastic-support critical speed: the first mode with a much greater separation margin from the operating speed than required.
- The third-mode natural frequency is not a critical speed, since the amplification factor is < 2.5.

- The rotor stiffness ratio is 17% higher than required by the gas density; so the rotor is not prone to subsynchronous vibrations.
- In spite of the relatively high pressure level there is no need to incorporate cross-coupling forces; the basic log dec calculation is sufficient and is four times the required minimum of 0.3.

The 1st, 2nd and 3rd rotor-mode vibration amplitudes versus rotational speed curves are shown in Fig. 6.45, and the log dec is shown in Fig. 6.46.

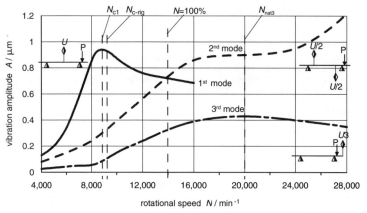

Fig. 6.45 Case study: calculated vibration amplitudes (major semi-axes of vibration ellipse at probe location). U, $U/2$, $U3$ unbalance locations for 1st, 2nd and 3rd rotor mode, respectively; P vibration probe location at coupling side.

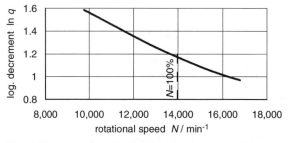

Fig. 6.46 Case study: calculated rotor logarithmic decrement.

6.8 Performance Map

The dimensionless so-called φ_3-performance curve of the single stage similar as in Fig. 5.19 serves as the basis for predicting the performance curves of the compressor. Once the impeller exit flow coefficient φ_3 at the best efficiency point is fixed (see Fig. 6.10), the work input factor can be calculated from Eq. (6.50).

Since the relative exit flow angle β_3 is constant for off-design operation, the impeller slip s^* is a linear function of φ_3:

$$s^* = 1 - \frac{\varphi_3}{\varphi_{3opt}}(1 - s^*_{opt}).\qquad(6.167)$$

The disk friction and shroud leakage work are calculated from Eqs. (6.52) to (6.57) giving the $s = f(\varphi_3)$ function a slight parabolic shape (Fig. 5.19), since the disk friction and shroud leakage constitute a growing percentage at partload.

Once the stage polytropic efficiency at the best efficiency point is fixed, the course of the efficiency curve at partload and overload is dictated by the Mach number. This can be calculated in detail by computing the various losses as described in the text books. In practice the Mach number dependency has been determined empirically.

So, for each flow coefficient (or for each distinctive standard stage within the impeller family, as shown in Fig. 3.4) a φ_3-curve or a set of curves exist for the stage, as shown in Fig. 6.47. In the case of installed adjustable inlet guide vanes an empirically determined dimensionless IGV φ_3-map serves as the basis for the prediction of the compressor IGV map (Lüdtke (1992)), as shown in Fig. 6.48.

The surge point of each individual curve has normally been determined experimentally and converted to a critical diffuser inlet angle α_{3c} to be used as calibration for Senoo's model of "critical flow angle for reverse flow" (Fig. 5.28).

The non-dimensional stage curves are aerodynamically superimposed by the following calculation steps: select the actual volume flow and IGV angle; calculate the Mach number, impeller volume ratio and exit flow coefficient of the first impeller; read from the relevant graph s, $\eta = f(\varphi_3)$ the work input factor and stage efficiency; calculate the head, gas power, pressure and temperature ratio; the resulting actual volume flow at the second stage permits us to repeat this procedure for the second stage, and so forth. Repeat for as many volume flows and IGV angles as necessary.

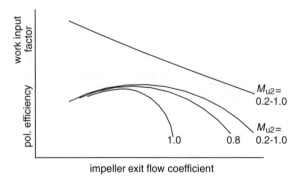

Fig. 6.47 Non-dimensional stage performance curve(s); variable Mach number.

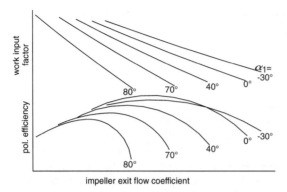

Fig. 6.48 Non-dimensional stage performance curve(s); variable IGV angle, constant Mach number.

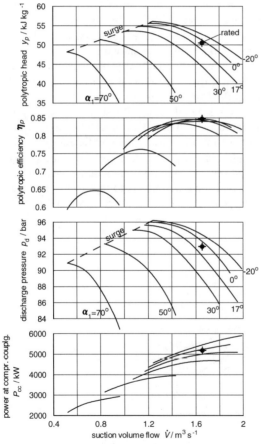

Fig. 6.49 Case study: predicted performance map. $M_{u2} = 0.553$, α_1 adjustable inlet guide vane angle.

Compressor surge will be reached either when an individual stage arrives at the critical flow angle ($\alpha_3 = \alpha_{3c}$, stage surge) or when the compound discharge pressure curve arrives at the horizontal tangent ($dp_d/d\dot{V} = 0$, system surge), whichever comes first. Adding up the stage gas powers plus the mechanical losses of bearings, seals and gear will result in the compressor power at the drive coupling.

Figure 6.49 displays the results of the described performance-map calculation in the form of the head, efficiency, discharge pressure and power at the compressor coupling. Note that in order to avoid a negative head tolerance the head at the specified volume flow has a reserve of 2%, normally achieved by an eventual speed increase of approximately 1%, which has to be in accordance with the exact gear teeth ratio. According to API 617 (1995), para 4.3.6.1.4, the actual head measured in a performance test has to be in the range of 100% to 105%.

6.9 Performance Test

The natural gas compressor according to the case study, can be shop performance-tested according to the "Type 2" criteria of the "Performance Test Code on Compressors and Exhausters", ASME PTC 10-1997. This class allows us to test with another gas, another speed at a different (usually lower) pressure and power input level. The common denominator is the adherence to the laws of aerodynamic similarity in the form of limited deviations of volume ratio, flow coefficient, Mach and Reynolds number (Table 6.20).

Table 6.20 Permissible deviations from specified non-dimensional similarity parameters for Type 2 tests for centrifugal compressors (t test, sp specification)

		Min. limit	Max. limit
Volume ratio across compressor	$(\dot{V}_d/\dot{V}_s)_t / (\dot{V}_d/\dot{V}_s)_{sp}$	0.95	1.05
Flow coefficient, 1st stage	φ_t/φ_{sp}	0.96	1.04
Machine Mach number, 1st stage, depending on M_{u2sp}	$M_{u2t} - M_{u2sp}$	–0.20 to –0.04	0.07–0.25
Machine Reynolds number, 1st stage, depending on Re_{u2sp}	Re_{u2t}/Re_{u2sp}	0.3–0.1	3–100

Table 6.21 Case study: performance test planning; comparison test/specification

		Test	Specification
Volume ratio	$\dot{V}_d/\dot{V}_s$	0.735	0.768
Flow coefficient	φ	0.1024	0.1024
Mach number	M_{u2s}	0.605	0.553
Reynolds number	Re_{u2s}	2.51×10^6	20.8×10^6

The compressor can be shop tested in a closed loop with nitrogen at a suction pressure of 9 bar at the design speed of 100%. The following deviations from aerodynamic similarity will occur:

$(\dot{V}_d/\dot{V}_s)_t / (\dot{V}_d/\dot{V}_s)_{sp} = 0.957$ (i.e. 4.3% deviation ; permissible max. 5%),
$\varphi_t/\varphi_{sp} = 1.0$ (no deviation; exactly on the point),
$M_{u2t} - M_{u2sp} = 0.052$ (permissible maximum 0.16),
$Re_t/Re_{sp} = 0.12$ (permissible minimum 0.10).

The shop test parameters are shown in Table 6.21. So for the above envisioned test conditions the deviations for the guarantee point are all in the allowable range. The calculated deviations warrant that the aerodynamic stage mismatching at the guarantee point is negligible and the Mach number increase tolerable. The low test Reynolds number reduces the test efficiency by only 0.003, so the Reynolds number corrections to adjust the test results remain very low.

If no heat insulation material is used on the compressor body during the shop test, the efficiency is measured as too high because the measured discharge temperature is too low. The resulting gas power, according to the first law of thermodynamics has to be increased by the radiation loss:

$$P_{icorr} = P_{it} + \dot{Q} = \left(\frac{\dot{m} y_p}{\eta_p}\right)_t + \dot{Q}, \tag{6.168}$$

where the first term on the right-hand side constitutes the measured power according to the heat-balance method (PTC 10, para. 4.15). According to VDI 2045 (1993) the radiation loss by calculation is

$$\dot{Q} = kA(\bar{t}_{body} - t_{amb}), \tag{6.169}$$

where the mean body temperature and an approximate heat transfer coefficient are given by

$$\bar{t}_{body} \approx \frac{t_s + t_d}{2} \quad \text{and:} \quad k \approx 14 \text{ W m}^{-2} \text{ K}^{-1}, \tag{6.170}$$

where A is the body surface including nozzles (in m^2) and t_{amb} (in °C) is the ambient temperature.

Thus the measured efficiency can be reduced by the heat loss, in order to be used for the guarantee comparison:

$$\eta_{p\,corr} = \frac{\dot{m}_t y_{pt}}{\dot{Q} + \left(\dfrac{\dot{m} y_p}{\eta_p}\right)_t}. \tag{6.171}$$

References

API Standard 617 (1995) Centrifugal Compressors for Petroleum, Chemical, and Gas Service Industries. 6th edn, American Petroleum Institute, Washington, D.C.

Baskharone EA (1994) Perturbed flow structure in an annular seal due to synchronous whirl. J. Fluids Eng., vol 116, pp 564–569; The American Society of Mechanical Engineers, New York, N. Y.

Benckert H, Wachter J (1978) Studies on vibrations stimulated by lateral forces in sealing gaps, AGARD-CP-237, pp 9.1–9.11; AGARD Conf., Meeting on Seal Technology in Gas Turbine Engines, 1978, London

BORSIG Pocket Book (1994), Borsig GmbH Berlin, 7th edn

Busemann A (1928) Das Förderhöhenverhältnis radialer Kreiselpumpen mit logarithmisch-spiraligen Schaufeln. Zeitschrift für Angewandte Mathematik und Mechanik, vol 8, pp 372–384, (in German)

Daily JW, Nece RE (1960) Chamber dimension effects on induced flow and frictional resistance of enclosed rotating disks. Trans. ASME, J. Basic Eng., pp 217–232; Hydraulic Conf., 1959, Ann Arbor, Michigan; The American Society of Mechanical Engineers, New York

Dallenbach F (1961) The aerodynamic design and performance of centrifugal and mixed-flow compressors. SAE Technical Progress Series, vol 3, pp 2–30; The American Society of Mechanical Engineers, New York

Ehrich R (1990) Instabilities with turborotors, MAN Turbomaschinen AG., Berlin, in-house publication

Lempart A (1992) Das Axialschubverhalten von industriellen Turboverdichtern. VDI Berichte 947 pp 177–194; Conference Turbocompressors in Industrial Use, 1992, Hannover, Germany; Verein Deutscher Ingenieure, VDI-Verlag, Düsseldorf, Germany (in German)

Lindner P (1983) Aerodynamic tests on centrifugal compressors – influence of diffuser diameter ratio, axial stage pitch, and impeller cutback. Trans. ASME, J. Eng. for Power, vol 105 pp 910–919; The American Society of Mechanical Engineers, New York

Lüdtke K (1985) Centrifugal process compressors – radial vs. tangential suction nozzles. ASME Paper 85-GT-80; Gas Turbine Conf. 1985, Houston, Texas; The American Society of Mechanical Engineers, New York

Lüdtke K (1992) The influence of adjustable guide vanes on the performance of multistage industrial centrifugal compressors. ASME Paper 92-GT-17; Gas turbine and Aeroengine Congress, 1992, Cologne, Germany; The American Society of Mechanical Engineers, New York

Lüdtke K (1999) Process centrifugal compressors – latest improvements of efficiency and operating range. Inst. Mech. Eng. Conf. Transactions pp 219–234, Paper No. C556/014; 7th European Congress on Fluid Machinery for the Oil, Petrochemical, and Related Industries, 1999, The Hague, Netherlands

McDonald AT, Fox RW (1965) An experimental investigation of incompressible flow in conical diffusers. ASME Paper 65-FE-25; Applied Mechanics and Fluids Engineering Conf., Washington, DC, 1965; The American Society of Mechanical Engineers, New York

Mishina H, Nishida H (1983) Effect of relative velocity distribution on efficiency and exit flow of centrifugal impellers. ASME Paper 83-GT-74; Gas Turbine Conf., Phoenix, Arizona,1983; The American Society of Mechanical Engineers, New York

Miskovish RS, Brennen CE (1992) Some unsteady fluid forces on pump impellers. J. Fluids Eng., vol 114, pp 632–637;

Nelson LC, Obert EF (1954) Trans ASME 76, p 1057; The American Society of Mechanical Engineers, New York

Nielsen KK, Van den Braembussche RA (1997) Rotordynamic impact of swirl brakes. Project Report 1997-28, von Karman Institute, Rhode Saint Genese, Belgium

Performance Test Code on Compressors and Exhausters (1997), ASME PTC 10, The American Society of Mechanical Engineers, New York

Runstadler PW Jr, Dolan FX, Dean RC Jr (1975) Diffuser Data Book. Creare TN-186, Creare, Hanover, New Hampshire

Strub RA, Bonciani L, Borer CJ, Casey MV, Cole SL, Cook BB, Kotzur J, Simon H, Strite MA (1987) Influence of the Reynolds number on the performance of centrifugal compressors. Final Report Working Group Process Compressor Sub-Committee International Compressed Air and Allied Machinery Committee (ICAAMC), ASME Paper 87-GT-10 and Trans. ASME, J. Turbomachinery 109, pp 541–544; The American Society of Mechanical Engineers, New York.

Traupel W (1962) Die Theorie der Strömung durch Radialmaschinen. G. Braun, Karlsruhe, Germany (in German)

VDI 2045 (1993) Acceptance and Performance Tests on Turbo Compressors and Displacement Compressors, Part 1 and 2 ; Verein Deutscher Ingenieure, VDI-Verlag, Düsseldorf, Germany

VDI-Wärmeatlas (1994), 7th edn.; VDI-Verlag Düsseldorf, Germany (in German)

Wiesner F J (1966) A review of slip factors for centrifugal impellers, ASME Paper 66-WA/FE-18, 1966, Equations (3), (12); Winter Annual Meeting of Fluids Engineering Division, New York 1966; The American Society of Mechanical Engineers, New York

Wu C H (1952) A general theory of three-dimensional flow in subsonic and supersonic turbomachines of axial, radial, and mixed flow types. NACA TN 2604

Zimmermann H, Radtke F, Ziemann M (1983) Scheibenreibung – Experimentelle und theoretische Untersuchungen des Reibungseinflusses an rotierenden Scheiben, FVV-Forschungsberichte, Forschungsvereinigung Verbrennungskraftmaschinen, Frankfurt, Germany, Heft 331; restricted (in German)

7 Application Examples

Low capital and low operating costs and high operational flexibility – these are the usual expectations the compressor user has when sending out enquiries to process compressor manufacturers. Operational flexibility is the ability of the compressor to cope with changing external conditions such as volume flow, pressure ratio, suction temperature and gas composition, as is frequently the case in the oil and gas industry. Needless to say, availability is just as essential. However, since this, for the most part, involves the reliability of bearings, shaft seals, couplings and gear boxes, it is outside the (aerodynamic) scope of this book. The only parameter that may reduce availability is the (excessive) rotational speed that is specified by the aerodynamic design results.

The starting point is to design a compressor with the highest possible flow coefficient level, because this would warrant not only a small frame size (i.e. low capital costs) but also a high efficiency (i.e. low operating costs). The head, however, which translates into the number of stages and shaft length, adds a rotordynamic dimension, so that the compressor design becomes much more than just the flow coefficient. The minimum number of stages is a function of the material yield strength and the maximum possible Mach number. If the number of stages turns out to be rotordynamically incompatible with the flow coefficient, this parameter must be reduced, leading to larger impeller diameters and a lower efficiency. If the Mach number turns out to be incompatible with the required flexibility, M_{u2} must be decreased, resulting in a still larger number of stages which, in turn, may bring about another flow coefficient decrease. A wide curve from surge to choke with a comparatively high head rise to surge and a moderate downturn to choke is synonymous with "flexibility".

The maximum rotational speed may constitute a powerful constraint in many cases:

1. Approximately 18,000 1/min is an established limit for single-shaft process compressors, which may be reduced for conservatively oriented users.
2. A usually lower speed may be dictated by a selected gas turbine drive.
3. In case of a two-casing train the speed is usually determined by the low-pressure rotor, leading to comparatively low speeds for the high-pressure rotor; this results in very low flow coefficients and low efficiencies in the HP casing.
4. In the case of high-density gases the shaft stiffness ratio has to be increased accordingly (to avoid sub-synchronous vibrations), which can be brought about by lowering the maximum continuous speed.

Temperature limits are important for compressors handling hydrocarbon gases, which tend to polymerisation, and for chlorine compressors, where temperatures above 150°C would, in the case of a rub, lead to a chlorine fire. This necessarily leads to three intercoolers per casing in many applications.

So it is essential that the compressor is within the framework of all the characteristics envelopes with strict observance of the constraints imposed by the particular application.

It is more than obvious that the design of a process centrifugal compressor is a multi-dimensional engineering task, since the compressor is subject to a multitude of aerodynamic, thermodynamic, rotordynamic and mechanical parameters that are mutually interconnected and whose constraints determine the machine design to a large extent. This makes the final solution a compromise between the requirements of the capital and operating costs and the operating flexibility.

The application examples bear witness to this multi-dimensional field of interdependent characteristics, resulting in very different manifestations of the compressor aero-package. So there are really striking differences between, say, a platformer and a chlorine compressor or between a gas lift and a propylene refrigeration compressor, as will be shown in the following figures, which show the peculiarities in each case (all photos by courtesy of MAN Turbomaschinen).

The following tour d'horizon covers six distinct groups of process compressors (made by MAN Turbomaschinen, Berlin; formerly Deutsche Babcock-Borsig) for hydrogen-rich gases, natural gas, air, refrigerants, chlorine and methanol synthesis gas. Although there are many more, it is thought that the compressor species described below are good examples to the diverse configurations of these machines.

7.1 Hydrogen-Rich Gas Compressors

This group of compressors (design data in Table 7.1) is characterized by very low molar masses translating into: very low tip-speed Mach numbers, nearly incompressible gas, very low stage pressure ratios, small temperature rises, all impellers nearly identical and very wide performance curves from surge to choke in spite of relatively high impeller tip speeds and heads. Usually all of them have a vertically (radially) split casing, even if the partial pressure of hydrogen is lower than 15 bar, which is the maximum for horizontally split casings according to API Standard 617.

A typical example is shown in Figs. 7.1–7.3: the four-stage compressor could be equipped with high flow coefficient impellers, since there were neither rotordynamic constraints nor limitations of tip and rotational speed (high volume flow!).

7.1 Hydrogen-Rich Gas Compressors

Table 7.1 Examples: hydrogen-rich gas, design data

Figure	Application	Size/i	p_d/p_s	$\dot{V}$ [m³/s]	y_p [kJ/kg]	u_2 [m/s]	H_2 [%]	M	φ	M_{u2}
7.1–7.3	Refinery, platforming	71/4	9.6/6.5 =1.48	8.5	160	290	88	7	0.09	0.4
7.4–7.5	Petrochemical	50/2	64/61 =1.048	5.4	60	270	93	2.2	0.12	0.2
7.6–7.7	Refinery, hydrotreating	42/6	68/51 =1.33	0.6	170	240	85	5	0.02	0.3
7.8	Refinery, hydrocracking	35/4	165/144 =1.14	0.7	80	200	85	5	0.04	0.2

All casings vertically split; drives: Figs. 7.1–7.5: electric motor; Figs. 7.6–7.8: steam turbine.
size frame size (max impeller OD in cm), *i* number of impellers, *c* number of casings, *s* number of sections, p_d/p_s pressure ratio, $\dot{V}$ suction volume flow, M molar mass, u_2 tip-speed of 1st impeller, y_p total compressor polytropic head, φ flow coefficient range of compressor, M_{u2} tip speed Mach number of 1st impeller, t_s suction temperature; $M, u_2, y_p, \varphi, M_{u2}$ are strongly rounded off.

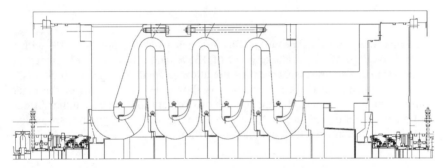

Fig. 7.1 Four-stage hydrogen-rich gas compressor (platforming process) with high flow coefficient impellers; 2,700 kW; 9,150 rpm.

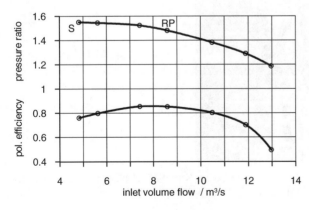

Fig. 7.2 Four-stage hydrogen rich gas compressor (Fig. 7.1); performance curve, RP rated, S surge.

Fig. 7.3 Four-stage hydrogen-rich gas recycle compressor (Fig. 7.1); barrel casing with shear-ring end covers.

The measured performance curve (Fig. 7.2) has a stability point of 57% with a head rise to surge of 9%, a choke point of 155%, and a best-point efficiency of 85% polytropic. The low discharge pressure allows a small casing wall thickness.

Another example of the same category is shown in Fig. 7.4 (part of the inner casing), having two identical impellers; the extremely low Mach number brings about a stage volume reduction of only 1% (see Fig. 4.6). The compressor is displayed in Fig. 7.5, again with a common base frame.

Fig. 7.4 Two-stage hydrogen-rich gas recycle compressor (93% H_2); horizontally split inner casing showing one end cover; identical impellers; 1,900 kW; 11,250 rpm.

Fig. 7.5 Two-stage hydrogen-rich gas recycle compressor (Fig. 7.4); compressor train.

Figure 7.6 shows a compressor with a considerably lower volume flow. Due to the required six stages, and in order to prevent an excessive rotational speed, a low flow coefficient design had to be selected (Fig. 7.7).

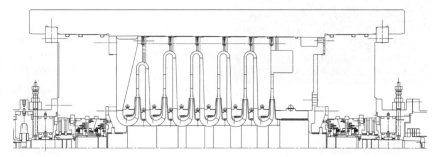

Fig. 7.6 Six-stage hydrogen-rich gas compressor (diesel hydro-treating process).

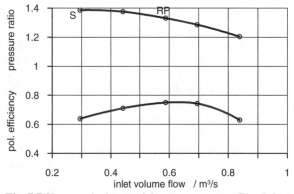

Fig. 7.7 Six-stage hydrogen-rich gas compressor (Fig. 7.6); performance curve, RP rated, S surge; 1,300 kW; 12,700 rpm.

Figure 7.8 shows a typical refinery compressor for hydrocracking service with bolted-on end covers. The metal-to-metal contact between cover and outer casing is sometimes preferred over the shear-ring design, which enables easier access to the rotor but requires O-rings between the cover and the casing wall.

Fig. 7.8 Four-stage hydrogen-rich gas recycle compressor (hydrocracking process), barrel casing with bolted-on end covers; 2,150 kW; 11,200 rpm.

7.2 Natural-Gas Compressors

The large group of natural gas compressors (some design data is given in Table 7.2) have discharge pressures of between 5 and 500 bar and higher, leading to an array of configurations from a single stage to 20 impellers and more, depending on a multitude of applications, especially in the oil and gas industry.

The combustion chamber feed compressor in Fig. 7.9 as part of a gas turbine power plant has nine stages and is thus at the upper limit of rotordynamic and rotational speed feasibility. The flow coefficient level (and thus efficiency level) is about the maximum that could be achieved for this head with a single casing design.

The gas gathering compressor with a fabricated casing shown in Fig. 7.10 has still smaller flow coefficients (at the very low end of the range) due to the extremely low inlet volume flow and the comparatively high head with stage efficiencies in the upper sixties. In order to keep the impeller exit widths at the compressor rear end from falling below 3 mm the diameters had to be continuously stepped down, which, in turn, increased the number of impellers. However, through this concept a solution was made possible without violating rotordynamic constraints.

Table 7.2 Examples: natural gas, design data

Fig.	Application	Size/i	c/s	p_d/p_s	$\dot{V}$ [m³/s]	y_p [kJ/kg]	u_2 [m/s]	nat. gas [%]	M	φ	M_{u2}
7.9	Gas turbine fuel supply	35/9	1/1	21.1/4 = 5.3	1.0	320	280	98	16	0.04 to 0.014	0.6
7.10 7.11	Gas gathering	35/8	1/1	32.5/6.6 = 4.9	0.46	170	240	49	29	0.02 to 0.012	0.7
7.12 to 7.16	Gas lift	60/8 60/9	2/4	77/2 = 38.5	6.7	480	270	70	24	0.1 to 0.011	0.8
7.17 to 7.19	Gas lift	42/9 42/8	2/3	177/9.1 = 19.5	1.35	440	300	80	19	0.04 to 0.011	0.7

Drives: Figs. 7.9–7.16: electric motor; 7.17–7.19: gas turbine "Solar Mars".

Fig. 7.9 Nine-stage natural gas compressor (fuelling combustion chamber of gas turbine); horizontally split cast casing; 1,300 kW; 15,800 rpm.

When comparing the test performance curves in Figs. 7.2, 7.7 and 7.11, it becomes clear how much the efficiency depends on the flow coefficient:

$\varphi = 0.090 \rightarrow \eta_p = 85\%$; $\varphi = 0.023 \rightarrow \eta_p = 75\%$; $\varphi = 0.016 \rightarrow \eta_p = 68\%$,
although this comparison is blurred somewhat by the different frame sizes.

Fig. 7.10 Eight-stage natural gas compressor for gas gathering service; low flow coefficient impellers, horizontally split fabricated casing; 960 kW; 13,600 rpm.

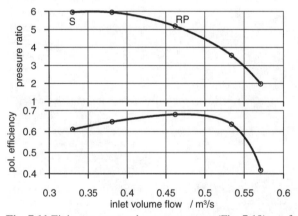

Fig. 7.11 Eight-stage natural gas compressor (Fig. 7.10); performance curve, RP rated, S surge.

On the other hand, the impact of Mach number and number of stages is evident, as far as surge and choke are concerned:

- $M_{u2} = 0.28/6$ stages: 52% surge and 165% choke,
- $M_{u2} = 0.40/4$ stages: 56% surge and 160% choke,
- $M_{u2} = 0.75/8$ stages: 72% surge and 126% choke.

A gas lifting compressor is presented in Figs. 7.12–7.16. It generates a 39:1 pressure ratio with 17 stages in four sections (i.e. three intercoolers) arranged in two casings at the same rotational speed. The horizontally split LP casing is fabricated; the HP casing is a forged barrel design with shear-ring-secured end covers. The spread of flow coefficients is tenfold. Again, due to the preset rotational speed (by the LP rotor), the HP rotor has to have very low flow-coefficient impellers. The section efficiencies are 83, 78, 72 and 67% polytropic from first to last. The horizontally split inner casing of the HP compressor with the rotor in place is shown in Fig. 7.13; note the considerably stepped-down impeller diameters. The relatively large diffuser ratios prevent the efficiency from dropping below 65%.

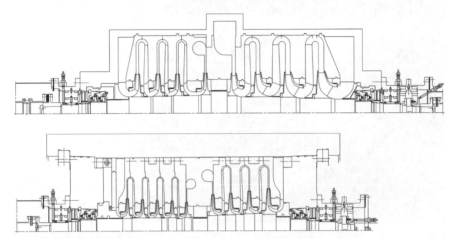

Fig. 7.12 Two-casing gas lift compressor train: 17 stages in four sections (3 intercoolers), fabricated horizontally split LP casing (*top*), forged vertically split HP casing (*bottom*); 7,900 kW; 9,560 rpm; motor drive with variable speed hydraulic coupling.

Fig. 7.13 Gas lift compressor: horizontally split inner casing of HP body (Fig. 7.12).

The complete compressor on the shop floor prior to shipment is shown in Fig. 7.14. The variable speed hydraulic coupling (between motor and LP casing), which comprises a two-stage step-up gear including a hydraulic compressor/turbine element, was selected to provide a no-load start-up of the train and can also be used for variable speed regulation during operation, although the compressor is bypass-regulated most of the time.

Fig. 7.14 Two-casing gas lift compressor train (Fig. 7.12). From *left*: electric motor, hydraulic coupling, LP compressor, HP compressor. *Above* LP: identical barrels and inner casing halves.

Fig. 7.15 Gas lift train (Fig. 7.12) at site in a Near East oil field.

The compressor at site is displayed in Fig. 7.15. The variable speed performance map is shown in Fig. 7.16. The four sections were shop performance-tested separately (PTC 10 type 2 test at reduced pressure and with substitute gas) and the total compressor map was arrived at by aerodynamic superposition. The turndown point is at 68% at a speed of 94%; the choke at 100% speed is a mere 104%. However, a 102% speed capability extends the maximum flow to 107% and the maximum choke to 109%.

The map also contains the 100% speed curve for a molar mass reduced by 5%. This demonstrates how sensitively this compressor responses to changes of the gas composition, which has to be accounted for when specifying a compressor for changing gas constituents.

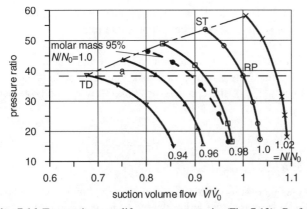

Fig. 7.16 Two-casing gas lift compressor train (Fig. 7.12). Performance map; RP rated point (6.7 m³/s), TD turndown point, ST stability point, a operating line, N/N_0 speed ratio, suction pressure $p_s = 2$ bar.

When comparing performance curves in Figs. 7.11 (8 stages) and 7.16 (17 stages) with regard to surge and choke the enormous influence of the number of stages becomes obvious:

- $M_{u2} = 0.75/08$ stages: 72% surge (stability) and 126% choke,
- $M_{u2} = 0.76/17$ stages: 92% surge (stability) and 104% choke.

The encroachment of the compressor (unregulated = stability) surge point $\dot{V}_{1-surge}$ on the rated point can be described best by the following basic equations. Since the compressor surge is determined by the last-stage surge volume flow $\dot{V}_{last-surge}$, the compressor surge volume flow is

$$\dot{V}_{1-surge} = \frac{\dot{V}_{last-surge}}{\left[\left(\dot{V}_{7t}/\dot{V}_{0t}\right)_{av-surge}\right]^{i-1}}, \tag{7.1}$$

where the denominator is the accumulated average stage volume ratio of all stages at surge except the last (see Eq. (4.9) and Fig. 4.6). So the compressor surge flow ratio (surge over rated) becomes

$$\frac{\dot{V}_{1-surge}}{\dot{V}_1} = \frac{\dot{V}_{last-surge}}{\dot{V}_{last}} \frac{\left[\left(\dot{V}_{7t}/\dot{V}_{0t}\right)_{av}\right]^{i-1}}{\left[\left(\dot{V}_{7t}/\dot{V}_{0t}\right)_{av-surge}\right]^{i-1}}. \qquad (7.2)$$

Regardless of the number of stages i, the last stage surge ratio $\dot{V}_{last-surge}/\dot{V}_{last}$ remains constant, with $\dot{V}_{last}$ being the last stage design point. If the number of stages increases, the accumulated stage volume ratio at surge (denominator of Eq. (7.2)) decreases to a larger degree than the accumulated stage volume ratio at rated (numerator of Eq. (7.2)); so the compressor surge encroaches on the rated point as the number of stages increases. This is all the more pronounced with a higher Mach number level, because this is the driving force behind these interstage compressibility changes.

The encroachment of the compressor choke point $\dot{V}_{1-choke}$ on the rated point can be described analogously, since the compressor choke is determined by the last-stage choke volume flow $\dot{V}_{last-choke}$, the compressor choke flow ratio (choke over rated) is

$$\frac{\dot{V}_{1-choke}}{\dot{V}_1} = \frac{\dot{V}_{last-choke}}{\dot{V}_{last}} \frac{\left[\left(\dot{V}_{7t}/\dot{V}_{0t}\right)_{av}\right]^{i-1}}{\left[\left(\dot{V}_{7t}/\dot{V}_{0t}\right)_{av-choke}\right]^{i-1}}. \qquad (7.3)$$

If the number of stages increases, the accumulated stage volume ratio at choke (denominator of Eq. (7.3)) increases to a larger degree than the accumulated stage volume ratio at rated (numerator of Eq. (7.3)); so the compressor choke encroaches on the rated point as the number of stages increases.

As was described in Chap. 5, aerodynamic stage mismatching, i.e. a disproportionate compressibility change between stages, is responsible for this typical behavior of multistage machines with tip-speed Mach numbers higher than approximately 0.5. Hydrogen-rich gas compressors, even with 6 stages and more, do not exhibit the dependency of surge and choke encroaching on the rated point, since there is (nearly) no interstage compressibility.

A gas turbine driven 17-stage off-shore gas lift compressor (installed on a "floating production storage and off loading vessel") is shown in Figs. 7.17–7.19.

Both casings are barrel designs. Consequently, when access to the LP rotor is required, the LP casing has to be decoupled and lifted off the base frame, since the inner casing has to be pulled axially out of the outer casing. Figure 7.17 shows the inner casings; twelve of the 17 impellers have flow coefficients of around 0.012, being at the very bottom of the line. The section efficiencies are 78%, 69% and 67% polytropic, reflecting the smaller volume flow in comparison with the compressor in Fig. 7.12. After four years of operation the train was uprated to 135% mass flow through larger impeller diameters and widths as well as through speed increase; the uprated LP body is shown in Fig. 7.19.

Fig. 7.17 Off-shore gas-lift compressor; two barrel casings, LP (*top*, two sections, back-to-back) and HP (*bottom*, one section); horizontally split inner casings with rotors in place. Sections 2 and 3 have very low flow coefficient impellers; 6,300 kW; 14,300 rpm.

Fig. 7.18 Off-shore gas lift compressor; two barrel casings, driven by 10 MW gas turbine via speed increasing gear (Fig. 7.17).

Fig. 7.19 Off-shore gas lift compressor; LP casing, uprated to 135% mass flow.

7.3 Air Compressors

Quite a number of specifications still call for single-shaft air compressors, although they are being gradually replaced by integrally geared compressors, which are more efficient and lower in capital cost, albeit rated lower in their availability. All of these single-shaft casings are horizontally split up to approximately 70 bar, which is about the maximum pressure of air compressors (e.g. for acetic acid plants). Example design data are given in Table 7.3.

Table 7.3 Examples: design data for air

Fig.	Application	Size/i	c/s	p_d/p_s	$\dot{V}$ [m³/s]	u_2 [m/s]	y_p [kJ/kg]	φ	M_{u2}
7.20	Refinery, FCC	85/5	1/1	3.85/0.98 = 3.9	21	270	150	0.12 to 0.08	0.8
7.21	Chemical, NH_3 – syngas	100/6 42/6	2/4	34.4/0.92 = 37.4	22	260	390	0.11 to 0.03	0.7
7.22	Chemical	35/4	1/1	17.6/4.8 = 3.7	1.1	270	150	0.07 to 0.03	0.8

All casings horizontally split; drives: Fig. 7.20: electric motor; Fig. 7.21: GE gas turbine "Frame 5"; Fig. 7.22: electric motor + exhaust gas expansion turbine. Molar mass 28.3–29.0 kg/kmol (humid air).

The air compressor for a Fluid Cat Cracking plant in Fig. 7.20 is equipped with optimum flow coefficient impellers, each with moderate pressure ratio, resulting in a high polytropic efficiency level of 84% from first to last stage.

Fig. 7.20 Large five-stage air compressor in FCC plant; fabricated casing, medium (= optimum) flow coefficient impellers; no intercooler; 4,400 kW; 6,000 rpm.

Fig. 7.21 Two-casing air compressor in an ammonia synthesis plant, gas turbine-driven; fabricated casings; *top* and *bottom right*: LP casing; *bottom left*: HP casing; 15,600 kW; 4,800 rpm (LP); 12,500 rpm (HP).

The large gas turbine-driven air compressor in Fig. 7.21 is the heart of a 1,350 t/d (tons per day) ammonia synthesis plant with twelve stages in two casings with an intermediate gear to achieve a satisfactory efficiency in the HP compressor as well (smaller impeller diameters → higher flow coefficients → increased efficiency and lower capital cost).

The four-stage air compressor shown in Fig. 7.22 is remarkable insofar as it has adjustable inlet guide vanes before the first stage, improving turndown and part-load power consumption, as was described in Chap. 5. And furthermore it has an integral overhung exhaust gas expansion turbine, utilizing the residual exergy (available energy) from the chemical process to reduce the actual motor power output. The turbine operates with nitrogen at 8 bar/200°C inlet conditions, generating 600 kW, i.e. 45% of the compressor power input.

Fig. 7.22 Four-stage air compressor in a chemical plant with adjustable inlet guide vanes and integral single-wheel exhaust gas turbine (600 kW); compressor power 1,300 kW; 17,200 rpm.

7.4 Refrigeration Compressors

The purpose of refrigeration compressors (some design data is given in Table 7.4) is to reduce the temperature of fluid flows below ambient rather than to achieve certain discharge pressures. The cooling takes place in the (suction side) evaporator and the condensing of the refrigerant takes place in the (discharge side) condenser, which is water- or air-cooled; the compressor discharge pressure is the saturation pressure related to the water or air temperature. The nomenclature in refrigeration engineering is "temperatures" instead of "pressures": "the compressor suction is –40°C, and the compressor discharge is +42°C". The mechanical engineer has to convert these into (saturated vapor) pressures, depending on the selected refrigerant. There can be up to two (in rare cases, three) evaporators at different temperature levels, corresponding to a suction and a side stream pressure, resulting in an additional side mass flow (a second "suction") of some 50% to 300% of suction mass flow. Thus the compressor can have, for instance, three suction nozzles and just one discharge nozzle. The pressure reduction of the liquid refrigerant from the condenser exit to the evaporator inlet is usually carried out in two steps: the first valve discharges into an economizer, releasing the vapor to another compressor sidestream inlet. This produces a mass flow increase in the following stage of approximately 5–15%. However, the mass flow in the first compressor section is thus reduced, decreasing the power consumption at the specified refrigerating capacity.

An example is shown in the pressure–enthalpy diagram in Fig. 7.23: two evaporators (e.g. at –40°C and –20°C) and one economizer require three compressor sections with different mass flows.

Table 7.4 Examples: refrigeration, design data

Fig.	Application	Size/i	c/s	p_d/p_s	$\dot{V}$ [m³/s]	u_2 [m/s]	y_p [kJ/kg]	φ	M_{u2}	Gas	t_s [°C]
7.24	Chemical	50/8	1/2	17/2 =8.5	2.9	322	337	0.05 to 0.015	0.8	NH_3	−5
7.25	Chemical	35/8	1/2	12/0.4 =30.0	1.6	280	240	0.07 to 0.011	1.1	C_2H_4	−101
7.26 7.27	Chemical	71/7	1/2	17.2/1.4 =12.3	4.1	200	130	0.1 to 0.014	0.9	C_3H_6	−40
7.28	Gas field	42/4	1/2	23.5/3.9 =6.0	2.4	210	90	0.09 to 0.04	1.0	C_3H_8	+2

All casings horizontally split; drives: Figs. 7.24 and 7.25: steam turbine; Figs. 7.26–7.28: electric motor.

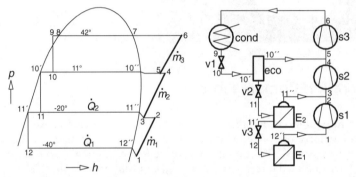

Fig. 7.23 Refrigeration process in pressure–enthalpy diagram: example. $\dot{Q}_{1,2}$ specified refrigeration capacities at evaporation temperatures −40°C and −20°C; $\dot{m}_{1,2,3}$ mass flows in compressor sections s1, s2, s3; E_1 evaporator, E_2 evaporator and economizer, eco economizer (at 11 °C), cond condenser (condensing temp. 42 °C); v1, v2, v3 expansion valves; h_6–h_7 desuperheating (separate or in one heat exchanger); h_7–h_8 condensing; h_8–h_9 subcooling; 12″–1 pressure loss and temperature heat-up between evaporator and compressor.

These are calculated as follows:

$$\dot{m}_1 = \frac{\dot{Q}_1}{h_{12''} - h_{12}}, \tag{7.4}$$

$$\dot{m}_2 = \dot{m}_1 + \dot{m}_1 \frac{h_{11} - h_{11'}}{h_{11''} - h_{11}} + \frac{\dot{Q}_2}{h_{11''} - h_{11}}, \tag{7.5}$$

$$\dot{m}_3 = \dot{m}_2 \frac{h_{10''} - h_{10'}}{h_{10''} - h_{10}}. \tag{7.6}$$

7.4 Refrigeration Compressors

In this example evaporator 2 serves as a second economizer, generating vapor described by the second term of Eq. (7.5).

The side streams can be inserted directly into the compressor or the mixing can be carried out in an external pipe, especially with side streams larger than approximately 300%.

The most common refrigerants continuously circulating in these closed systems are ammonia, propane, propylene and ethylene. R11 and R12 have been banned from the market due to their ozone-depletion potential (ODP) and global warming potential (GWP). R134a was introduced as a replacement.

An eight-stage ammonia compressor is shown in Fig. 7.24 with a suction temperature of –5 °C and one water-intercooler enabling two back-to-back sections. A single-casing solution was made possible by a high impeller tip speed of 322 m/s (the turbine drive with N_{max} = 105% increases this to 340 m/s); the required eight stages allow a first-stage maximum flow coefficient of approximately 0.05 (see Fig. 4.4 at M_{u2} = 0.8). The total compressor polytropic head is 337 kJ/kg at 100% speed, which is about the highest in one single casing. The first stage head is y_p = 56 kJ/kg at 100% speed, which is the approximate maximum for a single process compressor stage with a shrouded impeller.

Fig. 7.24 Eight-stage steam-turbine-driven ammonia refrigeration compressor in a chemical plant; two back-to-back sections, with one water cooler; 2,200 kW; 13,400 rpm.

The ethylene compressor displayed in Fig. 7.25 operates between –101°C saturation (suction) and –50°C saturation (discharge) corresponding to a pressure ratio of 30 : 1 in one casing achieved by eight impellers. These had to be stepped down substantially towards the rear end to arrive at aerodynamically feasible impeller widths. The ethylene compressor serves as the low-stage of a cascade system whose condenser (at –50°C) is the evaporator of the high-stage, which is usually a propylene compressor discharging to a normal water-cooled condenser at about +45 °C saturation (about 17 bar).

Fig. 7.25 Eight-stage steam-turbine-driven ethylene refrigeration compressor in a chemical plant; suction temperature –101°C, pressure ratio 30 : 1; 600 kW; 15,700 rpm; 23% side stream after 1st section.

Figures 7.26 and 7.27 show a seven-stage propylene compressor in a hydrogen chloride synthesis plant. The limit of a maximum of seven stages (by rotordynamics) required the stepping-up of the impeller diameters in the second section (required to meet the specified side-stream pressure). This, in turn, reduced the flow coefficients of the last stages substantially, resulting in a comparatively low second-section efficiency of 74%, compared to 82% in the first section.

This is an application example showing a satisfactory compromise between capital and operating costs and conforming to the aerodynamic and rotordynamic constraints: first-stage flow coefficient as high as possible (efficiency!), Mach number level as high as possible (to limit the number of stages and assure an acceptable performance curve shape!), last-stage flow coefficient above minimum. The high hub/tip ratio and the small axial stage pitch in the second section comply with rotor stiffness requirements. A higher flow coefficient level (to improve efficiency) would violate the law of shaft stiffness, making the rotor prone to subsynchronous vibrations.

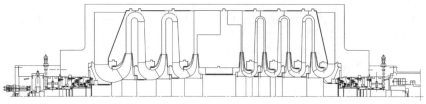

Fig. 7.26 Seven-stage propylene compressor in a chemical plant; suction temperature –40°C, pressure ratio 12 : 1; 2,400 kW; 7,100 rpm; 7% (economizer) side stream.

Fig. 7.27 Seven-stage propylene compressor (Fig. 7.26); fabricated horizontally split casing.

The cross-sectional drawing of a propane compressor operating in an oil field (Fig. 7.28) has an internal side stream of 50% (two suction nozzles, one discharge nozzle). Inserting the side mass flow at the last third of the return vane channel, where the blades are radial already, warrants an even distribution in the inlet annulus and a smooth admixture.

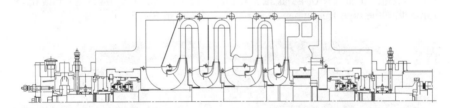

Fig. 7.28 Four-stage propane compressor for gas liquefaction in an oil field; 50% side stream before third stage with internal insertion; 2,850 kW; 10,500 rpm.

7.5 Chlorine Compressors

Until about 1994 chlorine compressors (some design data is given in Table 7.5) used to have discharge pressures of approximately 5 bar, resulting in three-stage machines. Due to a major change of the chlorine production process, the pressures were increased to between 13 and 17 bar, often combined with a suction-temperature reduction from ambient to $-30°C$ (i.e. setting up a classical refrigeration cycle with evaporator(s), economizer and condenser) bringing the number of compressor stages up to seven and eight. Together with the conventional temperature constraint of conservatively 110 °C at the rated point, this leads to a four-section machine with eight nozzles on one horizontally split casing. In order to be able to easily dismount the upper casing half, all nozzles have to be located in the lower half, which is a real challenge for the pattern-maker and the foundry.

Table 7.5 Examples: chlorine, design data

Figure	Application	size/i	c/s	p_d/p_s	$\dot{V}$	u_2	y_p	φ	M_{u2}	t_s
					[m³/s]	[m/s]	[kJ/kg]			[°C]
7.29–7.31	Chemical	50/8	1/4	12.5/1 = 12.5	3.4	170	100	0.11 to 0.02	0.8	25
7.32	Chemical	50/7	1/4	13.2/1 = 13.2	1.8	160	90	0.11 to 0.02	0.8	−32

Casings horizontally split; drives: electric motor; Fig. 7.32: 400 % side stream.

In Fig. 7.29 the lower casing half is shown with the eight-stage rotor in place. Whereas in many applications the back-to-back design is an option, it is (almost) mandatory with chlorine compressors in order to minimize the recirculation losses and the risk of rubbing in the balance piston labyrinth. An in-line impeller arrangement with a huge balance piston at the end would have a three- to fourfold balance-piston loss, because, in addition to what was said in Chap. 4, the piston labyrinth tip clearance would have to double to exclude any rub, which would instantly cause chlorine fire. So the multiple back-to-back arrangement not only minimizes the recirculation loss but also reduces the danger of rubbing of metal parts also, since the centrifugal strain of the interstage shaft labyrinth diameters is much smaller than that of a balance-piston diameter. The train on a common base frame with the pipe-free upper casing half is shown in Fig. 7.30.

Fig. 7.29 Eight-stage chlorine compressor in a chemical plant; four multiple back-to-back sections, eight nozzles in lower casing half, cast casing; stage numbers from *right* to *left*: 1,2,5,6,8,7,4,3; 1,300 kW; 7,100 rpm.

Fig. 7.30 Eight-stage chlorine compressor (Fig. 7.29).

Figure 7.31 demonstrates the maze of connecting pipework between the compressor and three vertically arranged coolers, including the bypass lines around the sections to prevent surging. This is a real challenge for systems engineering as well. Figure 7.32 shows the casing nozzles of a (different) seven-stage compressor.

The seven- to eight-stage chlorine compressor rotor is comparatively long and thin, although fully complying with the shaft stiffness requirements. This is because the impeller tip speed is low (170 m/s for the 1st stage, as shown in Fig. 7.29) due to Mach number constraints for this high-molar mass gas. So, since the rotational speed N_{max} is some 35% lower than for natural gas applications, the rigid-support critical speed $N_{c1\text{-}rig}$ can be relaxed by the same percentage, i.e. the bearing span may be longer (more impellers!) and/or the shaft diameter may be smaller (higher flow coefficient impellers!) without compromising the shaft stiffness ratio $RSR = N_{c1\text{-}rig}/N_{max}$.

Three elements are required to start a chlorine fire: "oxidant" (chlorine), fuel (iron), and heat. The heat can be generated by rubbing of the labyrinth tips on the mating part if the clearance is improperly sized or by dislodging of the seal strips or by excessive rotor imbalance or by ingested foreign particles (e.g. FeO or $FeCl_3$ depositing in the labyrinth passages).

A rise in temperature to approximately 150°C will initiate the run-away reaction. The gaseous oxidant Cl_2 (in the absence of oxygen) reacts with iron to form (solid) iron dichloride, which is highly exothermal. This is converted into solid and then into gaseous iron trichloride:

$Fe + Cl_2 \rightarrow FeCl_2$ (solid, exothermal),

$2FeCl_2$ (solid) $+ Cl_2$ (gas) $\rightarrow FeCl_3$ (solid, endothermal), (7.7)

$FeCl_3$ (solid) $+$ heat $\rightarrow FeCl_3$ (gas).

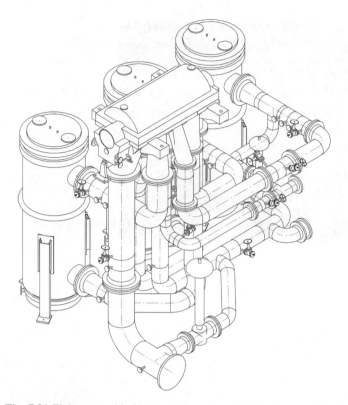

Fig. 7.31 Eight-stage chlorine compressor (Fig. 7.29); pipework between compressor and intercoolers, table foundation not shown.

Fig. 7.32 Seven-stage chlorine compressor; casing lower half with eight nozzles; 2,300 kW; 6,700 rpm; S suction, D discharge.

The fire will burn down the complete compressor unless immediate action is taken to extinguish it by removing the chlorine through purging with nitrogen. It is essential that during commissioning of the compressor all solid particles are removed from the flow paths by purging with N_2 for a sufficiently long time. It has to be kept in mind, though, that particles with a diameter smaller than the boundary-layer thickness cannot be removed by N_2 purging. When changing over from N_2 to Cl_2 operation, the boundary-layer thickness decreases due to the higher density. Consequently, many small particles on the surface are broken loose, because their diameter has now become larger than the boundary layer thickness and because the gas momentum ("brake-away force") is greater by a factor of 2.5 due to the higher molar mass. So, with the beginning of Cl_2 operation, it is probable that there are free (FeO) particles in the gas that can accumulate on the labyrinth strip surfaces eventually leading to blockage of the clearances and heat-generating rubbing.

7.6 Methanol Synthesis Gas Compressors

Methanol is used as fuel, solvent and feedstock in the manufacture of formaldehyde, fibers and paints. It is synthesized at 250 °C between 50 and 90 bar, chiefly from carbon monoxide and hydrogen: $CO + 2H_2 \rightarrow CH_3OH$ (exothermal). The synthesis gas, consisting of approximately 20% CO, 70% H_2 and 10% CO_2, produced from natural gas, petroleum residue or coal gasification, is compressed by the multistage syngas compressor (suction pressure: 7 to 28 bar) to a discharge pressure of 50 to 90 bar. It is then mixed with the recycle gas, which is circulated through the reactor, a heat exchanger, cooler and separator by a single-stage compressor (having a pressure ratio of 1.07 to 1.1). This compressor is either an integral part of the main compressor or a separate unit. Depending on the pressure ratio, the methanol syngas train may consist of three (Fig. 7.33), or two casings (Fig. 7.34), or one casing (Figs. 7.35–7.37). The exothermal synthesis in the reactor is used to generate steam for the compressor drive turbine. Design data is given in Table 7.6.

The relatively high pressure ratio of 6.6 : 1 and the low suction volume flow of 0.8 m³/s of the methanol syngas train in Fig. 7.33 resulted in three casings and a low flow coefficient level (i.e. a comparatively low efficiency) throughout the 15 stages. For maintenance purposes the low and the medium pressure casing were designed as horizontally split casings, although the MP casing has a hydrogen partial pressure of 21 bar, which is 6 bar higher than allowed in API Standard 617 for horizontally split casings.

A specified pressure ratio of 3 : 1 allowed us to reduce the number of casings to two: Fig. 7.34 shows the train during assembly for the closed-loop shop performance test. The recycle stage is integral with the HP barrel casing in the foreground. A horizontally split LP casing was selected to ease maintenance.

Table 7.6 Examples: design data for methanol synthesis gas

Fig.	Application	$size/i$	p_d/p_s	$\dot{V}$ [m³/s]	y_p [kJ/kg]	u_2 [m/s]	gas [%]	M	φ	M_{u2}
7.33	Petro-chemical	3×35/15	49/7.4 = 6.6	0.8	480	280	H_2 70 CO 21 CO_2 12	11.6	0.03 to 0.01	0.5
7.34	Petro-chemical	2×35/9	50/16.8 = 3	1.0	334	290	H_2 71 CO 14 CO_2 8	10.2	0.04 to 0.02	0.5
7.35 to 7.37	Petro-chemical	1×71								
	Syn →	$i = 7$	82.3/27 = 3	3.2	320	310	H_2 69 CO 21	11.3	0.03 to 0.02	0.5
	Rec →	$i = 1$	83/76.3 = 1.09	4.4		240	H_2 67 CH_4 23	8.9	0.06	0.4

Syn seven-stage synthesis gas section, *Rec* single-stage recycle stage. Drives: steam turbine.

Fig. 7.33 Three-casing methanol syngas compressor; two horizontally split casings, one barrel casing; pressure ratio 6.6 : 1; two intercoolers; drive: back-pressure extraction turbine; 2,000 kW; 15,200 rpm.

Fig. 7.34 Two-casing methanol syngas compressor; LP horizontally split casing, HP barrel casing; pressure ratio 3 : 1; nine syngas stages, no intercooler; drive: back-pressure extraction turbine; 2,500 kW; 15,600 rpm.

With the advent of mechanical-drive condensing-steam turbines in the suitable power/speed class, the compressor stage head could be increased by 25% due to an impeller tip speed increase from 280 to over 300 m/s, and, additionally, due to an efficiency increase of some 4% for higher volume flows. Both effects enabled the designer to reduce the number of syngas stages from nine to seven for a pressure ratio of 3 : 1. This makes a single-casing solution feasible, including the integrated recycle stage, as shown in Figs. 7.35–7.37.

Whereas the front wall of the plenum inlet to the syngas section is formed by the end cover (on the left in Fig. 7.35), the back wall of the recycle wheel, including the diffuser and scroll, is formed by the end cover (on the right). The casing is locked by massive end covers held in place by shear rings. The impellers as well as the diaphragms (return vane channels and volutes) are milled from solid forgings.

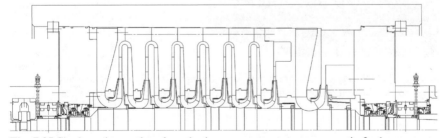

Fig. 7.35 Single-casing methanol synthesis gas compressor; pressure ratio 3 : 1; seven syngas stages, one recycle stage; barrel casing; 19,800 kW; 8,600 rpm.

The pressure ratio depends on the specialities of the different methanol processes developed by the licensors. Once the pressure ratio is markedly above 3 : 1, the compressor becomes a two-casing design with the option of a two-shaft-end-drive steam turbine, so that both compressors can be barrel types fully complying with the requirements of the API Standard 617.

Fig. 7.36 Single-casing methanol syngas compressor; (Fig. 7.35), inner casing with rotor in place, seven syngas and one recycle stage (*left*).

Fig. 7.37 20 MW single-casing methanol syngas compressor (Fig. 7.35); no intercooler; drive: condensing-extraction turbine. *Bottom*: syngas nozzles; *top*: recycle nozzles.

8 Improving Efficiency and Operating Range

Since energy consumption and operational flexibility have a high priority, revolutionary inventions and evolutionary improvements have been continuously introduced over the last 100 years to enhance compressor aero-thermodynamics in order to increase both the efficiency and operating range of multistage single-shaft centrifugal compressors. This is especially so in the hydrocarbon processing industry, where such compressors are predominantly used.[1]

8.1 Increase of Efficiency

The use of high flow coefficient impellers, the optimization of blade shape, diffuser ratios and volutes, the use of blade cut-back, the reduction of surface roughness and nozzle Mach number, and the limitation of recirculation losses have been the means to increase the efficiency to a maximum of 87% polytropic for medium flow coefficient impellers and to raise efficiencies of low flow coefficient impellers ($\varphi = 0.012$) from 63% to 68% for medium impeller diameters of 400–500 mm.

8.1.1 Optimization of Aero-Thermodynamics

Centrifugal compressor stages achieve the highest possible efficiency when designed with a flow coefficient φ of around 0.07 to 0.10, assuming the use of impellers with twisted (so-called 3D-)blades.

As a matter of fact the volume flow reduces towards the rear end of the machine, and the diameters of subsequent impellers on one single shaft cannot be stepped down in diameter in order to obtain optimum flow coefficients for all stages. So, necessarily, the efficiency reduces for each subsequent stage for single-shaft compressors.

---

[1] Reproduced (with supplements) with permission of Professional Engineering Publishing Limited, Bury St Edmunds, Suffolk, UK. From Inst. Mech. Eng. Conference Transactions: 7th European Congress on Fluid Machinery for the Oil, Petrochemical, and Related Industries, 1999, The Hague, Netherlands. Lüdtke K: Process centrifugal compressors – latest improvements of efficiency and operating range. Inst. Mech. Eng. (1999), C556/014, pp. 219–234

Figure 8.1 contains the η–φ- ranges of two multistage compressors from the first to the last stage; the first is a conventional and the second is a high flow coefficient design resulting in 20% smaller impeller diameters, 25% higher rotational speed and 5% lower power consumption. The prerequisite, of course, is that high flow coefficient impellers (say, up to 0.15) for process compressors are available, which was not the case until after the mid 1980s.

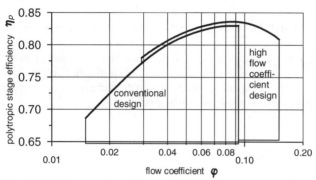

Fig. 8.1 Efficiency ranges of stages in a conventionally designed compressor and in a compressor with high flow coefficient impellers.

Prototypes of very high flow coefficient impellers with $\varphi \approx 0.2$ were developed by university institutes earlier, but for overhung impellers only. However, it is not too easy to design through-shaft (i.e. beam-type shaft) impellers with $\varphi = 0.15$ and hub/tip ratios of around 0.33 to enable multistage process compressors. This is because these stages tend to develop considerable secondary flows within the meridional flow path, which, in turn, may distort the meridional velocity profile at the impeller exit by accumulating low energy material in the shroud/suction side corner. This is not only detrimental to the impeller and diffuser efficiency (especially since the diffuser requires a smooth profile at the inlet to be efficient) but diminishes the operating range by moving the rotating stall point closer to the rated point.

During the last two decades high flow coefficient impellers with high hub/tip ratios have been developed. Three constraints have to be observed, though. Rotordynamic criteria for these multistage units are of the utmost importance. In order to avoid subsynchronous vibrations the shaft stiffness ratio, i.e rigid-support critical speed over maximum speed $N_{c1\text{-rig}}/N_{max}$, must be greater than 0.4 to 0.6, depending on the average gas density, as stated by Fulton (1984). Therefore, since high flow coefficient impellers have smaller shaft diameters and greater axial lengths, the maximum number of impellers to be accommodated in one casing is smaller than for conventional designs; guidelines are shown in Fig. 4.4. The increase of the flow coefficient by 0.02 reduces the potential of the maximum number of impellers per rotor by one. So it might well be that one ends up with a high efficiency requiring two casings as opposed to a single casing with medium efficiency. The second constraint is the increased rotational speed, which may exceed acceptable limits dictated by maximum driver or gear speeds or by experience or by customary practice, especially for small high-pressure compressors. So far, the maximum multistage single-shaft compressor speed is normally around 20,000 rpm. The third constraint of high flow coefficient impellers is the increased stress level at the given tip speed of approximately 115% compared to conventional impellers. This leads to a higher material yield strength to maintain the stage head or, in the case of a limited material yield strength (e.g. if H_2S is present), to a reduced stage head.

8.1.2 Increased Diffuser Diameter Ratio

In Sect. 6.2.3 it was described how the stage efficiency is affected by the radius ratio of the diffuser/return vane package (Figs. 6.18b, 6.19, 6.20). The increase of the radius ratio is an effective measure of the increase of the stage efficiency, not only for higher but also for very low flow coefficient stages.

Table 8.1 shows geometric and aerodynamic parameters, as well as shop test results of two similar three-stage contract compressor sections with medium flow coefficients and high Mach numbers. The increase of the average diffuser ratio from 1.52 to 1.67 resulted in an efficiency increase of 2.3%.

Table 8.1 Two compressor sections with different diffuser ratios and medium flow coefficient impellers

Version	A	B
Number of stages	3	3
Impeller diameters d_2	460; 460; 430	562; 552; 517
Width/diameter ratio b_2/d_2	0.058; 0.040; 0.034	0.05; 0.037; 0.034
Diffuser ratio r_4/r_2	1.49; 1.49	1.66; 1.69
Volute type/area	internal; var. diff. ratio, volute 10 in Fig. 3.14	external; circular/oval, volute 4 in Fig. 3.14
CG radius/impeller radius r_s/r_2	1.57	1.67
Average diffuser ratio	1.52	1.67
Flow coefficient φ, optimum	0.092–0.045	0.076–0.042
Tip speed Mach number M_{u2}	0.98–0.88	1.00–0.80
Average test Reynolds number Re_t	2.2×10^6	0.13×10^6
Pol. stage efficiency η_p, measured	0.810	0.836
Adjusted for radiation, d_2, b_2/d_2, Re and scroll shape: $\eta_{p\,adj}$		0.833
efficiency difference $\Delta\eta_p$, B vs. A		0.023

In order to be able to compare efficiencies, they were adjusted for the different heat radiation losses, diameters and impeller exit width/diameter ratios according to proven empirical experience and also for Reynolds numbers according to the method, introduced by the "International Compressed Air and Allied Machinery Committee" (Strub et al., ICAAMC 1987). The cross-sections of the relevant compressors are shown in Fig. 8.2.

262   8 Improving Efficiency and Operating Range

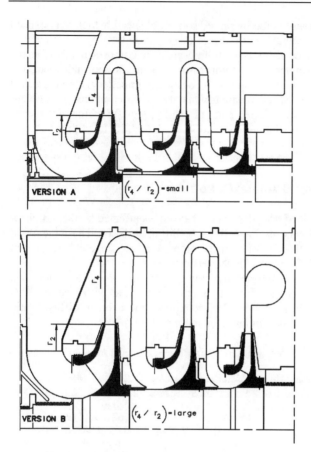

Fig. 8.2 Two compressor sections with different diffuser ratios and medium flow coefficient impellers.

Table 8.2 Two performance-tested compressor sections with different diffuser ratios, low flow coefficient impellers

Version	C	D
Number of stages	3	4
Impeller diameter d_2	615; 560; 515	515; 505; 480; 455
Width/dia ratio b_2/d_2	0.015/0.0148/0.015	0.015/0.0133/0.013/0.0136
Diffuser ratio r_4/r_2	1.51; 1.46	1.59; 1.62; 1.70
Volute type/area	external; circular volute 1 in Fig. 3.14	external; circular/oval volute 4 in Fig. 3.14
CG radius/imp. radius r_3/r_2	1.46	1.69
Average diffuser ratio	1.48	1.65
Flow coefficient φ, optimum	0.0125	0.012
Impeller/diaphragm gap a/r_2	0.035	0.015
Tip speed Mach number M_{u2}	0.85 – 0.61	0.68 – 0.55
Avg. test Reynolds number Re_t	1.5×10^5	3.8×10^5
Pol. stage eff. η_p, measured	0.639	0.673
η_p, adjusted for d_2, b_2/d_2, Re	-	0.663
Efficiency difference $\Delta\eta_p$, C vs. D		0.024

Table 8.2 shows that low flow coefficient stages can also be improved in efficiency by applying longer vaneless diffusers. The very similar three- and four-stage contract-compressor sections, having the same standard impeller types, differ in their average diffuser ratios (1.65 versus 1.48), bringing about an efficiency difference of 2.4% in favor of the long diffuser section. Again, the measured efficiencies were empirically adjusted to make them comparable. It should be added that version D had a smaller axial clearance between the impellers and diaphragms of $a/r_2 = 0.015$ versus 0.035 (version C), which contributed somewhat to the efficiency improvement. The cross-sections are shown in Fig. 8.3. As can be derived from the test results, the findings of Lindner's development tests are basically confirmed: some 2–3% difference in the efficiency for stages with increased diffuser radius ratios.

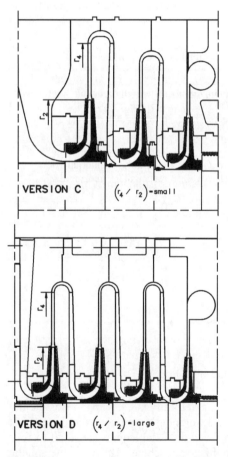

Fig. 8.3 Two compressor sections with different diffuser ratios and low flow coefficient impellers.

8.1.3 Improved Volute Geometry

The "diaphragm package" of the final stage at the end of a compressor section consists of an (normally vaneless) annular diffuser, a volute and a conical exit diffuser. The volute collects the gas circumferentially at the annular diffuser exit and guides it into the conical diffuser and discharge nozzle.

All three elements serve the purpose of efficiently converting a great deal of the kinetic energy at the impeller exit into static pressure. One casing can have up to four of these packages, i.e. four sections with three intercoolers.

In Table 8.3 three shop-performance-tested single-stage contract compressors with different volute configurations are compared with each other. They have the same scalable standard impellers, two of which were slightly modified regarding the exit width. All three tip-speed Mach numbers are in the range where Mach number differences do not affect the efficiency. To make them comparable, the efficiency of version E was adjusted to a radial inlet and version F was adjusted to zero cutback.

Table 8.3 Three performance tested single-stage compressors with different volutes

Version	E	F	G
Rotor type/suction nozzle	overhung/axial	beam/radial	beam/radial
Impeller disk/blade diameter d_{2o}/d_2	815/815	475/460	486/486
Width/diameter ratio b_2/d_2	0.0664[a]	0.080[b]	0.0631c[c]
Annular diffuser type	parallel walled		vaneless
Annular diffuser ratio d_4/d_2	1.4	1.61	1.94–1.5, avg.1.72
Volute type/volute no. in Fig. 3.14	external/4	internal/6	internal/10
Volute area shape	circular/oval	rectangular	rectangular
Volute outer radius/imp. radius d_a/d_2	2.2	2.06	2.06
Volute axial length/impeller dia. l/d_2	0.31	0.42	0.66
Effective diffuser ratio r_s/r_2	1.72	1.61	1.77
Exit diffuser L/d_1, area ratio AR	4.04/1.84	1.55/1.37	3.98/2.10
Flow coefficient φ, optimum	0.084	0.094	0.096
Tip speed Mach number M_{u2}	0.35	0.56	0.52
Pol. stage efficiency η_p, measured	0.87	0.825	0.855
adjusted to radial inlet and $d_2 = 460$	0.83		
adjusted to zero cutback		0.815	

[a] original standard impeller
[b] standard impeller exit width increased
[c] standard impeller exit width decreased.
The impeller is shrouded and has backward curved 3D-blades.

The performance curves are directly compared in Fig. 8.4. Version E is the traditional textbook volute, one-sided offset, developing externally from the diffuser exit. The stage efficiencies are satisfactory and the axial lengths are small to cope with the rotordynamic requirements of multistage machines. However, the annular diffuser ratio is relatively low and the volute tongue is close to the impeller exit, causing losses due to an uneven circumferential static-pressure distribution. The radial space requirements are excessive and the circular cross-sections allow casting as the sole manufacturing method. Version F has an acceptable annular diffuser ratio and a reduced outer radius. However, the reduced effective diffuser ratio results in additional losses through reacceleration of the flow. Version G combines the advantages of E and F: placement of the tongue further away from the impeller (increase of d_4/d_2) and increase of the effective diffuser ratio. The efficiency of version G is increased because:

- The circumferential static pressure distribution at the impeller exit is excellent in spite of the variable ratio of the annular diffuser (as revealed by a viscous three dimensional CFD calculation).
- The average throughflow velocity level in the volute is reduced.
- The swirl velocity (responsible for the bulk of the losses) does not play a significant role due to the diminished meridional velocity in the long diffuser.

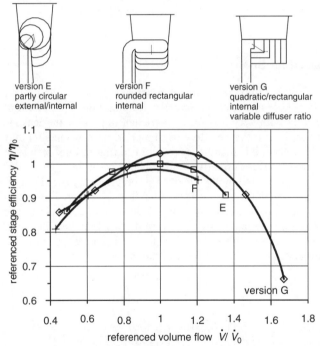

Fig. 8.4 Three single-stage compressors with different volutes: versions E, F, G; measured curves (see also Fig. 3.14).

The vaneless diffuser ahead of the volute should preferably be parallel walled, because pinching would increase the meridional velocity at the volute inlet, thus increasing the secondary vortex across the through-flow area which in turn may reduce the efficiency considerably. LSD vanes, however, are allowed.

The design of the conical exit diffuser has to be in accordance with Fig. 6.27. In order to achieve pressure recovery factors $c_p \geq 0.45$, which assures a satisfactory efficiency, the diffuser area ratio should preferably be ≥ 3 and the length/inlet diameter ratio should be ≥ 4.

8.1.4 Impeller-Blade Cut-Back

As explained in Chap. 6 (and Figs. 6.17 and 6.18), the stage efficiency of medium flow coefficient impellers can be increased by up to 1% if the blades are moderately cut-back at the exit, whereby the disks are left unchanged. The favorable effect derives from reduced friction in the remaining "rotating diffuser" and by re-accelerating the stagnating boundary layer at the impeller exit, thus generating favorable diffuser inlet conditions. If the blades are trimmed by more than approximately 5%, the efficiency gain is lost.

Blade trimming is restricted to medium flow coefficient impellers ($\varphi = 0.04$–0.1), because with narrow impellers the head-reducing effect diminishes and the efficiency cannot be improved.

8.1.5 Cover-Disk Labyrinth Seal

The cover-disk leakage flow has to be recirculated into the main flow in an orderly way so that the velocity profile ahead of the impeller leading edge is distorted as little as possible. Especially with very low flow coefficient impellers, the leakage may amount to as much as 7%, so that the flow in an a priori narrow channel may be severely disturbed. Therefore, it is mandatory to avoid configurations such as in Fig. 8.5a where the leakage bounces directly on the main flow at an angle of 90° and where a vortex may be generated by the uncovered shroud face. Since both effects are detrimental to the efficiency, the face should at least be covered by an extension of the labyrinth seal carrier, as shown in Fig. 8.5b, so that the leakage can be guided in a better way into the main flow. For very narrow impellers the solution shown in Fig. 8.5c is recommended, because this nearly tangential blend-in warrants the least disturbance at this crucial point of the impeller inlet.

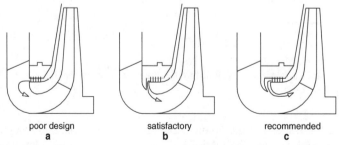

Fig. 8.5a-c Low flow coefficient impellers and shroud face configurations of cover-disk labyrinth seal. **a** uncovered face, **b** straight face cover, **c** curved face cover; arrows indicate path of leakage flow.

8.1.6 Reduction of Plenum Inlet Loss

As explained in detail in Chap. 6, it is recommended that the nozzles at compressor inlet and downstream of intercoolers should be designed in such a way that the head loss as given by Fig. 6.15 or Eq. (6.86) from the inlet flange to the impeller eye should normally not exceed 0.01, i.e. 1%. In some cases 2% would still be tolerable. If there is only one suction nozzle followed by more than five stages, the plenum inlet loss can be increased beyond 2% to come up with a smaller nozzle diameter, because the weight of this loss is low.

However, it should be kept in mind that empirically designing suction nozzles by more or less arbitrary values of gas velocities may lead to excessively high efficiency penalties. An example is: air compressor, stage pressure ratio downstream of nozzle 1.5, selected velocity 30 m/s → nozzle Mach number 0.086, related inlet head loss 0.045, efficiency decrease 3.8% (Fig. 6.15)! This would be a poor nozzle design, reducing the high stage efficiency of 84% to 80%!

Inlet nozzles ought to be sized by loss rather than by velocity!

8.1.7 Reduction of Flow-Channel Surface Roughness

Casey (1984) suggested the definition of an equivalent pipe flow having a friction factor similar to the compressor flow. Using the well-known Moody diagram, which is a plot of the friction coefficient λ versus the Reynolds number $Re = u_2 b_2 / v_1$ with the relative arithmetic average roughness Ra/b_2 as a parameter, an approximation of the influence of the surface roughness on the efficiency can be calculated as follows.

The efficiency change with friction coefficient (Eq. (1.92)) is

$$\Delta \eta_p = -\frac{k}{s} \Delta \lambda, \qquad (8.1)$$

where k is a proportionality factor, s is the work input factor, and λ is the friction coefficient.

The Colebrook–White formula (Eq. (1.91)) for the friction factor was first published by Moody (1944):

$$\lambda = \frac{1}{\left[1.74 - 2\log_{10}\left(2\frac{Ra}{b_2} + \frac{18.7}{Re\sqrt{\lambda}}\right)\right]^2}. \qquad (8.2)$$

The Reynolds number is given by

$$Re = \frac{M_{u2} a_0 \frac{b_2}{d_2} d_2}{v_1}, \qquad (8.3)$$

where a_0 is the sonic velocity at the impeller inlet, and v_1 is the kinematic viscosity at the impeller inlet.

Equation (8.1) was calibrated by utilizing test results that were systematically conducted on various stages with known roughness and Reynolds number. Therefore the dimensionless proportionality factor k as a function of b_2/d_2 and M_{u2} is defined as follows:

$$k = \frac{(1-\eta_p)s}{\lambda}. \tag{8.4}$$

where k is between 6.0 and 8.0 for wide impellers and between 8.0 and 12.0 for narrow impellers (low values at low Mach numbers). As can be seen from Fig. 8.6, which shows as an example how the efficiency increases when the flow-path roughness is reduced to 50%, the efficiency picks up between 1% and 3%, depending on the relative impeller exit width and the tip-speed Mach number.

So there is a great potential for improving low flow coefficient stages, which are typical for many applications in the oil and gas industry.

Surface smoothing is carried out conventionally and also by chemical or electropolishing. Needless to say, that it does not make sense to treat compressors handling gases with contaminating constituents.

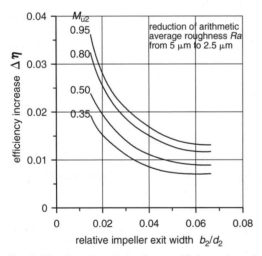

Fig. 8.6 Predicted increase of stage efficiency by reducing flow-channel surface roughness. M_{u2} is the tip speed Mach number.

8.1.8 Minimizing Impeller/Diaphragm Clearance

The magnitude of the clearances between the impeller hub and cover disks and the opposite diaphragm walls has an influence on the losses caused by disk friction, as revealed by Daily and Nece (1960). The frictional power of a two-sided disk is

$$P_F = C_M \frac{\rho_3}{2} u_2^3 r_2^2 .\tag{8.5}$$

The torque coefficient and Reynolds number are

$$C_M = \frac{0.102(a/r_2)^{0.1}}{Re^{0.2}}; \quad Re = \frac{\rho_3 u_2 r_2}{\eta_3}.\tag{8.6}$$

C_M is valid for turbulent flow and separate boundary layers on both sides of the clearance.

The disk friction loss per unit of gas power of a compressor stage is

$$\frac{P_F}{P_i} = \frac{C_M \rho_3 u_2^3 (d_2^2/4)}{2 \dot{m} s u_2^2} \tag{8.7}$$

Substituting the mass flow $\dot{m}$ with $\dot{V}_3 \rho_3$ and introducing the flow coefficient φ leads to a very simple formula for the impeller disk-friction loss:

$$\frac{P_F}{P_i} = \frac{C_M}{4\pi s \varphi(\dot{V}_3 / \dot{V}_{0t})} \tag{8.8}$$

The "feeling" that a disk, rotating in a denser gas and at a higher speed, will consume more frictional power P_F is not betrayed by this formula because the gas power P_i increases accordingly.

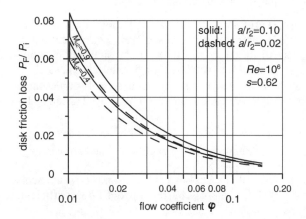

Fig. 8.7 Impeller disk friction loss as a function of flow coefficient and axial gap impeller/diaphragm. *a* average gap impeller/diaphragm, r_2 impeller radius, *Re* Reynolds number, *s* work input factor.

The torque coefficient C_M, depending on the Reynolds number, is between 0.002 and 0.008 for most applications. The total to static impeller volume ratio $\dot{V}_3/\dot{V}_{0t}$, according to Figs. 6.13 and 6.14, is a strong function of the Mach number and is between 0.7 for heavy gases and ≈1.0 for H_2-rich gases.

Figure 8.7 clearly shows the influences of the various parameters and the possible improvement potential of a reduced clearance ($a/r_2 = 0.02$ versus 0.10). So the efficiencies of very low flow coefficient stages ($\varphi = 0.01$) can be improved by an impressive 1% by reducing the axial gap from 10% to 2% of the impeller radius. An improvement of 0.5% is possible for impellers with a flow coefficient of $\varphi = 0.025$. For very high flow coefficient impellers of $\varphi > 0.1$ the improvement is less than 0.1%.

8.1.9 Reduction of Recirculation Losses

The parasitic leakage flows, occurring throughout the compressor, have to be continuously compressed but do not contribute to the process mass flow. The leakage flow at each cover-disk seal is recirculated from the impeller exit to the inlet. The interstage shaft leakage flow is either centrifugally recirculated from the stage discharge to the impeller exit or, in the case of an adjoining balance piston, centripetally recirculated from the impeller exit via the piston labyrinth to the connected impeller inlet. These leakages account for 3% to as high as 12% of the overall power, depending on the flow coefficient level, pressure ratio, and impeller arrangement.

In particular, compressors with narrow impellers arranged in-line can be made more efficient if the conventional steel labyrinths are replaced by "plastic" seals with clearances of only 30 to 40% of those of steel seals. This "plastic engineering" is being introduced more and more into present-day process compressors. There are basically two types of plastic seal material:

- Abradable materials, like polytetrafluoroethylene (PTFE) combined with mica and silicon–aluminum (AlSi) combined with polyester (PE), permit rubbing of rotating parts by giving way to intruding tips through abrasion, i.e. adapting their geometry locally. Labyrinth tips are of conventional material (e.g. steel).
- Rub-resistant materials forming the labyrinth tips, such as polyamide-imide (PAI) and poly-ether-ether-ketone (PEEK), tolerate rubbing by deforming locally and resuming nearly all their former shape. AlSi+PE can be used on stationary as well as on rotating parts; all others are used as stationary parts.

In Fig. 8.8 an example is shown of an impeller cover-disk seal and an interstage shaft seal.

The overall energy savings attainable by applying rub-resistant labyrinth seals is between 2% and 8% and varies with compressor size, impeller flow coefficient, impeller arrangement and pressure ratio.

The best method of reducing the balance piston recirculation with intercooled compressors is the back-to-back impeller arrangement, as described in Sect. 4.2.5. Compressors with in-line arranged impellers have balance piston losses of between 1% and 3.5%, depending on the flow coefficient and pressure ratio. The losses are reduced to approximately one third in the case where the impellers of an intercooled compressor are arranged back-to-back (Figs. 4.9 and 4.10). This is because the piston diameter, labyrinth clearance and pressure ratio across the piston are considerably decreased and because the leakage flow is circulated in one compressor section only. So, the power savings of back-to-back compressors, compared to in-line compressors, is between 1% with high flow coefficient and 3% with low flow coefficient compressors.

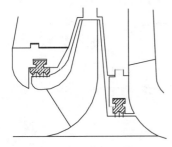

Fig. 8.8 Efficiency increase by using abradable plastic seals. The sealing tips are of conventional material.

8.2 Increase of Operating Range

The following methods have become established for achieving a greater operating range: steepening of the characteristic curve by partload efficiency increase or overload design or impeller blade angle reduction; increasing of the flow angle at the impeller exit by diffuser pinching or LSD vanes; smoothing of the velocity profiles and shaping the diffuser inlet; and placing IGVs with multistage compressors.

As described in Sect. 5.5.1, there are two aerodynamically different surge-triggering mechanisms that determine the compressor operating range:

- System surge starts when, upon flow reduction, the negatively sloped pressure–volume curve leads into the horizontal tangent, which is about the limit that the machine/system interaction can tolerate. A positively sloped curve is a priori excluded as a possible stable operating area.
- Stage surge starts when, upon flow reduction, the flow begins to reverse locally at the diffuser inlet, causing rotating stall in the stage with the eventual surging of the machine. The primary cause is the diffuser inlet flow angle at one of the walls becoming zero, with the result that the radial velocity component does not have enough kinetic energy to maintain its forward movement against the radial pressure gradient.

8.2.1 Delay of System Surge

Since the steepness of the pressure–volume curve is the criterion for this stability limit, it suffices to increase the curve slope to attain a surge delay, i.e. a shift to smaller volume flow. System surge is not a compressor-intrinsic fluid-dynamic phenomenon but solely determined by the dynamic interdependence of the system compressor and its associated volume; so, not the inner working mechanism, but rather the plain curve slope is the culprit for an unsatisfactorily narrow operating envelope.

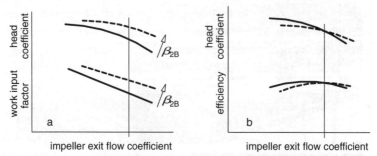

Fig. 8.9a, b Slope increase of pressure–volume curve. a impeller blade exit angle reduction, β_{2B} blade angle, b partload efficiency increase.

Slope increase can be brought about by choosing a smaller impeller-blade exit angle or by reducing the losses at partload operation, as shown in Fig. 8.9.

It is one of the basic textbook wisdoms that a smaller blade angle (from tangent) increases the slope of the work input factor curve, which automatically produces a steeper pressure–volume curve, as shown in Fig. 8.9a. At the same time, however, it reduces the head accordingly, necessitating a tip-speed increase or, if impossible, adding one more stage. And, as a matter of fact, manufacturers seldom have the option of blade-angle variation, since their experienced impeller families are based on one particular blade shape per flow coefficient.

Increase of Partload Efficiency by Overload Design

The second option of curve steepening by efficiency increase at partload (see Fig. 8.9b) is more realistic. Volumetric overload design is explained in Fig. 5.2: the compressor as a whole is designed in such a way that the optimum point is located to the left of the rated point (specified point). This means that the partload efficiency increases and the operating range is $\dot{V}_0 - \dot{V}_{TD}$ instead of $\dot{V}_{opt} - \dot{V}_{TD'}$. The consequences are that the head decrease at the rated point has to be compensated by a higher tip speed and hence a higher Mach number; and of course the choke point is placed closer to rated. The latter is often tolerable, but in some cases the specification for the particular application has to be adhered to. A positive effect is the increase of the flow angle α_3 at rated point, which pushes the critical flow angle for reverse flow α_{3c} to a smaller volume flow, thus delaying rotating stall. So, what is good for system surge delay is good for stage surge delay as well. By overload design the surge can be delayed by 7–12%, corresponding to $\dot{V}_0 / \dot{V}_{opt} = 1.07 - 1.12$.

8.2 Increase of Operating Range   273

Increase of Partload Efficiency by a Longer Diffuser

It has been observed during the experiments with diffuser ratio variation that the longer diffuser not only resulted in a higher stage efficiency but also in a more moderate efficiency decrease at partload. Thus, the head rise to surge became greater with a surge limit improved by 16% for this medium flow coefficient impeller, as can clearly be seen in Fig. 6.19 (reported by Lindner (1983)). So, a higher diffuser radius ratio adds to the compressor cost or, for a given frame size, reduces the maximum capacity (by having to apply smaller impeller outer diameters (OD)), but improves the operational flexibility.

Increase of Partload Efficiency by Diffuser Pinching or LSD Vanes

Narrowing of axial diffuser widths and placing of low-solidity diffuser vanes do increase the partload efficiency, definitely steepening the curve because in both cases the efficiency maximum is shifted to a lower volume flow. Since the aerodynamic effects of these two measures go beyond this, by decisively delaying rotating stall, their functioning is explained below.

8.2.2 Delay of Stage Stall

Stage stall, also known as rotating stall (RS), will occur, upon volume flow reduction, if the critical flow angle α_{3c} at the diffuser inlet of one particular stage (mostly one of the rear stages) in the compressor is reached. This can only be the case if system surge has not been reached yet, i.e. normally at the negative slope of the pressure–volume curve, as can be seen in Fig. 5.23.

As a trigger for stage surge, the rotating stall can hardly be suppressed but it can be delayed under certain conditions. Suppression in the pragmatic sense is possible only by limiting the operating envelope by intentionally opening a bypass or blow-off valve to prevent the compressor from operating in the unstable region. The bypass line is normally 5% to 10% away from either system surge or the rotating-stall line.

Delay of RS is usually appropriate when the specified operating points are far apart from each other and when the highest operational flexibility is required, for example, variation of suction and recooling temperatures, of mass flow, and of gas composition at constant discharge pressure, as is often the case in the oil and gas industry. RS delay may also be necessary if two or more "identical" compressors have to operate in parallel whose performance curves might be incompatible due to excessive manufacturing tolerances.

The sine qua non for delay of RS is: *step up the flow angle at the diffuser inlet!* The delay of RS can be obtained by:

(1) flow-angle increase through:
   – axial width reduction of the vaneless diffuser ("pinching")
   – equipping the parallel-walled diffuser with vanes of high pitch/length ratio (low-solidity diffuser vanes, "LSD")
   – geometric shape of diffuser inlet

– a priori compressor design with volumetric overload
– adjustable inlet guide vanes with multistage machines
(2) straightening of the velocity profile at the impeller exit.

Diffuser Pinching

The underlying principle is to stabilize the diffuser flow right after the impeller.

Abrupt or gradual diffuser pinching increases the flow angle at the rated point, shifting the critical flow angle to lower flow and thus widening the operation range by warranting stable flow conditions over a wider capacity domain, as can be seen from Fig. 8.10. The experiments conducted with a medium flow coefficient single stage, having a relative impeller exit width of $b_2/d_2 = 0.06$, reported by Lüdtke (1983), yielded a surge-flow reduction of 27% when the parallel-walled diffuser was replaced by a highly tapered diffuser. For this extreme case of diffuser pinching the head reduction and hence the efficiency trade off is some 5%. So, in principle turndown improvement is not free of charge. However, a very good compromise can be achieved by a constant-area diffuser, which improves the surge point by approximately 10% at a cost of only 1% efficiency decrease.

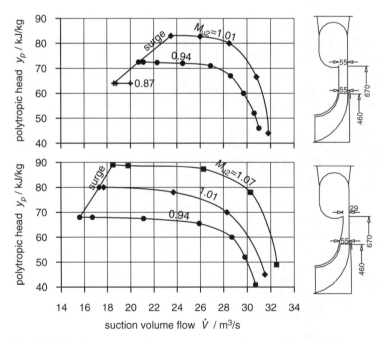

Fig. 8.10 Single-stage compressor operating range: parallel-walled (*top*) versus highly tapered (*bottom*) diffuser. $d_2 = 460$ mm, $\varphi = 0.11$, tip speed Mach number $M_{u2} = 0.87–1.07$; impeller semi-open, R-type, through-shaft.

Also a mild diffuser taper of 10% to 15% downstream of the medium-to-high flow-coefficient impeller, and then continuing on with a parallel-walled diffuser is quite effective with no appreciable efficiency decline.

If the diffuser is pinched, i.e. $b_3 < b_2$, the flow angle will be steeper and the onset of RS will be delayed. However, according to Nishida et al. (1988), the critical flow angle α_{3c} as per Fig. 5.28 is no longer valid and has to be modified according to Eq. (5.24).

The experiments with the various degrees of diffuser pinching have shown that the diffuser is the cause of the stage surge triggered by RS. The flow-angle increase for the pinch displayed in Fig. 8.10 is approximately 15°. This extreme measure clearly shows which extreme effects can be reached with pinching.

Nishida also reported on two different degrees of diffuser pinching. However with a low flow coefficient stage, having a relative exit width of $b_2/d_2 = 0.0183$, a pinch of $b_3/b_2 = 0.69$ places the RS point at a volume flow that is 50% greater than the system surge. A pinch of $b_3/b_2 = 0.38$ has the RS totally disappear to the left of the system surge, which then remains the only performance-map limitation. Here again, this "success" costs a 12% efficiency decrease!

Low-Solidity Diffuser Vanes

The function of LSD vanes is described in Sect. 3.5. Figure 8.11 shows the results of measurements on a medium-Mach-number six-stage process compressor with and without LSD vanes on each stage. The surge flow was improved by 28%, the head rise to surge was improved from 110% to 117% (beneficial for molecular mass changes and controllability), the rated point efficiency remained constant, the partload efficiency was improved and the choke flow was reduced from 142% to 133%. The overall curve range $\dot{V}_{choke}/\dot{V}_{surge}$ was substantially increased from 2.2 to 2.9 (thus partload efficiency increase also delays the system surge).

The effect of the LSD vanes is basically the same as diffuser pinching: increase of flow angles (from tangent) at partload so that no surge-triggering reverse flow can take place where the vaneless diffuser normally stalls. After a thorough calculation of flow angles along the performance curve, the vanes must be positioned in such a way that, at the desired part load and at the rated point, the incidence angle $i = (\alpha_{chord} - \alpha_{flow})$ is positive, assuring a positive lift coefficient and a sufficiently small drag coefficient. Thus, the diffuser flow angles are increased by the vanes. This is also the case from rated to an appropriate volumetric overload. At higher volumetric overloads the incidence angle and the lift coefficient become negative (i.e. the flow angles are reduced by the vanes) and the drag coefficient increases dramatically, so that the LSD vanes shift the vaneless choke point to smaller volume flows. The chord angle is between the logarithmic spiral chord (resulting from conformal mapping from linear to angular) and the circumferential direction of the airfoil profile.

It should be mentioned, though, that the application of LSD vanes has been limited so far to maximum tip-speed Mach numbers of approximately 0.75, because the LSD effect on the entire performance curve of multistage compressors diminishes as the Mach number increases.

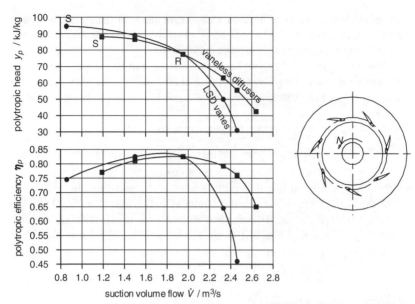

Fig. 8.11 Six-stage compressor operating range: measured curves. Parallel-walled vaneless diffusers versus diffusers with LSD vanes. R rated point, S surge point, N rotational speed, $d_2 = 425$ mm, $\varphi = 0.065$, $M_{u2} = 0.51$; all stages with LSD vanes.

As far as the location of the LSD leading edge is concerned, the following has to be observed: the leading edge must be placed at a radius which has to be smaller than the one where flow separation sets in. Nishida et al. (1991) reported tests with the above-mentioned impeller with $b_2/d_2 = 0.0183$: (a) at a pinch ratio of $b_3/b_2 = 0.64$ without LSD the RS point was delayed to 50% to the right of system surge; (b) at a pinch ratio of $b_3/b_2 = 0.66$ plus LSD with a leading edge at $r_{LE}/r_2 = 1.2$ the RS point was delayed to 30% to the right of system surge; (c) the same with $r_{LE}/r_2 = 1.07$ caused the RS to totally disappear. Accoording to Fig. 5.27, the RS onset is predicted to be at around $r/r_2 = 1.12$ for this impeller; so, a leading edge at $r_{LE}/r_2 = 1.2$ is already in the separated zone, which diminishes the effect of the LSD vanes. The full effect of totally suppressing RS is achieved when the leading edge is positioned at $r_{LE}/r_2 = 1.07$, where the flow is still healthy.

Geometric Shape of Diffuser Inlet

Kobayashi et al. (1990) reported tests with a very low flow coefficient impeller with $d_2 = 300$ mm and $b_2/d_2 = 0.0116$, with equal b/r_2 in impeller and diffuser. The diffuser inlet geometry was varied as shown in Fig. 8.12:
- Shape T: impeller discharges into a wide annulus; a pinch radius of 5 mm reduces the annulus to $b_3 = 3.5$ mm which is identical with the impeller exit width; parallel-walled diffuser inlet at $1.05 \times r_2$.
- Shape B: pinch radius 2.5 mm, parallel-walled diffuser inlet at $1.024 \times r_2$.
- Shape A: pinch radius 0.5 mm, parallel-walled diffuser inlet at $1.01 \times r_2$.

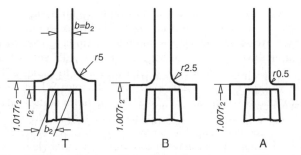

Fig. 8.12 Low flow coefficient impeller: shape of diffuser inlet (from Kobayashi et al. (1990) redrawn). T, B poor designs, A recommended.

Since with shape T the flow angle decreases abruptly due to the sudden width increase, the RS point is extremely to the right of system surge; B improves marginally; but A delays the RS to half-way between the RS of shape T and system surge.

Therefore, with low flow coefficient impellers, it is mandatory to properly shape the diffuser inlet and to avoid large round-off radii. Wider impellers and diffusers are less sensitive to the inlet geometry of the diffuser, since the RS starts at larger ratios of r/r_2.

Volumetric Overload Design

The method is described in Sect. 8.2.1. By designing compressors with the rated point placed at a volume flow that is 5–12% greater than optimum, the flow angle α_3 is increased by some 3° at the rated point so that the RS flow is delayed by the same amount.

Therefore, it is always advisable to use this design approach, because it delays system as well as stage surge. But it also reduces the choke flow ratio $\dot{V}_{choke}/\dot{V}_{rated}$, which may, for some applications with high molar mass gases, not be appropriate or not even permitted.

Adjustable Inlet Guide Vanes

In Sect. 5.2.3 it is explained how an improvement of turndown with multistage machines, in comparison with suction throttling or speed regulation, is brought about by placing just one row of adjustable IGVs ahead of the first stage. This becomes effective for tip-speed Mach numbers above approximately 0.6–0.7. The fact that IGVs ahead of the first stage can improve turndown even for multistage machines is widely ignored by the turbomachinery community because it sounds so unrealistic. However, just one simple formula proves the basic fact; therefore, Eq. (5.11) is repeated here. The minimum volume flow of the first stage is dictated by the surge volume flow of the second stage of a two-stage "model" compressor

$\dot{V}_{\min\_1\,\text{stg}} = \dot{V}_{\text{surge}\_2\,\text{stg}} \dfrac{1}{(\dot{V}_2/\dot{V}_1)_{1\,\text{stg}}}$, where $(\dot{V}_2/\dot{V}_1)_{1\,\text{stg}}$ is the volume ratio across the first stage (as shown in Fig. 4.6.). By closing the IGVs, the low volume ratio (at high M_{u2}) is markedly increased. This, in turn, reduces the minimum flow of the first stage (which is the compressor surge!) by the same percentage, because the second-stage surge volume flow remains the same. Actually, the "second stage" of the model is normally a stage group, which does not change the basic truth of the above equation. It has to be kept in mind, though, that only the turndown point is improved, not the stability point on the unregulated curve, because the IGVs are open at that point!

Smoothing of Velocity Profiles

According to Senoo and Kinoshita (1977), the critical flow angle, according to Fig. 5.28, is valid for undistorted velocity profiles only; if there are distortions, the values of α_{3c} must be upward corrected, as shown in Table 8.4.

Table 8.4 Critical flow-angle correction due to distortion of velocity profiles at the impeller exit

Relative impeller exit width	b_2/d_2	0.025	0.05	0.10
Radial distortion	$\Delta c_m/c_m$		0.3	
Correction of critical flow angle	$\Delta \alpha_{3c}$	+3.5°	+4.0°	+5.5°
Circumferential distortion	$\Delta c_u/c_u$		0.3	
Correction of critical flow angle	$\Delta \alpha_{3c}$		+1.0°	

Δc_m, Δc_u difference of main flow velocities, front wall/back wall.

A 4° difference of the angle α_3 represents a 12% difference of the volume flow. Thus, if the radial velocity profile is distorted by 30% (which is not much for this critical area), the rotating stall flow is already increased by 12%. Because, in addition to this, a distorted profile reduces the efficient energy transfer in the diffuser and hence the head rise to surge, it is essential, in impeller design to achieve a velocity profile that is as smooth as possible, especially in the radial direction. It should be noted that a distorted profile at the rated point gets all the more pronounced as the flow is reduced. An example of extremely warped measured profiles is shown in Fig. 8.13. The flow angle is 7° at the front wall and 30° at the back wall, with associated radial velocities of 50 m/s and 230 m/s respectively. The reason for this large difference was that the impeller exit width was excessive for the recorded operating point, with the rotating stall being just 8% away. The radial velocity c_m at the diffuser front wall has not quite reached the value zero, which would be the condition for the onset of RS. The tests were carried out by Hass et al. (1974), within the FVV, the joint research association of German and Swiss turbomachinery companies. As pointed out earlier, a smooth velocity profile is not easily achievable for high flow coefficient impellers with the hub/tip ratios required for multistage process compressors. Normally a viscous, three-dimensional CFD calculation is carried out to verify the profile shape.

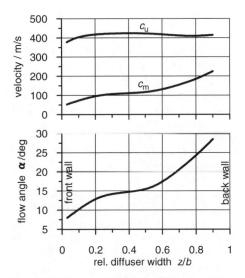

Fig. 8.13 Measured velocity profiles at diffuser inlet (from FVV, Hass et al. (1974), redrawn). c_u absolute velocity, c_m meridional velocity, α absolute flow angle from tangent, z/b fraction of axial diffuser width, $d_2 = 400$ mm, $b_2 = 26$ mm, $N = 22{,}000$ 1/min, $\dot{m} = 8.6$ kg/s, 8% away from rotating stall.

References

Casey MV (1984) The effects of Reynolds number on the efficiency of centrifugal compressor stages. ASME Paper No. 84-GT-247, Gas Turbine Conf. Amsterdam, The Netherlands, 1984; The American Society of Mechanical Engineers, New York

Daily JW and Nece RE (1960) Chamber dimension effects on induced flow and frictional resistance of enclosed rotating disks. Trans. ASME, J. Basic Eng., pp 217–232; The American Society of Mechanical Engineers, New York

Fulton JW (1984) The decision to full load test a high pressure centrifugal compressor in its module prior to tow-out. IMechE Conference Publications, C45/84. In: Fluid Machinery for the Oil, Petrochemical, and Related Industries; Proc. Inst. Mech. Eng. pp 133–138, Conference Transactions, European Congress on Fluid Machinery, 1984, The Hague, Netherlands

Hass U, Kassens K, Knapp P, Mobarak A, Siegmann W, Wittekind W (1974) Untersuchung der Laufradstömung in hochbelasteten Radialverdichterrädern. Forschungsberichte Verbrennungskraftmaschinen (FVV), Heft 155, Abschlussbericht, restricted, in German

Kobayashi H, Nishida H, Takagi T, Fukushima Y, (1990) A study on the rotating stall of centrifugal compressors. 2nd Report, effect of vaneless diffuser inlet shape on rotating stall. Trans. Japan Soc. Mech. Eng. 56 (529), pp 98–103

Lindner P (1983) Aerodynamic tests on centrifugal process compressors – influence of diffusor diameter ratio, axial stage pitch and impeller cutback. Trans. ASME, J. Eng.

Power, 105, No. 4, pp 910–919; The American Society of Mechanical Engineers, New York

Lüdtke K (1983) Aerodynamic tests on centrifugal process compressors – the influence of the vaneless diffusor shape. Trans. ASME, J. Eng. Power, 105, No.4, pp 902–909; The American Society of Mechanical Engineers, New York

Moody LF (1944) Friction factors for pipe flow. Trans. ASME, Nov. 1944, pp 671–684; The American Society of Mechanical Engineers, New York

Nishida H, Kobayashi H, Takagi T, Fukushima Y (1988) A study on the rotating stall of centrifugal compressors. 1st Report, effect of vaneless diffuser width on rotating stall. Trans. Japan Soc. Mech. Eng. 54 (499), pp 589–594

Nishida H, Kobayashi H, Fukushima Y (1991) A study on the rotating stall of centrifugal compressors. 3rd Report, rotating stall suppression method. Trans. Japan Soc. Mech. Eng. 57 (543), pp 3794–3800

Senoo Y and Kinoshita Y (1977) Influence of inlet flow conditions and geometries of centrifugal vaneless diffusers on critical flow angle for reverse flow. Trans. ASME, J. Fluids Eng. 1977, pp 98–103; The American Society of Mechanical Engineers, New York

Strub RA, Bonciani L, Borer CJ, Casey MV, Cole SL, Cook BB, Kotzur J, Simon H, Strite MA (1987) Influence of the Reynolds number on the performance of centrifugal compressors. Final Report Working Group Process Compressor Sub-Committee International Compressed Air and Allied Machinery Committee (ICAAMC), ASME Paper 87-GT-10 and Trans. ASME, J. Turbomachinery 109, pp 541–544; The American Society of Mechanical Engineers, New York.

9 Rerate of Process Centrifugal Compressors

9.1 Introduction[1]

Quite a number of users wish to increase production as early as 3, or as late as 30 years after commissioning their plant, even if they did not envision such an extension at the time of the initial design. Two questions are almost inevitable:

1. Can the compressor be uprated without replacing the casing, its connecting pipework and auxiliary equipment?
2. Can the exchange of parts be accomplished during one of the short normal planned shutdowns?

Almost every compressor has the potential of being uprated by means of one of the following approaches:

1. wider impellers,
2. higher speed,
3. suction side boosting.

Experience has shown that all parts and new equipment to be exchanged can be ready at the beginning of the shutdown, including the installation of a booster compressor. Needless to say, rerates are lower in cost than new builds, although efficiencies may be somewhat lower than for a completely new compressor.

9.1.1 Definition of Rerates

A rerate is defined as the geometric modification of a compressor already in operation to change its aero-thermodynamic behavior. The modification may cover flow, head, efficiency, and operating range and results in a performance curve change.

---

[1]Reproduced (with supplements) with permission of the Turbomachinery Laboratory (http://turbolab.tamu.edu). From Proceedings of the Twenty-Sixth Turbomachinery Symposium, 1997, Turbomachinery Laboratory, Texas A&M University, College Station, Texas, pp. 43–55, Copyright 1997.

"Rerate" should not be confused with "retrofit", which is defined as an exchange of functionally enhanced components, such as bearings, shaft seals and couplings, which improve the mechanical behavior, maintainability, operational safety, availability, etc. of a compressor already in operation. Retrofitting will not alter the pressure–volume curve. The term "revamp" is not used in this book because its definition appears to be blurred.

The majority of rerates encompass uprates (increased flow and/or head) rather than downrates (reduced flow and/or head). Accomplishing compressor uprates is a demanding engineering task, since the degrees of design freedom are less than in designing new compressors from scratch.

9.1.2 Increase of Mass Flow

The options of increasing the mass flow through any flow channel (e.g. impeller, diffuser) are basically described by the continuity equation:

$$\dot{m} = A\,w\,\rho = A\,w\,\frac{p}{RT}, \tag{9.1}$$

where A is the through-flow area, p is the pressure, w is the average flow velocity (see Fig. 9.1), R is the gas constant, ρ is the density, and T is the temperature.

There are three customary options:

- Compressor with wider impellers and diaphragms enlarges area A.
- Higher rotational speed increases gas velocity w.
- Upstream-booster increases compressor suction-density ρ.

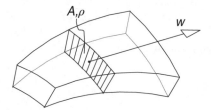

Fig. 9.1 Flow channel. A through-flow area, ρ density, w velocity.

In principle, a reduction of the inlet temperature would also increase the density. However, this is not commonplace, because it is comparatively too expensive to install and operate a refrigerating system that achieves the same effect as a booster compressor. In the literature only one example could be found where a favorable situation justified suction cooling. Runge (1996) reports that for rerating an ammonia plant the excess refrigerating capacity of the ammonia compressor was used to chill the inlet and the intercooler exit temperature of the syngas compressor, thus achieving an increase of mass flow at constant discharge pressure.

Reducing the gas constant R (increasing molar mass) excludes itself due to process reasons.

9.2 New Wider Impellers and Diaphragms

The increase of the suction volume flow of an existing casing by installing wider impellers with wider stationary flow channels is feasible if there is some additional space to accommodate them. It is often desired not to alter the casing with its inlet and discharge nozzles, scroll(s), bearings, bearing span and rotational speed. The impeller flow-path width can generally be increased through higher flow coefficients or through larger impeller diameters, as described by the mass flow ratio of the uprated to the existing compressor (at constant suction density and constant speed).

The following equation reveals the mass flow relationship between the original and the rerated compressor.

Rearranging the defining equation for the flow coefficient, we obtain

$$\dot{m} = (\pi^2/4)\rho_s N \varphi d_2^3, \tag{9.2}$$

where N is the rotational speed and ρ_s is the suction density. The mass flow increase due to the flow coefficient and diameter increase is given by

$$\dot{m}/\dot{m}_0 = \varphi/\varphi_0 \times (d_2/d_{20})^3. \tag{9.3}$$

Since the flow benefits from the diameter-induced head increase, the flow ratio is to be corrected by the slope effect:

$$S = \frac{(y_p/y_{p0})-1}{(\dot{m}/\dot{m}_0)-1}. \tag{9.4}$$

Since for multistage compressors with compressible flow (i.e. medium to high Mach numbers) the head ratio is $y_p/y_{p0} > (d_2/d_{20})^2$, a diameter-mismatch correction factor has to be accounted for (valid for uprate only):

$$F_{d_2} = \frac{(y_p/y_{p0})-1}{(d_2/d_{20})^2-1}. \tag{9.5}$$

So the mass flow increase due to the flow coefficient and diameter increase is

$$\frac{\dot{m}}{\dot{m}_0} = \frac{\varphi}{\varphi_0} \times \left(\frac{d_2}{d_{20}}\right)^3 \times \left[\left(\left(\frac{d_2}{d_{20}}\right)^2 - 1\right)\frac{F_{d_2}}{S} + 1\right], \tag{9.6}$$

where subscript "0" refers to the existing compressor, no subscript refers to the uprated compressor.

The first term represents the extension of the flow coefficient related to the first-stage flow coefficient of the existing machine (Fig. 9.2). During the last 40 years the maximum flow coefficient of centrifugal impellers has doubled from approximately 0.075 to 0.15, made possible by developing efficient backward-leaning 3D-bladed impellers with high hub/tip ratios to be used in multistage single-shaft compressors. Therefore, the older compressors still in operation have a comparatively very high uprating potential through the use of higher flow coefficient impellers.

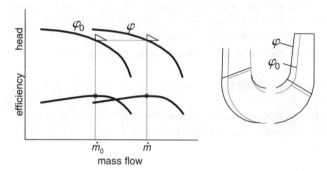

Fig. 9.2 Uprating by using a higher flow coefficient impeller with area increase, $d_2 =$ const, $N =$ const, $\rho =$ const.

Wider impellers can also be designed by just scaling up existing wheels, as described by the second term of Eq. (9.6). The increased head, caused by the higher tip speed, is converted into a higher mass flow at the specified head via the performance-curve slope represented by the third term of the equation, as shown in Fig. 9.3.

This means that an actual increase of the flow coefficient is forcefully brought about by a volumetric overload at some efficiency sacrifice. The diameter-mismatch factor constitutes the fan-law deviation, i.e. the overproportional head increase in relation to d_2^2 due to progressive aerodynamic stage-mismatching, especially for multistage compressors with high-Mach-number impellers.

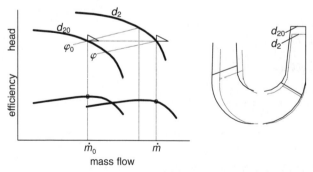

Fig. 9.3 Uprating by using scaled up impeller with area increase, $N =$ const, $\rho =$ const.

Typical diameter-mismatch factors F_{d_2} are: 1.0 for low, 1.5 for medium, and up to 2.0 for high pressure ratio applications.

Guide lines for curve slopes S at the design point are: 0.8 for pressure ratios of 2 : 1 (flat), 2.0 for pressure ratios of 20 : 1 (normal), and 2.8 for pressure ratios of 40 : 1 (steep).

As can be seen from Eq. (9.6), each 1% flow coefficient increase yields a 1% mass flow increase, and each 1% impeller-diameter increase results in some 5% mass flow increase. So even if, due to technological reasons, higher flow coefficient impellers are not available, an efficient uprating is feasible through the scaling-up of existing impellers.

The consequences of installing wider impellers in an existing casing are as follows:

- The impeller stress increases due to the wider flow channels. The separation margin to material yield strength of the wheel material should be checked.
- The shaft diameter may have to be increased to offset higher impeller weights to assure critical speed separation margin and stability against subsynchronous vibrations.
- If the rerate power input exceeds the maximum power capacity of the existing shafts and power train components, the shaft journals, couplings, gear and the driver should be checked and uprated or replaced if necessary.
- Higher efficiencies, inherent with higher flow coefficient stages, are reduced by the higher losses in the unchanged piping, inlet nozzles, casing openings, inlet duct(s) and scroll(s), so that the original efficiency level can normally be maintained.
- Larger impeller diameters reduce the diffuser radius ratios, thus diminishing the efficiencies.

9.3 Higher Rotational Speed

In the case of a motor drive, the volume flow can, in many cases, be uprated by increasing the rotational speed through exchanging the gear elements. This can be less complicated and less costly for the compressor. In the case of a turbine drive, however, a new turbine rotor and guide blade carrier(s) are generally required. The compressor casing with nozzles, scroll(s), bearings and bearing span remains constant.

The mass flow ratio of the uprated compressor referenced to the existing compressor can be calculated as follows. Since for multistage compressors with compressible flow (i.e. medium-to-high Mach numbers) the head ratio is $y_p/y_{p0} > (N/N_0)^2$, a speed mismatch correction factor has to be accounted for (valid for uprate only):

$$F_N = \frac{(y_p/y_{p0})-1}{(N/N_0)^2 - 1}. \tag{9.7}$$

With slope S the mass flow increase caused by the speed increase is

$$\frac{\dot{m}}{\dot{m}_0} = \frac{N}{N_0} \times \left[\left(\left(\frac{N}{N_0}\right)^2 - 1\right)\frac{F_N}{S} + 1\right]. \qquad (9.8)$$

The first term is the speed increase of the rerated compressor at constant flow coefficient, and the second term is, again, the increased head caused by the higher speed and converted into increased mass flow at constant head via the curve slope (Fig. 9.4). This also means that an actual flow coefficient increase takes place by volumetrically overloading the stage with some efficiency decrease.

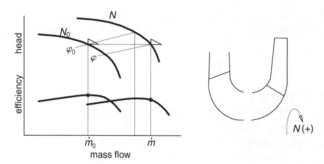

Fig. 9.4 Uprating by using higher rotational speed with area = constant, d_2 = const, ρ = const.

Although theoretically different, the curve slope is similar for the increased impeller diameter. The same is true for the speed-mismatch factor, representing the overproportional head increase in relation to N^2 due to stage mismatching. Each 1% speed increase yields some 3% mass flow increase. The consequences of a speed increase are as follows:

- A higher machine Mach number brings about a higher reduction of volume flows for each consecutive stage, leading to an aerodynamic stage mismatch for constant pressure ratio, which necessarily reduces the efficiency (Fig. 9.4). The higher the Mach number, the greater is the reduction in the efficiency.
- The operating flow range decreases, since the surge and choke point (maximum flow at zero head) approach the best efficiency point.
- The impeller stress increases. The separation margin to impeller material yield strength should be checked.
- The rotordynamics should be checked with regard to critical speed separation margin and stability.
- The impeller and coupling shrink fits should be checked.
- A new overspeed test may be required.

- If the rerate power input exceeds the maximum power capacity of the existing shafts and power train components, then the shaft journals, couplings, gear and driver should be checked and uprated if necessary.

9.3.1 Additional Losses

For both wider impellers and a speed increase, the friction losses caused by higher flow velocities in the casing nozzles, casing openings and stationary ducts will increase.

The increased inlet loss from the inlet flange via the plenum inlet up to the eye of the first impeller, referenced to the gas power of the pertinent downstream stage, can be calculated via the increased Mach number at each inlet flange from Eqs. (6.86)–(6.90) and from Fig. 6.15.

9.3.2 Case Study 1: Wider Impellers and Speed Increase Combined

In 1968, the author's company was awarded a contract by a major oil company in Germany for a turbine-driven five-stage hydrocarbon compressor in a fluid cat-cracking plant. Its design specifications are shown in Table 9.1, designated FCC 68; the suction volume flow is 6,220 m^3/h. In 1986, this compressor was uprated to 200% of the original volume flow, i.e. 12,500 m^3/h, designated FCC 86. In 1989, another uprate to 243%, i.e. 15,088 m^3/h, designated FCC 89, was required. The uprate pressure ratio was approximately constant; however, due to molar mass changes the head requirements rose by 5% for the first rerate and fell by 8% for the second rerate. How was this extreme uprating of an existing casing accomplished?

Since an increase of impeller diameters was not deemed appropriate due to the short diffusers, a combination of flow coefficient and speed change was the only way.

Table 9.1 Case study 1: Original design and two rerates

Specification			FCC 68	FCC 86	FCC 89
Mass flow	$\dot{m}$	kg/h	16,445	32,206	44,010
Volume flow	$\dot{V}$	m^3/h	6,220	12,500	15,088
Gas			hydrocarbons	+ H$_2$	+ H$_2$S
Molar mass	M	kg/kmol	36.58	34.98	39.64
Suct. pressure	p_s	bar	1.84	1.88	1.86
Suct. temp.	t_s	°C	38	38	37
Disch. pressure	p_d	bar	9.31	9.4	9.7
Pressure ratio	p_d/p_s		5.05	5.0	5.2

Table 9.1 (cont.)

Geometry					
Number stages	i		5	5	4
Impeller dia.	d_2	mm	5×425	425/400	3×425/400
Rel. impeller exit width	b_2/d_2		0.036–0.02	0.060–0.026	0.09–0.04
Diffuser ratio	d_4/d_2		1.41	1.55	1.51/1.55
Shaft diameter	d_w	mm	143	162	164
Shaft dia. at: cplg./brg./seal		mm	64/70/90	70/85/90	70/85/95
Bearing span	L	mm	1,480	1,480	1,480
Bearing type			fixed lobe	tilting pad	tilting pad
Seal type			floatg. ring	mech.contact	dry gas
Mechanics					
Rotor mass	m_R	kg	312	330	327
Rotor stiffness ratio	N_{1c}/N_{max}		0.41	0.41	0.45
Max. impeller stress/yield	σ_v/Rp_{02}		0.47	0.62	0.73
Aero data					
Flow coeff. 1st/last stage	φ_1/φ_{last}		0.053/ 0.017	0.094/ 0.036	0.119/ 0.045
Mach no., 1st	M_{u2}		0.815	0.90	0.93
Tip speed, 1st	u_2	m/s	229	261	245
Suction nozzle Mach number	M_s		0.06	0.12	0.16
Inlet loss	$\Delta\eta$		0.02	0.07	0.11
Pol. efficiency	η_p		0.76	0.74	0.78
Polytropic head	y_p	kJ/kg	127.4	133.9	117.8
Speed	N	1/min	10,300	11,710	11,140
Maximum cont. speed	N_{max}	1/min	10,820 (105%)	12,650 (108%)	12,500 (112%)
Power input	P_c	kW	825	1,712	1,930
Maximum turbine power	P_{Tmax}	kW	1,300	2,500	2,500
Comparison (relative)					
Mass flow	$\dot{m}/\dot{m}_0$		1.0	1.96	2.68
Volume flow	$\dot{V}/\dot{V}_0$		1.0	2.0	2.43
Pol. head	y_p/y_{p0}		1.0	1.05	0.93
Speed	N/N_0		1.0	1.14	1.08
Power input	P_c/P_{c0}		1.0	2.08	2.34

First Rerate

Since the maximum flow coefficient, which is regarded as a technological limit, increased considerably between 1968 and 1986 as a result of R&D activities, the greatest uprate potential was applying the ratio of flow coefficients φ/φ_0 (first term in Eq. (9.2)), which amounted to a whopping 162%. The second largest potential was speed increase. In spite of the low impeller material yield strength (630 N/mm^2 = 90,000 psi) there was enough margin for a hefty speed increase of 14% bringing the flow up to 185% (first term of Eq. (9.3)). The curve-slope effect according to the second term of Eq. (9.3) increased the flow to 200%; see Fig. 9.5.

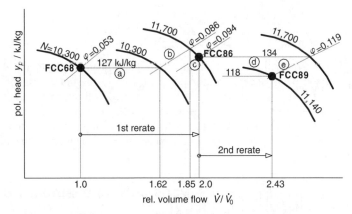

Fig. 9.5 Case study 1. a wider impellers, b speed increase, c slope effect, d wider impellers, e speed reduction.

As can be seen from Table 9.1 and from a comparison of the cross-sectional drawings for FCC 68 and FCC 86 in Figs. 9.6 and 9.7, the prerequisites and consequences of this large flow increase were the following:

- The complete internals, including rotor and diaphragms, were completely replaced within the short plant shutdown time.
- No attempt was made to salvage any of the old components. It was decided that this possible cost saving was not a reasonable risk compared to potential losses for each day's production.
- The original compressor had enough internal voids to accomodate wider impellers and wider stationary flow paths.
- The diffuser radius ratios could be increased from 1.41 to 1.55 benefitting the efficiency.
- The scroll-through flow area could be increased by 25%.
- The specification not only called for 200% flow of the original design but also for increased operational flexibility. This was accomplished by a so-called volumetric overload design which resulted in an efficiency penalty of 5% at the design point of 12,500 m^3/h. Thus, the turndown could be improved from 30% to more than 40%.
- The increased turndown was responsible for the efficiency degradation from 0.76 to 0.74. If the turndown could have been maintained at 30%, the efficiency would have gone up to 0.79.

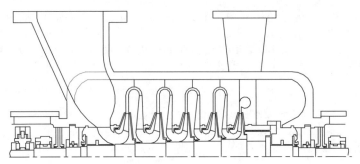

Fig. 9.6 Case study 1: original compressor of 1968, 100% flow.

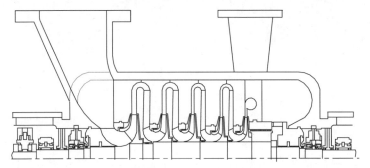

Fig. 9.7 Case study 1: first compressor rerate of 1986, 200% flow.

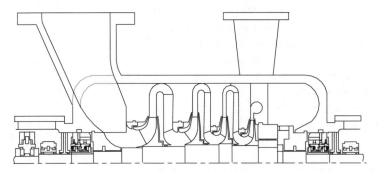

Fig. 9.8 Case study 1: second compressor rerate of 1989, 243% flow.

- The flow coefficient increase from 0.053 to 0.094 was possible through a first-stage impeller with backward-curved (twisted) 3D-blades. These were developed around 1970 and had accumulated a good record of experience by 1986.
- The stages used for FCC 86 represent the scalable fixed-geometry component system, the so-called "standard impeller family" introduced in 1971 and described in detail by Lüdtke (1992). As can be seen, this comprehensive system with its flexibility can be successfully used for extreme rerates as well, without any compromises with the philosophy of the system.

- The inlet loss from the suction flange to the impeller eye increased from 2% (FCC 68) to 7% (FCC 86), due to the considerably increased velocity. Therefore, the difference of 5% had to be charged to the first stage. In a similar way, the additional losses in the scroll and the adjoining diffuser and discharge nozzle were charged to the last stage.
- Since the maximum continuous speed increased by 17%, the shaft diameter had to be increased in order to maintain the shaft stiffness ratio, i.e. the ratio of the first rigid support critical speed to the maximum speed N_{1c}/N_{max}, which can be a cause for the onset of subsynchronous vibrations.
- In spite of the higher rotational speed, the actual impeller stress at maximum continuous speed is acceptable.
- The maximum turbine power increased from 1,300 to 2,500 kW. This required a thickening of the shaft diameter under the coupling and an increase of the compressor journal-bearing diameter from 70 to 85 mm.
- Doubling the power output necessitated a new rotor and guide-blade carrier for the back-pressure reaction-type turbine with partial admission-control stage, as shown in Figs. 9.9 and 9.10.

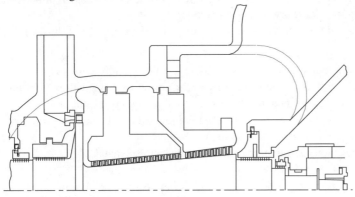

Fig. 9.9 Case study 1: original drive turbine of 1968.

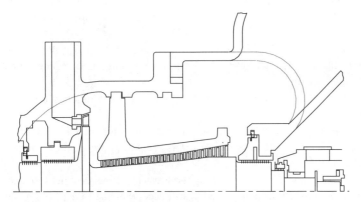

Fig. 9.10 Case study 1: rerated turbine, 234% power output.

The consequences of higher steam consumption and speed were as follows:
- The number of stages was reduced from 25 to 20.
- Longer, more efficient blades with better profiles were used.
- Increasing the first-stage exit pressure (wheel-chamber pressure) permitted a more favorable load distribution between the less efficient control stage and the highly efficient reaction-type stages.
- The end result of these changes was an increase of the turbine efficiency by 7%, offsetting part of the increased compressor power requirements.

Second Rerate

In 1989 the same compressor got another uprate (designated FCC 89), this time by 22%, so that the total uprate accumulated to an unprecedented 243% of the original volume flow. However, the head dropped by 12%, so that 7% was chopped off the original head because the gas became more dense. By 1989 another extension of flow coefficients to more than 0.12 was available and already well established, which led to the application of such a high flow coefficient first-stage impeller being the prerequisite for this extreme capacity uprate.

In Table 9.1 and on the cross-sectional drawing (Fig. 9.8), the following facts are revealed:

- The number of compressor stages could be reduced from five to four.
- This, in turn, was a prerequisite for increasing the first-stage flow coefficient from 0.086 to 0.119, since the axial stage pitch increased substantially.
- Needless to say, this new radical change required the replacement of the rotor and diaphragms with the casing being untouched.
- No old superfluous aerodynamic components were salvaged.
- The scroll was the only flow-path component to be retained.
- A large turndown was not required anymore, so that the design point came close to the best efficiency point.
- Standard impeller family components were applied exclusively. The second impeller had to be blade-trimmed at the exit. The first two impellers had 3D-blades.
- Inlet and exit losses due to the higher velocities in the nozzles, casing openings, inlet, and exit ducts were charged to the first and last stage and reduced their efficiencies by 9% each.
- In spite of these additional losses, the original compressor efficiency could be raised by 2%.
- The rotor weight was reduced and the mean shaft diameter increased slightly, thus increasing the rotor stiffness ratio to 0.45, safeguarding against subsynchronous vibration.
- The wider impellers have an inherently higher stress level; nevertheless, the actual stress at maximum continuous speed is still acceptable at 73% of the material yield strength.

- Although the power consumption increased, the maximum turbine power was still sufficient. Thus the turbine did not need another rerate.

It is thought to be a significant achievement that, in spite of the massive capacity increase, the efficiency suffered a minor loss of only 2% for the first rerate and actually gained 2% for the second rerate. Performance maps of FCC 68, FCC 86, and FCC 89, displaying polytropic head, polytropic efficiency, and power input versus actual volume flow, are shown in Fig. 9.11.

The first rerate incorporated a retrofit from floating-ring oil-shaft seals to mechanical contact seals to minimize the inner-seal oil leakage, which had to be disposed of; see Figs. 9.6 and 9.7.

The second rerate comprised a retrofit to dry gas seals in order to totally eliminate the seal oil system.

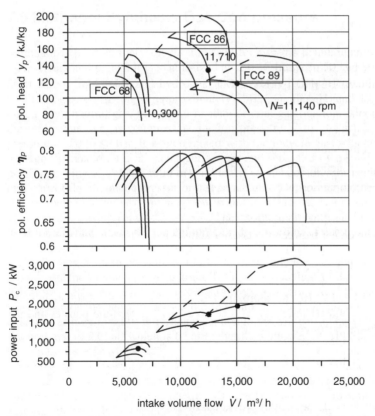

Fig. 9.11 Case study 1: compressor performance maps; original, first and second rerate; dots constitute "rated points".

9.4 Supercharging with a Booster Compressor

In certain cases the application of a single-stage booster upstream of the existing compressor is advantageous compared to new impellers or a speed increase. The advantages are:

- The flow-increasing potential is as high as 100% and more, even if the compressor is already equipped with high-flow-coefficient impellers.
- The main compressor remains (nearly) unchanged and can be operated with almost constant power input.

The approximate mass-flow increase of the main compressor with an aftercooled booster in operation is (all following equations are based on perfect-gas behavior):

$$\frac{\dot{m}}{\dot{m}_0} = \pi_B \frac{T_{s0}}{T_R} \left(\frac{\dot{V}}{\dot{V}_0}\right)_{MC} ; \qquad (9.9)$$

the symbols are defined as follows:

π_B booster pressure ratio
π total pressure ratio (booster + MC)
T_{s0} booster suction temperature
T_R recooling temperature
T_s suction temperature main compressor
$\dot{V}$ suction volume flow main compressor with booster in operation
$\dot{V}_0$ original suction volume flow main compressor
MC   main compressor
B   booster.

If there is no cooler between the booster and the main compressor, we obtain

$$T_{s0}/T_R = T_{s0}/T_s = \pi_B^{\frac{\kappa-1}{\kappa \eta_{pB}}} . \qquad (9.10)$$

From Eqs. (9.9) and (9.10) the approximate mass-flow increase of the main compressor with the booster in operation (and no cooler between the booster and the main compressor) becomes

$$\frac{\dot{m}}{\dot{m}_0} = \pi_B^{\frac{1+k\eta_{pB}-k}{k\eta_{pB}}} \left(\frac{\dot{V}}{\dot{V}_0}\right)_{MC} . \qquad (9.11)$$

As previously derived, the booster pressure ratio is

$$\pi_B = \left[s(\kappa-1)M_{u2}^2 + 1\right]^{\frac{\kappa \eta_{pB}}{\kappa-1}} \qquad (9.12)$$

9.4 Supercharging with a Booster Compressor

The approximate polytropic head of an intercooled compressor (from Eq. (6.7)) is

$$y_p \approx [T_s + cT_R(1+h_c)]R\frac{\kappa \eta_p}{\kappa-1}\left[\pi^{\frac{\kappa-1}{\kappa \eta_p(c+1)}} - 1\right], \quad (9.13)$$

where c is the number of intercoolers and h_c is the cooler head loss.

The original gas power input of the MC is

$$P_0 = \dot{m}_0 y_{p0}/\eta_{p0} = \dot{m}_0 T_{s0}\left(1+c\frac{T_R}{T_{s0}}\right)R\frac{\kappa}{\kappa-1}\left[\pi^{\frac{\kappa-1}{\kappa \eta_{p0\text{MC}}(c+1)}} - 1\right]. \quad (9.14)$$

The gas power input of the MC with booster in operation is

$$P = \dot{m}T_R(1+c)R\frac{\kappa}{\kappa-1}\left[\left(\frac{\pi}{\pi_B}\right)^{\frac{\kappa-1}{\kappa \eta_{p\text{MC}}(c+1)}} - 1\right]. \quad (9.15)$$

The MC power referenced to the original power and combined with Eq. (9.9) is

$$\frac{P}{P_0} = \left(\frac{\dot{V}}{\dot{V}_0}\right)_{\text{MC}}\frac{1+c}{1+c\frac{T_R}{T_{s0}}}\pi_B \frac{\left(\frac{\pi}{\pi_B}\right)^{\frac{\kappa-1}{\kappa \eta_{p\text{MC}}(c+1)}} - 1}{\pi^{\frac{\kappa-1}{\kappa \eta_{p0\text{MC}}(c+1)}} - 1}. \quad (9.16)$$

The approximate referenced MC volume flow with aftercooled booster in operation is

$$\left(\frac{\dot{V}}{\dot{V}_0}\right)_{\text{MC}} = \frac{P}{P_0}\frac{\left(1+c\frac{T_R}{T_{s0}}\right)}{(1+c)\pi_B}\frac{\pi^{\frac{\kappa-1}{\kappa \eta_{p0\text{MC}}(c+1)}} - 1}{\left(\frac{\pi}{\pi_B}\right)^{\frac{\kappa-1}{\kappa \eta_{p\text{MC}}(c+1)}} - 1}. \quad (9.17)$$

The approximate suction volume-flow ratio of the main compressor with the booster in operation (and no intercoolers, i.e. $c = 0$) is

$$\left(\frac{\dot{V}}{\dot{V}_0}\right)_{\text{MC}} = \frac{P}{P_0}\frac{\pi^{\frac{\kappa-1}{\kappa \eta_{p0\text{MC}}}} - 1}{\pi_B\left[\left(\frac{\pi}{\pi_B}\right)^{\frac{\kappa-1}{\kappa \eta_{p\text{MC}}}} - 1\right]}. \quad (9.18)$$

In Eqs. (9.10)–(9.18) the symbols are defined as follows:

- s work input factor booster, κ isentropic exponent,
- M_{u2} tip speed Mach number booster,
- $\dot{m}_0$ original mass flow, $\dot{m}$ uprated mass flow,
- π total pressure ratio (MC+ booster), π_B booster pressure ratio,
- η_{pOMC} original polytropic efficiency MC, $_{pB}$ polytropic efficiency booster,
- η_{pMC} polytropic efficiency MC with booster in operation,
- P power input MC with operating booster,
- P_0 original power input of the main compressor.

As can be seen from the formulae, there are many parameters that influence the booster-induced mass-flow increase. Simplified calculation results are displayed in Figs. 9.12 and 9.13 for the conditions indicated.

These curves also contain lines of constant main compressor volume ratios, indicating how much the suction volume flow necessarily has to change when the booster is put in operation. Higher pressure ratios require volumetric partload, and lower pressure ratios require overload operation of the main compressor. Overload is feasible, because the main compressor head reduces! For H_2-rich gas applications the overload can be as high as 35% due to the very flat curve between rated and choke point.

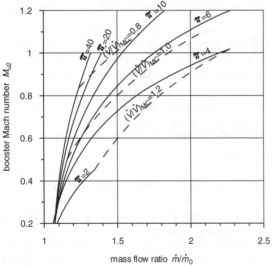

Fig. 9.12 Uprating by suction boosting for pressure ratio 2 to 40; *solid lines* total pressure ratio, *dashed lines* main compressor volume ratio, booster with aftercooler. Power ratio main compressor $(P/P_0) = 1.1$, isentropic exponent $\kappa = 1.4$, work input factor booster $s = 0.65$, polytropic efficiency booster $\eta_p = 0.8$.

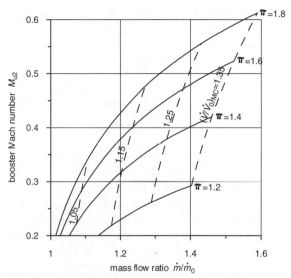

Fig. 9.13 Uprating by suction boosting: pressure ratio 1.2 to 1.8; *solid lines* total pressure ratio, *dashed lines* main compressor volume ratio, no intercoolers. Power ratio of main compressor $(P/P_0) = 1.0$, isentropic exponent $\kappa = 1.35$, work input factor booster $s = 0.62$, polytropic efficiency booster $\eta_p = 0.8$.

In general the following facts can be deduced:
- Maximum mass-flow increase caused by boosting reduces sharply with increasing pressure ratio.
- Full increasing potential can only be utilized if the booster operates at high Mach numbers.
- Compressors for light gases (e.g. H_2) with inherently low Mach numbers (i.e. very low pressure ratios) can only with effort be uprated by boosting.

Note: the maximum attainable booster Mach number is subjected to the maximum impeller tip speed: $M_{u2\,max} = u_{2\,max}/\sqrt{\kappa_s Z_s R T_s}$ with $u_{2max} = 280$ to 320 m/s. For example: a compressor with a 1.4 : 1 pressure ratio and a maximum booster Mach number of 0.3 can only be uprated to 117%.

At constant overall pressure ratio the main compressor excess head must be eliminated. This can be accomplished by:

- removal of impeller(s),
- impeller blade trimming,
- volumetric overload operation,
- speed reduction.

The latter is easy with a turbine drive. However, a motor drive will require the replacement of the gear box or, if possible, the replacement of the gear elements. For a turbine drive, in the event of a booster trip, the variable-speed main compressor can stay in operation at the original discharge pressure and mass flow by resuming the original 100% speed. The main compressor operating at constant speed, however, has to be shut down in the case of a booster trip, since it is no longer able to develop the original head.

All head-lowering measures widen the operating range, thus adding more flexibility for the joint operation with the booster.

Case Study 2: Boosting of a Two-Casing Air Compressor

In 1979 the author's company was awarded a contract by a major petrochemical company in the Far East for two two-casing steam and expansion turbine-driven 17 MW air compressors in a terephthalic acid plant (PTA), whose design specifications are shown in column 4 of Table 9.2. In 1994 one of the turbo trains was uprated to 164% volume flow, corresponding to 148% mass flow by means of a single-stage integrally geared compressor boosting the main compressor suction pressure to 2.0 bar. Booster data are shown in column 6 and the data for the main compressor jointly operating with the booster are exhibited in column 5.

For such an uprate, from the very beginning, it was a common understanding between the user and the manufacturer that the capital cost for new rotors and stationary internals for LP, HP compressor, steam and expansion turbine (i.e. a total of eight rotors including spares and four stationary aero-packages) would be comparatively expensive, and disassembly and reassembly time would be excessively long causing high losses of production.

A speed increase was out of the question for such a high uprate due to the high tip speeds used in the original designs.

The planned PTA train rerate lent itself to suction-side boosting, since it required no geometrical changes to the original compressor itself due to its turbine drive, since space for placing the booster, cooler, and pipes was available and since it was the least costly solution.

There were further advantages for suction side boosting:

- erection of the skid-mounted booster with cooler and the new piping system could be accomplished without disturbing normal train operation;
- the integration of the additional pipes, valves, control instrumentation, and monitoring system, and the booster commissioning and tying to the main compressor would increase the normal plant shutdown time insignificantly;
- it offered the possibility, in the case of a booster trip, to continue operating the main compressor at the original capacity and specified discharge pressure, as if the booster had never existed and would not interrupt production.

From Table 9.2 and the performance maps in Figs. 9.14 (base: mass flow) and 9.15 (base: volume flow), the following is deduced for the operating mode "booster + main compressor":

Table 9.2 Case study 2: technical data for original compressor with and without booster

			Original compressor rated point	MC with booster operating at rated point	Booster rated point
Mass flow, suction	$\dot{m}$	kg/h	107,150	153,790	159,084
Volume flow, suction	$\dot{V}$	m³/h	93,500	71,880	153,150
Gas			humid air	humid air	humid air
Molar mass	M	kg/kmol	28.6	28.6	28.0
Suction conditions	p_s/t_s	bar/°C	1.0/28	1.95/40	0.975/43
	rel. H.	%	85	100	100
Discharge pressure	p_d	bar	26.0	26.0	2.0
Pressure ratio	p_d/p_s		26.0	13.3	2.05
Design				single-shaft 2-casing no gear box	integrally geared, single stg.
Drive				steam turbine + expansion turbine	electric motor
Number of stages	i		5+6	5+6	1
Impeller diameter	d_2	mm	880–515	880–515	1,120
Flow coeff. 1st stage	φ		0.128	0.108	0.11
Tip speed	u_2	m/s	335	304	394
Mach number 1st stage	M_{u2}		0.96	0.85	1.09
Polytropic head	y_p	kJ/kg	349	270	77
Polytropic efficiency	η_p		0.803	0.78	0.82
Speed	N	1/min	7,270	6,600	6,710
Power input	P_c	kW	13,275	15,040	4,250
Maximum drive power	P_{max}	kW	17,500	17,500	4,800
Surge at p_d = const	$\dot{V}_{SL}$	m³/h	63,000	55,000	–
Turndown		%	33	23	–
Mass flow ratio	$\dot{m}/\dot{m}_0$		1.0	1.44	1.48
Volume flow ratio	$\dot{V}/\dot{V}_0$		1.0	0.77	1.64
Head ratio	y_p/y_{p0}		1.0	0.772	0.221
Speed ratio	N/N_0		1.0	0.91	–
Power input ratio	P_c/P_{c0}		1.0	1.13	–
Power input ratio, total	P_{cT}/P_{c0}		1.0	1.45	

- The specified mass-flow increase requires raising the main compressor suction pressure from 1.0 to 1.95 bar in order to keep the main compressor power input below the maximum driver power (mass-for-head-swap). The power input of 15,040 kW remains well below the maximum power. The difference between the volume increase (164%) and mass increase (148%) derives from a redefinition of the suction conditions (lower pressure, higher temperature, higher relative humidity).
- The main compressor speed dropped to 91% of the original speed.
- The booster pressure ratio of 2.05 requires a high head/high Mach-number machine, which can only be achieved with an unshrouded impeller. So, the single-stage booster is responsible for 22% of the total head, three times the main compressor average stage head.

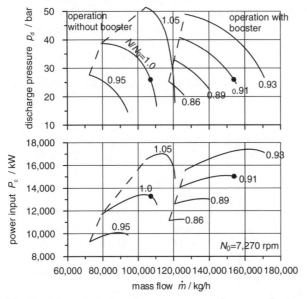

Fig. 9.14 Case study 2: main compressor performance. Basis: mass flow; original and with booster in operation; dots constitute "rated points".

- The main compressor suction-volume flow necessarily has to drop to 71,880 m³/h or 77% of the original value, also depicted by the flow coefficient drop from 0.128 to 0.108. This small flow can be handled without blow-off, since the surge limit at the reduced speed falls to a comfortable 55,000 m³/h at constant discharge pressure versus 63,000 m³/h at 100% speed.
- The main compressor performance maps with and without booster are shown in Figs. 9.14 and 9.15. The surge limit shift is clearly illustrated. The efficiency drops some 2% with boosting, because at 9% lower speed the latter stages tend

to volumetrically overload, thus sacrificing efficiency through aerodynamic mismatching.
- As can be seen in Fig. 9.15, comparing the heads of both operating modes, the mismatch, on the other hand, benefits the surge limit, thus enabling a safe surge-free operation with the booster in operation.
- The overall efficiency is kept approximately constant, since the booster compensates for the efficiency drop.
- The single-stage booster is a compact integrally geared design with axial inlet. The common base frame for the compressor, gear and motor, houses the complete oil system, which enables an easy installation on a table foundation or on the ground next to the main compressor.

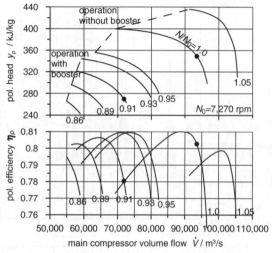

Fig. 9.15 Case study 2: main compressor performance. Basis: volume flow; original and with booster in operation; dots constitute "rated points".

- The normally unshrouded overhung impeller is mounted on the extended-pinion shaft. Impeller blade channels are five-axis-NC-milled from the solid-hub forging, even for such large impellers, as shown in Fig. 9.16. The axial inlet assures a high efficiency and can be equipped with adjustable inlet guide vanes that serve as an efficient means to adapt both compressors to each other during seasonal changes of operating conditions and to bring about load sharing if required. The booster shaft can be equipped with a choice of a carbon ring seal, a mechanical contact seal or a dry gas seal, depending on the gas handled (see drawing in Fig. 9.17, photo in Fig. 9.18).
- The booster, with its aftercooler, was placed on a ground-level foundation beside the main compressor.

Fig. 9.16 Case study 2: booster impeller, 1,120 mm diameter, unshrouded, backswept blades with axial inducer (S-type impeller, definition see Sect. 3.1.4), five-axis-NC-milled.

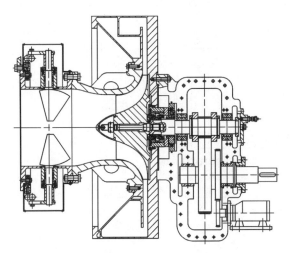

Fig. 9.17 Case study 2: single-stage booster, compact integrally geared, with IGVs, overhung unshrouded high-head impeller.

- Refer to Fig. 9.19 for a schematic of the piping arrangement for the booster and main compressor. During normal operation the booster/main compressor hookup valve V1 is open, while the main compressor atmospheric-suction valve V2 and the booster blow-off valve V3 are closed. During the sequential start-up of the two compressors, the valve positions are vice versa.
- In 1996 the expansion turbine was uprated to 150% mass flow and power output by means of a new rotor and new nozzles.

Fig. 9.18 Case study 2: single-stage booster with adjustable IGVs, skid mounted.

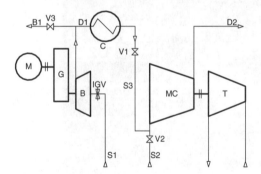

Fig. 9.19 Case study 2. Main compressor with booster (schematic diagram). MC main compressor, T steam turbine, B booster, G gear box, M motor, C intercooler, IGV adjustable inlet guide vanes, S1 booster suction, S2 MC atmospheric suction, S3 booster hook-up line, V1 booster/MC hook-up valve, V2 MC atmospheric suction valve, V3 blow-off valve, D1 booster discharge, D2 MC discharge line, B1 blowoff line.

- Another very similar example of an air compressor in an FCC Plant is shown in Fig. 9.20, whose volume flow was increased by 83% through boosting (a photo of the booster and MC at site for case study 2 is not available).

Fig. 9.20 Example: Main compressor with booster and intercooler. MC main compressor, B booster with noise hood, G gear box with noise hood, M motor, C booster aftercooler, H header, F1 table foundation, F2 ground level foundation booster, D1 booster discharge, D2 MC discharge line, S1 booster suction, S3 booster hook-up line, B1 blow-off line.

9.5 Rerate Methods: Advantages/Disadvantages

A comparison of the rerate methods is compiled in Table 9.3.

Table 9.3 Advantages and disadvantages of the three rerate modes

Wider impellers	Speed increase	Boosting
Advantages		
casing unchanged	gear-wheel set change easy, if feasible	main compressor train with drive cooler(s) nearly unchanged
older compressors: high uprate potential	light gases: high capacity increment per speed increment	turbine drive: original operation resumed in case booster trips
drive-turbine casing unchanged		booster switch-off: favorable bypass-free turndown
		low-cost compact single-stage booster, skid-mounted
Disadvantages		
internal space must be available	new turbine rotor	space required for booster unit and aftercooler
new rotor, diaphragms	motor, gear, cooler(s) uprate or new	new connecting pipes, valves, aftercooler
new turbine rotor, new stator	check foundation, skid	new comprehensive control and monitoring system
motor, gear, cooler(s) uprate or new	high pressure ratios: limited uprates (steep performance curves)	start-up more complicated
check foundation, skid	no uprates if original speed is maximum	light gases: limited uprate potential
newer compressors: lower uprate potential		high pressure ratios: lower uprate potential

9.6 Anticipating Rerates

The specification for a new compressor should be carefully scrutinized with regard to any anticipated capacity increases. If there are any plans to increase the flow sometime in the future, even if it is only a faint idea, it is highly advisable to take that into account when specifying and ordering the new compressor. Not only nozzles, shaft diameters and bearings, but also the flow volutes, which are often the bottlenecks, should have appropriate reserves for these future extensions. These provisions do not normally carry an additional price tag. However, it would also be advisable in such a case that couplings, gears and main drives be selected on the basis of the required extra power reserve exceeding the API 617 minimum reserve of 10%. This, of course, does increase the capital costs, but to a much smaller extent than would otherwise be required later on, when these items have to be uprated or even totally replaced.

9.7 Conclusions

Quite an array of options are available for centrifugal compressor rerating:

- Method 1: wider impellers, i.e. increase of the flow channel through-flow area that can be brought about by either higher flow coefficient impellers or impeller scale-up.
- Method 2: increasing the flow path gas velocity, which is achieved by raising the rotational speed.
- Method 3: increasing the suction density, which is attainable by placing a separate booster compressor upstream of the existing machine.

The potential of all three approaches was demonstrated by means of two case studies from the daily life of an OEM.

In the first study, concerning a hydrocarbon compressor in an FCC plant built in 1968, method 1 was combined with method 2 to obtain 200% of the original flow rate in the first step and 243% of the original flow rate in the second step. Step 1 involved a complete replacement of the compressor rotor, including diaphragms, and of the drive turbine rotor, including the guide blade carrier. Step 2 comprised another new compressor rotor and diaphragms. The step-up of flow coefficients from 0.05 to 0.09 to 0.12 reflected the R&D efforts over the decades and demonstrated clearly that a 20-year-old compressor has an enormous uprate potential, possibly without changing speed. However, an uprate of the magnitude described required a speed increase as well: 14% for step 1 and 8% for step 2, related to the original speed. The inevitable head rise as a side effect of the speed increase was converted to a flow increase via the performance-curve slope.

The first flow extension of 62% was achieved by a flow-coefficient increase and 38% by a speed increase. The final extension to 243% flow would have rarely been feasible without the elimination of one stage made possible by a specified head reduction.

In general, an older compressor (say, 25 years old) has an uprate potential by flow coefficient of at least 50% and at most around 100%. In most cases, the uprate potential by speed increase is at least around 20%. Scaling up impellers normally yields around 20% increase also.

Newer compressors (say, 10 years old), equipped with high-flow-coefficient impellers with the maximum possible diameter in the particular casing and running at the highest permissible speed can only with difficulty be uprated by wider impellers or speed-up.

Even if the compressor aero-package is already congested, jam-packed and tightly overcrowded, suction-side boosting can be applied to bring about a substantial uprate. This is what the second case study is all about. It concerns a 17 MW two-casing air-compressor train in a PTA plant built in 1980. It was uprated to 148% mass flow by boosting with a single-stage compressor. The suction pressure of the existing machine was raised from 1.0 bar to 1.95 bar, reducing its head to 77%, its speed to 91%, and its suction volume flow to 77%, and increasing its power input to 113%, well below the maximum turbine power output. Geometrically, the main compressor remained unchanged. Hence, with the booster not in operation, it can instantly resume the old operating mode at atmospheric suction and specified discharge pressure. Boosting was, by far, the lowest-cost solution compared to new rotors and/or speed increase.

In general, the uprate potential through suction-side boosting depends on many parameters, especially on the overall pressure ratio and the feasible tip-speed Mach number of the booster. For pressure ratios of 2 up to 10, uprates of between 160 and 220% are possible. For pressure ratios up to 40, uprates of between 130 to 150 percent are possible. Low pressure ratio compressors (e.g. H_2-rich gases) have a lower uprate potential of between 110 to 140%, depending on how much the booster tip-speed Mach number can be maximized, e.g. through high yield strength impeller material.

In spite of additional losses occurring for any rerate, be it in the nozzles or by aerodynamic mismatch or by part load or overload operation, the efficiency level can, in most cases, be maintained. Efficiency levels of completely new compressors designed from scratch for the high flow rates must necessarily be higher than for rerated compressors. Efficiencies can be improved, though, by flow-channel surface smoothing and replacement of conventional interstage seals with rubresistent labyrinth-seal material.

When new compressors are specified and ordered, future uprates should be anticipated from the very beginning, even if a rerate is only a remote idea. Every compressor casing ought to have some space reserves internally, in the casing openings, or some oversized shaft diameters and couplings. Every gear box should have the possibility of accommodating a new gear-wheel set suitable for higher speed and power, in order to rerate at a later date with wider impellers and higher power inputs.

Compressor rerating is a favorable alternative for increasing the production, since the efficiency is maintained. The costs are lower than for new builds and the hardware exchanges can be accomplished in little more than the routine plant shutdown period.

References

Runge KM (1996) Increasing capacity and energy efficiency by revamp – an interesting option for any ammonia plant. Abu Qir Ammonia/Urea Technology Symposium, 1996, Alexandria, Egypt. Arabian Fertilizer Association, Cairo, Egypt

Lüdtke K (1992) Twenty years of experience with a modular design system for process centrifugal compressors. Proc. 21st Turbomachinery Symposium, 1992, Houston, Texas. Turbomachinery Laboratory, Texas A&M University, College Station, Texas

10 Standardization of Compressor Components

Process centrifugal compressors of the single-shaft type are characterized by an almost infinite number of variation options. They compress gases with molar masses of between 2 and 100, covering volume flows within the range of 0.1 to 200 m^3/s and discharge pressures from below atmosphere to 800 bar, many with side streams or extractions. From one impeller up to 25 impellers are arranged in-line or back-to-back in up to four casings. Compressors are either uncooled or have a maximum of three intercoolers per casing, resulting in a maximum number of eight nozzles per casing. Process compressors are driven by electric motors, steam or gas turbines either directly or via a gear box between the drive and compressor or between the casings.

Such a vast array of combinations excludes any package-unit standardization as is the case, for example, for plant-air or water-chilling compressors, where the gas never changes and the pressure ratios are within narrow limits.

In spite of all this, there were early attempts in the industry to standardize, i.e. to limit the design options of impellers and of stationary components; nobody wants to reinvent the wheel for every new compressor to be built, because extrapolating means reducing prediction accuracy.

There is, however, no doubt that any attempt to bring about a standardization has to concentrate on the stage as the smallest element common to all process compressor variants.

Basically, there are two philosophies for standardizing impellers: meridional impeller blade cutoff and scalable fixed-geometry impellers.

10.1 Meridional Impeller Blade Cutoff

A wide-spread conventional method is to work with a predetermined scalable impeller blade shape in space; from contract to contract, cut-outs from this singlecurved (2D surface) or twisted (3D surface) confined area are calculated that fit the special application. Once the impeller outer diameter is set, the meridional impeller contour is adapted in this way to arrive at the following necessary parameters:

- hub diameter (to attain the necessary shaft stiffness),
- eye diameter (to attain the specified volume flow),
- impeller exit width (to attain the envisioned head),
- impeller diffusion (to attain anticipated efficiency and operating range).

The method is straightforward for impellers having 2D-blades, where the meridional hub contour can be regarded as fixed. The blade cutoff is carried out on the shroud side only, giving the impeller the calculated inlet and exit widths: highly convergent with high molar masses and (almost) parallel with hydrogen. Of course, this requires certain base master models at different flow-coefficient levels, e.g. one for $\varphi = 0.07$–0.05 through shroud-side cutoff, the next one for $\varphi = 0.05$–0.03, and so on. This takes care of the different eye and hub diameters and lower blade inlet angles at lower flow coefficients. And it requires most likely a second master-model line with a different hub/tip ratio to adapt to varying shaft stiffness requirements; see Fig. 10.1. The cover disk is manufactured accordingly.

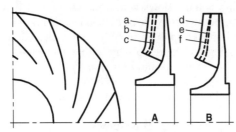

Fig. 10.1 Meridional blade cutoff for a 2D-impeller. **A** high hub/tip ratio, **B** low hub/tip ratio, a,d maximum blade surface, b medium flow coefficient, c low flow coefficient, high convergency, e medium flow coefficient, high convergency, f low flow coefficient, low convergency.

The meridional cutback approach is somewhat more complicated for impellers having 3D-blades, where the blade angle varies with the axial distance from the hub side. A mere shroud-side cutoff can aerodynamically be unacceptable because the blade inlet angle at the eye diameter increases excessively (at point b on the left-hand side of Fig. 10.2A) compromising the w_1-minimum criterion and blade loading. So, to convert a wide impeller to a narrow one, it does not suffice to just cut off a blade strip from the shroud side; instead, the blade leading edge must be extended towards the eye to assure the proper inlet angle. The cutoff for high-flow coefficient impellers is along line abc, and the cutoff for low flow coefficient impellers is along line def.

Separately manufacturing the impeller blades, e.g. through die forging, requires several dies of different size and angle gradient. A die whose surface is made up of logarithmic spiral elements offers several options: variation of outer diameter, flow coefficient, convergency, blade exit angle and hub diameter. The flat plate, prior to forging, has just to be placed at different locations of the die to realize these options (Fig. 10.2B), which have to be endorsed by computations of blade angle course, blade loading and diffusion. The log spiral stacking line LSSL is the "axis" where all log spirals are hung up. If this line coincides with d_2 the rake angle is 90°; if d_2 is smaller, the rake angle is < 90°; and if d_2 is larger, the rake angle is >90°, i.e. "overdrawn". The rake angle is between the blade trailing edge and the hub disk when looking right-angled at the shaft. The straight-line blade elements, required for flank milling, historically came to replace the die-forging concept; and they turned out to be a very good approximation for the log spirals without compromising the aerodynamics at all. Separately produced blades require welding to the hub and cover disks.

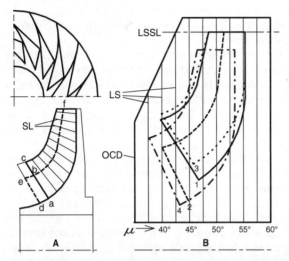

Fig. 10.2A, B Meridional blade cutoff for a 3D-impeller. **A** 5-axis NC blade milling, **B** die forging of blades, abc cutoff for wide impellers, def cutoff for narrow impellers, SL straight line blade elements, OCD outer die contour, LS logarithmic spirals, μ log spiral angle, LSSL log spiral stacking line, 1 high flow coefficient, 2 low flow coefficient, 3 high hub diameter, 4 small outer diameter and smaller blade exit angle.

The meridional blade cutoff as a semi-standard concept (working with a fixed and a variable dimension) gives the impression that the stage is designed exactly on the rated point. However, there are unavoidable compromises with regard to efficiency and prediction accuracy by extrapolating established experience limits.

10.2 Modular Design System[1]

The "modular design system", on the other hand, is rather unconventional, although it has been used for 30 years now. It is defined as a fixed but scalable geometry impeller family, whose members represent discrete flow coefficients between 0.01 and 0.15 (Fig. 3.4). "Scalable" means that the stage geometry is fixed

---

[1] Extracted (with modifications and supplements) from: K. Lüdtke: "Twenty years of experience with a modular design system for centrifugal process compressors". Reproduced with permission of the Turbomachinery Laboratory (http://turbolab.tamu.edu). From Proceedings of the Twenty-First Turbomachinery Symposium, 1992, Turbomachinery Laboratory, Texas A&M University, College Station, Texas, pp. 21-34, Copyright 1992.

in relative terms, and that all dimensions in relation to the impeller outer diameter d_2 remain constant when increasing or decreasing this diameter: b_2/d_2, b_1/d_2, d_H/d_2, d_E/d_2, and so forth; the blade angles are invariant.

Matching annular diffusers and return vane channels are assigned to each standard impeller, making up the complete modular stage. Impellers can also be combined with plenum inlets, when located downstream of a suction nozzle, or with volutes, when located upstream of a discharge nozzle. Standard side-stream arrangements are part of the system as well.

For any new job the stages are not designed but selected from the modular family, and it is then calculated how this stage performs under the anticipated conditions; that means that the stage is allowed to operate in a narrow limit to the left or right of the best efficiency point. If this performance does not meet the expectations, the adjacent impeller will be selected. The method is sometimes called "reverse engineering", i.e. calculating the aero-thermodynamic performance on the basis of an existing geometry.

The inlet flow coefficient mentioned above for the nomenclature of the line, however, is not sufficient to define the complete impeller geometry. The second parameter is the impeller volume ratio, which is a strong function of the molar mass of the gas (Figs. 6.13 and 6.14). This in turn influences the "convergency", i.e. the meridional area reduction from impeller inlet to exit. Each impeller at a given flow coefficient splits up into two convergencies, representing two discrete volume ratios, covering molar masses of approximately between 2 and 40 (wider exit) and between 35 and 100 (narrower exit), and the wider impellers split up into two hub/tip ratios to satisfy the rotordynamic requirements.

10.2.1 Aerodynamic Stage Rating Range

The starting point is the measured variable-speed map, Fig. 5.17, which is converted into a non-dimensional so-called φ_3-map in Fig. 5.19. The next step from there is to set up a "total performance map", Fig. 5.20, displaying the impeller volume ratio versus the flow coefficient, with efficiency, Mach number, work input factor and impeller exit flow coefficient as parameters.

With this information the rating range of this given measured stage can be established in such a way that it covers the optimum efficiency field (area enclosed by thick solid lines in Fig. 10.3).

The rating range limits are defined by a minimum φ_3-line, a maximum φ_3-line, and a maximum φ-line, leading to an efficiency drop-off of 0.5% in 90% of the range and a maximum of 1% near the borders of the range. So the philosophy of the system is "limited deviations from the aerodynamic similarity" (much in the same way as the power test code PTC 10 permits deviations of the flow coefficient, the volume ratio and the Mach number when reproducing the compressor rated point with a substitute gas on the shop test bed). The rating range is the φ/φ_3-range the individual stage is permitted to operate in at the rated point of the multistage compressor. So some stages may operate at the optimum, some at the range-border lines, and others right in the middle. So the modular design concept is based on stage rating ranges rather than on a stage rating point.

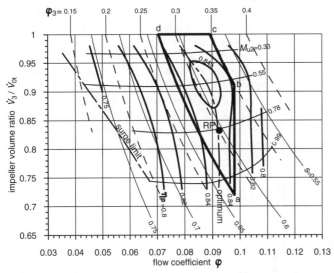

Fig. 10.3 Total performance map of a standard stage (example with 3D-impeller, $\varphi = 0.092$). abcda rating range, RP rated point, φ_3 exit flow coefficient, s work input factor, M_{u2} tip speed Mach number, η_p polytropic efficiency.

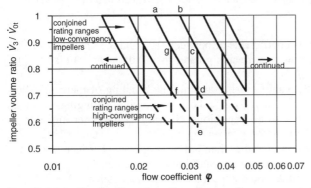

Fig. 10.4 Impeller family: schematic, selection diagram, consisting of conjoined stage rating ranges. abcda rating range of low-convergency impeller, defgd rating range of high-convergency impeller; both stages have identical rated-point φ-values.

10.2.2 Impeller Family

The neighboring rating range pertaining to the next higher or lower flow coefficient impeller can now be added, so that the maximum φ_3-line coincides with the mimimum φ_3-line of the next higher stage, and so that the minimum φ_3-line coincides with the maximum φ_3-line of the next lower stage. In this way all rating ranges are strung together, forming an impeller series with a convergency covering the lower molar mass gases (i.e. lower Mach number applications), as shown in Fig. 10.4. The lower voids resembling a sawtooth are filled in by a second impeller line operating at higher Mach numbers and, hence, with lower impeller volume ratios, leading to a more convergent meridional impeller shape. So the only difference between two impellers of constant inlet flow coefficient lies in the variation of the convergence of the meridional flow path, as shown in Fig. 10.5.

The conjunction of all these rating ranges leads to the impeller family; see Fig. 3.4. It remains to be added that besides two impeller convergencies there are two hub/tip-ratio options to cope with the shaft-stiffness requirements; see Fig. 6.8.

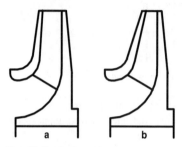

Fig. 10.5a, b Impeller convergency. **a** low-convergency impeller (low Mach number), **b** high-convergency impeller (high Mach number); flow coefficients of a and b identical.

Impellers with inlet flow coefficients of between 0.01 and 0.05 have 2D-blades, and impellers with inlet flow coefficients greater than 0.05 have 3D-blades (philosophy explained in Chap. 3).

With the advent of NC-manufacturing an unlimited scaling of these impellers within the complete outer-diameter range became possible, bringing about an enormous flexibility to adapt to all conditions. By varying the outer impeller diameter within very small limits for a given impeller type, any standard stage can be maneuvered into its optimum, rather than letting it operate near its rating range border lines. This leads sometimes to ups and downs of the outer impeller diameters in a compressor where one would expect a constant diameter.

10.2.3 Frame-Size Standardization

Whereas Fig. 3.4 shows the different impellers at constant outer diameter, Fig. 10.6 presents the approximate range chart of the complete modular system, covering actual volume flows from 1 to 60 m^3/s (approximate) with nine frame sizes geometrically stepped at 19% diameter increments, representing 42% volume flow increments at constant tip speed. The frame-size designations indicate the maximum possible impeller diameter per casing.

It has to be kept in mind, though, that the maximum volume flow per frame size can vary considerably. It can be as low as 17% and as high as 130%, related to the maximum value in Fig. 10.6 (valid for a maximum of four impellers and a normal material yield strength of 830 MPa) and depending on the gas, number of stages, material yield strength and casing type, as described in Chap. 4 (e.g. see Fig. 4.8).

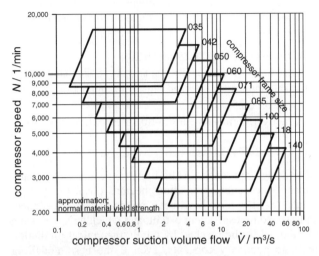

Fig. 10.6 Compressor range chart for single-flow horizontally split casings. Frame-size designations: maximum impeller outer diameter in cm. Maximum volume flow to be reduced for: barrel casings, number of stages > 4, molar mass > 35, cold gases, corrosive gas constituents.

10.2.4 Diaphragms

For each frame size two diaphragm lines are available: the 1.7-line and 1.5-line, representing vaneless diffuser radius ratios of 1.7 and 1.5, referenced to a nominal impeller diameter. However, since the impeller diameters vary in a given casing, because steps are necessary toward the rear end, depending on the Mach number, the diffuser radius ratios vary from approximately 1.4 to 1.8. The stage-efficiency prediction includes diffuser and return vane losses, thus accounting for diffuser radius ratio variations; see Fig. 6.20.

10.2.5 Discharge Volutes

The ten cast standard volutes per frame size that used to be available for selection, accounting for the enormously varying discharge volume flows of the compressor sections, were replaced by milled volutes (described in Chap. 6) with rectangular cross-sections and a large upstream diffuser radius ratio, placing the volute tongue sufficiently away from the impeller exit. The volute exit area range varies between 100% (largest) and 15% (smallest) per frame size. Conical exit diffusers provide a rectangular/circular transition to the pertaining standard discharge nozzles.

10.2.6 Systematic Tests of Modular Design Stages

To confirm the theory and to establish a reliable data base, numerous aerothermodynamic tests were systematically conducted on the development test stand. Basic air tests were carried out on every other impeller type of the family at a constant diameter of 400 mm and with a plenum inlet and a return vane channel outlet with the following variations:

- long and short vaneless diffusers (see Figs. 6.19 and 6.20),
- large and small stage pitches (axial distance between two successive stages),
- impeller blade cutbacks at the outer diameter (see Figs. 6.17 and 6.18a),
- large and small volutes (see Fig. 6.25),
- supplementary tests with the refrigerant R12 up to tip speed Mach numbers of 1.4 (see Fig. 4.1),
- supplementary tests with low-solidity diffuser (LSD) vanes (see Fig. 8.11).

The modular compressor design approach on the basis of detailed systematic aerodynamic stage tests makes theoretical prediction methods obsolete. A great but finite number of modules exist, which have been pre-engineered, designed, drafted and aerodynamically tested to a sufficient extent, and which have yielded a prediction accuracy that cannot so far be reached by purely theoretical prediction techniques with variable stage geometry concepts, even if the best modelling methods available today are applied.

The strength of the fixed-geometry system lies in the fact that only those discrete impellers are being used, together with their specific aerodynamic characteristics, that have been measured or have been derived from measurements.

Appendix

Unit Equations

Quantity	SI system of units		US system of units
Length	0.30480 m	=	1 ft
	25.400 mm	=	1 in
Mass	1 kg	=	2.2046 lb_m [1]
Pressure	1 bar	=	14.504 lb_f/in^2 [2] (=PSIA)
Pressure loss	1 mbar	=	0.014504 PSI
Temperature	°C	=	1.8×(°C+32) °F
	(°F-32)/1.8 °C	=	°F
Thermodynamic temperature	1 K	=	1.8 R
Volume flow (= capacity)	1 m^3/s	=	2118.88 ft^3/min (=CFM)
	1.699 m^3/h	=	1 CFM
Mass flow	1 kg/s	=	132.28 lb_m/min
Standard volume flow	1.6075 m^3/h [3]	=	1 ft^3/min (=SCFM) [4]
	1,116.3 m^3/h [3]	=	1 MMSCFD [4]
Head (= work)	1 kJ/kg	=	334.55 ft lb_f/lb_m
Specific enthapy	2.326 kJ/kg	=	1 BTU/lb_m
Specific entropy	4.1868 kJ/(kg K)	=	1 $BTU/(lb_m\ R)$
Isobaric specific heat capacity	4.1868 kJ/(kg K)	=	1 $BTU/(lb_m\ R)$
Isobaric molar heat capacity	4.1868 kJ/(kmol K)	=	1 $BTU/(lb_m$ mole R)
Molar mass	1 kg/kmol	=	1 lb_m/lb_m mole
Density	16.018 kg/m^3	=	1 lb_m/ft^3
Specific volume	1 m^3/kg	=	16.018 ft^3/lb_m
Power	1 kW	=	1.341 hp
Peripheral speed	1 m/s	=	3.2808 ft/s (=FPS)
Universal gas constant,	8.3145 kJ/(kmol K)	=	1545.3 ft $lb_f/(lb_m\ R)$
Acceleration of gravity, g (standard, sea level, latitude 45°)	9.80665 m/s^2	=	32.1740 ft/s^2
Basic equations			
Force, MKS system	1 kg_f	=	1 kg × 9.80665 m/s^2 = 9.80665 N
Force, SI system	1 N	=	1 kg × 1m/s^2
Force, US system	1 lb_f	=	1 lb_m × 32.174 ft/s^2
Energy, SI system	1 J	=	1 Nm = 1 Ws

[1] lb_m pound mass,
[2] lb_f pound force
[3] SI standard at 1.01325 bar/0°C
[4] US standard at 1.01325 bar/15.5 °C (14.7 PSIA/520 R)

K Kelvin, R Rankine, PSI pounds per square inch, PSIA pounds per square inch absolute, CFM cubic feet per minute, SCFM standard cubic feet per minute, MMSCFD million standard cubic feet per day, BTU British thermal unit, FPS feet per second

The unit equations numerically relate different units to each other, thus converting any unit of a physical quantity into another (desired) unit. The simple rules are:

(a) Write down the physical quantity as the numerical value times the unit.
(b) Select a suitable unit equation.
(c) Set up the two units as a ratio having a value of 1.0.
(d) Multiply the physical quantity with this unity brace.
(e) Cancel equal units in the numerator and denominator (as in an ordinary fraction) to arrive at the desired end unit.

Example: Calculate the head at an impeller tip speed of $u_2 = 300$ m/s with head coefficient $\psi = 1.0$.

Physical equation for the head:

$$y = \psi u_2^2 / 2 .$$

Relevant unity braces:

$$1 = \left\{ \frac{kJ}{1{,}000\,N\,m} \right\} \quad \text{and} \quad 1 = \left\{ \frac{N s^2}{kg\,m} \right\} .$$

Numerical equation (with N, s^2, m^2 being finally cancelled out):

$$y = \frac{\psi u_2^2}{2} = \frac{1.0 \times 300^2}{2} \frac{m^2}{s^2} \left\{ \frac{kJ}{10^3\,N\,m} \right\} \left\{ \frac{N s^2}{kg\,m} \right\} = 45 \frac{kJ}{kg} .$$

Corresponding tip speed in the US system of units using the {ft/m} unity brace:

$$u_2 = 300 \frac{m}{s} \left\{ \frac{ft}{0.3048\,m} \right\} = 984.25 \frac{ft}{s} .$$

Numerical equation for the head using the $\{lb_f / lb_m\}$ unity brace:

$$y = \frac{1.0 \times 984.25^2}{2} \frac{ft^2}{s^2} \left\{ \frac{lb_f\,s^2}{lb_m \times 32.174\,ft} \right\} = 15{,}054.9 \frac{ft\,lb_f}{lb_m} ,$$

(note: vernacularly "feet" for head is wrong (head is specific energy, not length; $lb_f \neq lb_m \rightarrow$ not cancelable!).

English–German Glossary on Compressors

acceptance test	Abnahmemessung
adjustable inlet guide vanes	bewegliche Eintrittsleitschaufeln, Dralldrossel
aftercooler	Nachkühler
amplification factor	Verstärkungsfaktor
anergy (unavailable energy)	Anergie
angular momentum	Drehimpuls, Drall
annular diffuser	Scheiben-Diffusor
application example	Anwendungsbeispiel
asynchronous motor	Asynchronmotor
axial compressor	Axialverdichter
axial thrust	Axialschub
back-to-back impellers	gegeneinander geschaltete Laufräder
backward curved	rückwärts gekrümmt
balance	auswuchten
balance piston recirculation loss	Ausgleichskolben-Verlust
barrel (radially split) casing	Topfgehäuse
beam type rotor	beidseitig gelagerter Läufer
bearing loss	Lagerverlust
blade angle	Schaufelwinkel
blade blockage	Schaufelversperrung
blade channel	Schaufelkanal
blade cutback	Schaufelausdrehung
blade design	Schaufel-Auslegung
blade loading diagram	Schaufelbelastungs-Diagramm
blade milling	Schaufel-Fräsung
blade thickness	Schaufelstärke
booster compressor	Vorverdichter
breakaway torque	Losbrech-Moment
casing	Gehäuse
casing design	Gehäusetyp, Gehäusebauart
cast impeller	Guss-Laufrad
center of gravity (CG)	Schwerpunkt
centrifugal compressor	Radialverdichter, Radialkompressor
centrifugal pump	Kreiselpumpe, radiale Turbopumpe
choke	Schluckgrenze, Sperrgrenze
clearance	Spiel
combustion chamber	Brennkammer
compressibility factor	Realgasfaktor
compressor characteristic	Verdichter-Kennzahl
compressor design	Verdichterbauart, Verdichterauslegung
compressor sizing	Verdichter-Dimensionierung
condenser	Kondensator
constraint	Begrenzung, Einschränkung
continuity equation	Kontinuitätsgleichung
core rotation factor	Kernrotations-Faktor
coupling thrust	Kupplungsschub
cover disk, shroud	Deckscheibe

critical speed	kritische Drehzahl
cross-coupling force	Querkraft
cross-coupling spring constant	Querfederzahl
damping coefficient	Dämpfungszahl
density	Dichte
desuperheating	Enthitzung
dew point	Taupunkt
diaphragm	Leitteil (Diffusor+Rückführstufe (Spirale))
diaphragm cooling	Innenkühlung
diffuser diameter (radius) ratio	Diffusor-Verhältnis
diffuser pinching	Diffusor-Verengung (-Einschnürung)
diffusion	Verzögerung
discharge nozzle	Druckstutzen
discharge pressure	Enddruck
disk friction	Scheibenreibung
double-arc blade	Zweikreisbogen-Schaufel
double-flow	doppelflutig
downstream	stromabwärts
dry gas seal	Gas geschmierte Gleitringdichtung
economizer	Zwischenentspannungs-Behälter
efficiency	Wirkungsgrad
energy transfer	Energieübertragung
equation	Gleichung
Euler head	wirkliche spezifische Arbeit, Enthalpie-Differenz
evaporator	Verdampfer
exergy (available energy)	Exergie
exit	Austritt
exit angle	Austritts-Winkel
expansion valve	Entspannungsventil
fabricated casing	Schweißgehäuse
fan law	Strömungsmaschinen-Gesetz
flow channel	Strömungskanal
flow coefficient	Volumenstromzahl
flow reversal	Stömungsumkehr
fluid-flow machine	Strömungsmaschine
frame size	Gehäusegröße
friction factor	Reibungsfaktor
gap	Spalt, Abstand
gasoline	Benzin
gas power	innere Leistung
gear box	Getriebe
gear loss	Getriebe-Verlust
ground level block foundation	Blockfundament
head	Förderhöhe, spezifische Arbeit
heat-up	Aufwärmung
high-temperature vacuum brazing	Hochtemperatur-Vakuumlötung
horizontally (axially) split casing	horizontal geteiltes Gehäuse
hub disk	Nabenscheibe

English–German Glossary on Compressors 321

humid gas	feuchtes Gas
hydraulic diameter	hydraulischer Durchmesser
hydrocarbon	Kohlenwasserstoff
hydrogen-rich gas compressor	Wasserstoff-Reichgas-Verdichter
impeller	Laufrad
impeller arrangement	Laufrad-Anordnung
impeller exit flow coefficient	Laufradaustritts-Volumenstromzahl
impeller exit width	Laufrad-Austrittsbreite
impeller eye	Saugmund
impeller outer diameter	Laufrad-Außendurchmesser
impeller slip	Laufrad-Schlupf (-Slip)
impeller tip speed	Laufrad-Umfangsgeschwindigkeit
impulse turbine	Gleichdruckturbine
inboard bearing	kupplungsnahes Lager
incipience of surge	Pumpbeginn
inlet	Eintritt
inlet angle	Eintritts-Winkel
inlet duct	Eintrittskanal
inline impellers	hintereinander geschaltete Laufräder
integrally geared compressor	Mehrwellen-Getriebeverdichter
intercooler	Zwischenkühler
interference fit, shrink fit	Schrumpfsitz
interlocking labyrinth seal	echte Labyrinth-Dichtung
isentropic exponent	Isentropenexponent
isentropic head	isentrope Förderhöhe (isentrope Arbeit)
isobaric heat capacity	isobare Wärmekapazität
jet engine	Düsentriebwerk
journal bearing	Radiallager
lateral critical speed	biegekritische Drehzahl
lateral vibration	Biegeschwingung
law of similarity	Ähnlichkeitsgesetz
law of thermodynamics	Hauptsatz der Thermodynamik
leakage flow	Leckmassenstrom
liquefaction	Verflüssigung
low-solidity diffuser vanes	Diffusorschaufeln großer Teilung
Mach number	Machzahl
machine Mach number	Umfangs-Machzahl
mass flow	Massenstrom
material yield strength	Streckgrenze
maximum continuous speed	maximale Dauerdrehzahl
modular design system	Baukasten-System
molar mass, molecular weight	molare Masse
momentum	Impuls
momentum thrust	Impulsschub
motor nameplate rating	Bestimmung der Motornennleistung
multistage compressor	mehrstufiger Verdichter
natural frequency	Eigenfrequenz
natural gas	Erdgas

negative prerotation (prewhirl)	Gegendrall
number of blades	Schaufelzahl
number of stages	Stufenzahl
numerical equation	Zahlenwert-Gleichung
operating range	Betriebsbereich
outboard bearing	kupplungsfernes Lager
overhung impeller	fliegend gelagertes Laufrad
overload	Überlast
parting joint	Teilfuge
partload	Teillast
perfect gas (ideal gas)	ideales Gas
performance curve	Kennlinie
performance map	Kennfeld
petroleum	Erdöl
physical equation	Größen-Gleichung
plenum inlet	Plenum-Einlauf
polytropic efficiency	polytroper Wirkungsgrad
polytropic head coefficient	polytrope Druckzahl
positive prerotation (prewhirl)	Mitdrall
power, power input	Leistung
pressure ratio	Druckverhältnis
pressure side	Druckseite
radial pressure equation	radiale Druckgleichung
rated point	Auslegungspunkt
rating range	Auslegungsbereich
reaction	Reaktionsgrad
reaction turbine	Reaktionsturbine
real-gas equation of state	Zustandsgleichung für reale Gase
refrigeration compressor	Kälteverdichter
regulation method	Regelungsart
relative humidity	relative Feuchte
rerate	Umbau
research and development (R & D)	Forschung und Entwicklung
residual unbalance	Restunwucht
return (crossover) bend	Umlenkkanal
return vane channel	beschaufelter Rückführkanal, Rückführstufe
reverse flow	Rückwärtsströmung
Reynolds number	Reynoldszahl
riveted impeller	Niet-Laufrad
rocket engine	Raketentriebwerk
rotating stall	Rotating Stall, rotierende Abreißströmung
rotational speed	Drehzahl
rotor	Läufer, Rotor
saturation pressure	Sattdampfdruck
scalable	skalierbar
section	Stufengruppe
see-through labyrinth seal	unechte Labyrinth-Dichtung
semi-open impeller	halboffenes Laufrad

separation margin	Abstand, Spielraum
shaft	Welle
shaft stiffness	Wellensteifigkeit
shear ring	Scherring
shear stress	Schubspannung
shop performance test	Werksprobelauf
shrouded impeller	Deckscheiben-Laufrad
side-stream inlet	Zwischenzuführung
similarity law	Ähnlichkeitsgesetz
single-flow	einflutig
single-shaft compressor	Einwellen-Verdichter
single-stage compressor	einstufiger Verdichter
slope	Steigung
sonic velocity, velocity of sound	Schallgeschwindigkeit
specific speed	spezifische Drehzahl
stability (performance curve)	ungeregelter Fahrbereich
stability point (performance curve)	Pumpgrenzpunkt, ungeregelt
stage	Stufe
stage head	Stufen-Förderhöhe, Stufenarbeit
stage pitch	axialer Stufen-Abstand
stage stall	Stufen-Abrissgrenze
stage surge	Stufen-Pumpgrenze
start-up time	Hochlaufzeit
state	Zustand
steady state	Beharrungszustand
steam turbine	Dampfturbine
stream tube	Stromröhre
stress analysis	Festigkeits-Berechnung
subcooling	Unterkühlung
substitute gas	Ersatzgas
subsynchronous vibration	subsynchrone Schwingung
suction nozzle	Saugstutzen
suction pressure	Ansaugdruck
suction side	Saugseite
suction throttle valve	Saugdrosselklappe (-ventil)
suction throttling	Saugdrosselung
surface roughness	Oberflächen-Rauheit
surge	Pumpgrenze
surge cycle	Pumpzyklus
system resistance curve	Anlagen-Widerstands-Kennlinie
system surge	System-Pumpgrenze
table foundation	Tischfundament
throat	engste Stelle
throttle ratio	Drosselverhältnis
thrust bearing	Axiallager
thrust collar	Druckscheibe
tilting pad (pivoted shoe) bearing	Kippsegment-Lager
tip speed (peripheral speed)	Umfangsgeschwindigkeit
tip speed Mach number	Umfangs-Machzahl
torque	Drehmoment
torsional critical speed	torsionskritische Drehzahl
trip speed	Schnellschluss-Drehzahl

turbocharger	Turbolader
turndown (perf. curve)	Konstant-Druck-Fahrbereich
turndown point (perf. curve)	Pumgrenzpunkt bei konstantem Enddruck
twisted (3D-) blade	räumlich gekrümmte Schaufel
unbalance response analysis	Unwucht-Rechnung
unit equation	Einheiten-Gleichung
uprate	Vergrößerung
upstream	stromaufwärts
vaned diffuser	beschaufelter Diffusor
variable speed	variable Drehzahl
velocity triangle	Geschwindigkeits-Dreieck
viscous flow	reibungsbehaftete Strömung
volume flow (capacity)	Volumenstrom
volume ratio	Volumenverhältnis
volute, scroll	Spirale
welded (fabricated) impeller	Schweiß-Laufrad
work input factor	Arbeitszahl
work, head	(spezifische) Arbeit
wrap angle, azimuth	Umschlingungswinkel, Azimuth

Index

adjustable inlet guide vanes 79
affine lines (suction throttling) 129
air compressor 244
ammonia refrigeration compressor 249
amplification factor 219
application examples 231
axial rotor thrust 210

balance piston thrust 214
bearing/gear losses 164
best efficiency point 115
binary interaction parameters 63
blade
   blockage 170
   exit angle 175
   inlet angle 171
   inlet width 171
   loading diagram 208
   pressure side 208
   suction side 208
bypass ratio 123

chlorine compressor 251
chlorine fire 253
choke limit 115, 149
Colebrook–White formula 267
compressibility factor 57
compressor design
   approximate 158
   detailed 165
   final 217
compressor range chart 315
condenser 248
conical exit diffuser 204
cooled bypassing 122
core rotation factor 212
coupling thrust 216
cover-disk leakage 168
cross-coupling force 221
cross-coupling spring constant 221
customary performance maps 123

damping coefficient 221
damping force 221
design constraints
   discharge temperature 97
   flow coefficient 93
   impeller exit width 98
   impeller tip speed 94
   impellers per rotor 100
   Mach number 95
   min. discharge volume flow 108
   rotational speed 94
   shaft stiffness 98
   stage head 99
   volume flow per frame size 104
desuperheating 248
diaphragm gap, pressure 211
diffuser
   aerodynamics 196
   friction factor 198
   increased diameter ratio 261
   inlet critical flow angle 147
   inlet Mach number 147
   inlet shape 276
   isentropic enthalpy recovery 83
   pressure recovery 83
   pinching 274
discharge temperature
   perfect gas 56
   real gas 56
disk friction efficiency, torque 178

economizer 248
equations, general 14
   numerical equation 14
   physical equation 14
   unit equation 14, 317
ethylene refrigeration compressor 250
Euler (actual) head 55
Euler turbomachinery equation 30
evaporator 248
exit flow angle 176
exit flow coefficient 175
expansion mode beyond choke 151

Index

fan law
  deviation 127
  parabola 127
fan laws 116
flow coefficient 37
flow reversal 145
frame size
  designations 339
  standardization 339

gas constant, universal 48, 317
gas field operations 9
gas gathering 238
gas lift compressor 239, 243
gas thrust map 215
gas thrust of the stage 215

head rise to surge 115
heat-up by recirculation 166
hub thrust 214
hub/tip ratio 171
humid gases
  absolute water content 67
  Clausius–Clapeyron equation 68
  dew point 68
  humid molar mass 66
  relative humidity 66
  saturation pressure 68
  water separation 68
hydrocarbon gases, abbreviations 164
hydrogen-rich gas compressors 232

impeller
  arrangement 108
  blade cutback 266
  blade design 208
  convergency 314
  disk friction loss 269
  exit 174
  eye diameter 169
  family 75, 313
  inlet 165
  inlet throat area 150
  slip factor 176
  stress analysis 84
  volume ratio 179, 181
impeller/diaphragm clearance 268
impeller manufacture
  blade milling 76
  casting 76
  electroerosion 77
  high-temperature vacuum brazing 76
  riveting 77
  stereolithography 77
  welding of cover disk 76
impeller, stage, section (definition) 13
impeller types
  semi-open impeller 73
  shrouded 2D-impeller 72
  shrouded 3D-impeller 72
inlet duct 78
inlet loss expansion 166
isentropic head, perfect, real gas 54, 55
isentropic temperature exponent 56
isentropic volume exponent 56
isobaric heat capacity 56
isothermal head, perfect, real gas 55

lateral critical speed 218
lateral vibration analysis 218
leakage flow, direction 212
leakage flow, labyrinth seals 217
lock-up thrust, toothed couplings 216
logarithmic decrement 222
low-solidity diffuser vanes 84, 275

machine (tip speed) Mach number 41
machine Reynolds number 43
material yield strength 95
methanol synthesis gas compressor 255
molar isobaric heat capacity 160
momentum thrust, impeller eye 214
motor nameplate rating 194

natural-gas compressor 236
negative slope (performance curve) 143
Nelson–Obert chart 158
number of blades 176

off-design operation 113
off-shore gas lift compressor 243
operating range, increase 271
optimization, aero-thermodynamics 259

partload efficiency, increase
  diffuser pinching 274
  LSD vanes 275
  longer diffuser 273
  overload design 272
performance
  curve limits 140
  curve slope 295
  dimensionless curve 137
  invariant curve 137

map 224
map, predicted 226
test 227
total map 138
variable conditions 133
variable Mach number 96
plastic seals
  abradable materials 270
  rub-resistant materials 270
plenum inlet loss 186
plenum inlet loss coefficient 168
polytropic diffuser efficiency 182
polytropic efficiency 38, 40
polytropic exponents 56
polytropic head 27
polytropic head coefficient 40
polytropic head, perfect gas 55
polytropic head, real gas 55
polytropic impeller efficiency 182
polytropic stage efficiency 182
positive slope 142
prerotation 118
pressure field, whirling rotor 220
processes
  chemical processing 9
  gas-field operations 9
  air 10
  ammonia refrigeration 9
  ammonia synthesis 9
  chlorine 10
  cracked gas 10
  ethylene refrigeration 11
  gas lifting 9
  gasoline production 9
  hydrocarbon processing 9
  hydrogen chloride 11
  methanol synthesis 9
  reinjection 11
  urea 11
propane compressor, liquefaction 251
propylene compressor 250
pseudocritical pressure 158
pseudocritical temperature 158
pseudocritical values 47

radial pressure equation 210
radiation loss 228
rated point 115
ratio of specific heats 52
reaction 135, 181
real gas equations of state
  Benedict, Webb, and Rubin 59
  Kammerlingh Onnes 59

Lee and Kesler 61
Peng and Robinson 62
Redlich and Kwong 61
van der Waals 59
recirculation flow paths 109
recirculation loss 110, 270
reduced pressure, temperature 47, 158
reference process
  isentropic 18
  isothermal 18
  polytropic 18
refinery
  hydrotreating 233
  hydrocracking 233
  platforming 233
refrigeration capacity 248
refrigeration compressor 247
refrigeration process 248
regulation modes
  bypassing 122
  diffuser vanes 120
  inlet guide vanes 118
  speed variation 116
  suction throttling 117
relative inlet Mach number 96
rerates, definition 281
rerates
  higher rotational speed 285
  suction boosting 294
  wider impellers, diaphragms 283
rerates, advantages/disadvantages 305
residual unbalance 219
return (crossover) bend 86
return vane channel 87, 201
return vane channel parameters 201
reverse engineering 312
reverse flow zone 146
rigid support critical speed 222
rotating stall 146
rotating stall cells 148
rotating stall excitation frequency 148
rotordynamics 218
rotor stiffness ratio 99

separation margins, critical speed 220
shroud face configurations 266
shroud thrust 214
side-stream inlet 88
similarity parameters, type 2 tests 227
single-stage booster 294
slip/work input factor relationship 179
sonic velocity 100, 150
specific speed 37

stability, performance curve 115
stability, rotordynamics 220
stage efficiency 184
   blade cutback correction 187
   diffuser radius ratio correction 188
   inlet loss correction 185
   Mach-number correction 190
   Reynolds-number correction 190
   size-effect correction 184
stage
   mismatching 126
   pitch 316
   volume ratio 104
stage stall 143
   abrupt or violent 144
   delay 273
   progressive or mild 144
standardization
   meridional impeller blade cutoff 309
   modular design system 311
static volume flow 168
static inlet thrust 214
stress, impellers 94
subsynchronous vibrations 220
supercharging, booster compressor 294
surge
   stage surge 143
   system surge 140
surge cycle 141
surface roughness 267
system resistance curves 114
system surge, delay 271

throat area (impeller inlet) 150
throttle ratio 129
total-to-static expansion (suction) 168
turndown (performance curve) 115

unbalance response analysis 219
unit conversion 317
uprating
   higher flow coefficient 284
   higher rotational speed 285
   scaled-up impeller 284
   suction boosting 294

vaned diffuser 84
vaneless diffuser hydraulic dia. 198
velocity profiles, smoothing 278
velocity triangles 34
volute (scroll) 202
   azimuth 202
   geometry, improvement 264
   parameters 207
   size, influence 203
   sizing 206
volute type
   circular 90
   elliptical 90
   one-sided offset 90
   quadratic 90
   rectangular 90
   symmetrical 90
volute wrap angle 202
volute wrapping
   central volute 90
   external volute 90
   internal volute 90

w_1-minimum criterion 170
wall shear stress 196
whirl frequency ratio 221
work input factor 40, 177

Druck: Strauss Offsetdruck, Mörlenbach
Verarbeitung: Schäffer, Grünstadt